Masonry
Level Three

Trainee Guide
Fourth Edition

PEARSON

Boston   Columbus   Indianapolis   New York   San Francisco   Upper Saddle River
Amsterdam   Cape Town   Dubai   London   Madrid   Milan   Munich   Paris   Montreal   Toronto
Delhi   Mexico City   São Paulo   Sydney   Hong Kong   Seoul   Singapore   Taipei   Tokyo

NCCER
President: Don Whyte
Director of Product Development: Daniele Dixon
Masonry Project Manager: Tim Davis
Senior Manager of Production: Tim Davis
Quality Assurance Coordinator: Debie Hicks

Desktop Publishing Coordinator: James McKay
Permissions Specialist: Megan Casey
Production Specialist: Adrienne Payne
Editor: Tanner Yea

Writing and development services provided by S4Carlisle Publishing Services, Dubuque, IA
Lead Writer/Project Manager: Michael B. Kopf
Writer: Paul Lagassse, Jack Klasey
Art Development: S4Carlisle Publishing Services

Permissions Specialist: Kim Schmidt, Karyn Morrison, Katherine Benzer
Media Specialist: Genevieve Brand
Copy Editor: Michael H. Toporek

Pearson Education, Inc.
Director, Global Employability Solutions: Jonell Sanchez
Head of Associations: Andrew Taylor
Editorial Assistant: Douglas Greive
Program Manager: Alexandrina B. Wolf
Project Manager: Janet Portisch
Operations Supervisor: Deidra M. Skahill
Art Director: Diane Ernsberger
Directors of Marketing: David Gesell, Margaret Waples
Field Marketers: Brian Hoehl, Stacey Martinez

Composition: NCCER
Printer/Binder: LSC Communications
Cover Printer: LSC Communications
Text Fonts: Palatino and Univers

Credits and acknowledgments for content borrowed from other sources and reproduced, with permission, in this textbook appear at the end of each module.

30 2023

PEARSON

Perfect bound   ISBN-13: 978-0-13-375045-4
                ISBN-10:   0-13-375045-0

Preface

To the Trainee

Masons are recognized as premier craftworkers on any construction site. Although masonry is one of the world's oldest crafts, masons also use 21st-century technology on the job. Using brick, block, or stone, and bound with mortar, masons build durable structures with optimized energy performance.

With the support of the Mason Contractor Association of America (MCAA), NCCER's program has been designed and revised by subject matter experts from across the nation and industry to update the curriculum with modern techniques. Our three levels present an apprentice approach to the masonry field and will help to keep you knowledgeable, safe, and effective on the job.

We wish you the best as you continue your training for an exciting and promising career. This newly revised masonry curriculum will help you enter the workforce with the knowledge and skills needed to perform productively in either the residential or commercial market.

New with *Masonry Level Three*

NCCER is proud to release the newest edition of *Masonry Level Three* in full color with updates to the curriculum that will engage you and give you the best training possible. In this edition, you will find that the layout has changed to better align with the learning objectives. There are also new end-of-section review questions to compliment the module review. The text, graphics, and special features have been enhanced to reflect advancements in masonry technology and techniques.

We invite you to visit the NCCER website at **www.nccer.org** for information on the latest product releases and training, as well as online versions of the *Cornerstone* magazine and Pearson's NCCER product catalog.

Your feedback is welcome. You may email your comments to **curriculum@nccer.org** or send general comments and inquiries to **info@nccer.org**.

NCCER Standardized Curricula

NCCER is a not-for-profit 501(c)(3) education foundation established in 1996 by the world's largest and most progressive construction companies and national construction associations. It was founded to address the severe workforce shortage facing the industry and to develop a standardized training process and curricula. Today, NCCER is supported by hundreds of leading construction and maintenance companies, manufacturers, and national associations. The NCCER Standardized Curricula was developed by NCCER in partnership with Pearson, the world's largest educational publisher.

Some features of the NCCER Standardized Curricula are as follows:

- An industry-proven record of success
- Curricula developed by the industry for the industry
- National standardization providing portability of learned job skills and educational credits
- Compliance with the Office of Apprenticeship requirements for related classroom training (*CFR 29:29*)
- Well-illustrated, up-to-date, and practical information

NCCER also maintains a Registry that provides transcripts, certificates, and wallet cards to individuals who have successfully completed a level of training within a craft in NCCER's Curricula. *Training programs must be delivered by an NCCER Accredited Training Sponsor in order to receive these credentials.*

Special Features

In an effort to provide a comprehensive, user-friendly training resource, we have incorporated many different features for your use. Whether you are a visual or hands-on learner, this book will provide you with the proper tools and information to orient you to the important skills and techniques of the masonry trade.

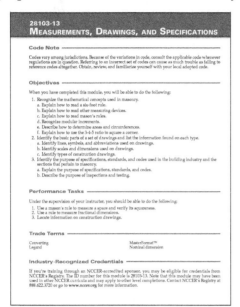

Introduction

This page is found at the beginning of each module and lists the Objectives, Performance Tasks, Trade Terms, and Required Trainee Materials for that module. The Objectives list the skills and knowledge you will need in order to complete the module successfully. The Performance Tasks give you an opportunity to apply your knowledge to real-world tasks you will undertake as a mason. The list of Trade Terms identifies important terms you will need to know by the end of the module. Required Trainee Materials list the materials and supplies needed for the module.

Special Features

Features provide a head start for those learning masonry by presenting technical tips and professional practices. These features often include real-life scenarios similar to those you might encounter on the job site.

Fall Protection

Most workers who die from falls were wearing harnesses but had failed to tie off properly. Always follow the manufacturer's instructions when wearing a harness. Know and follow your company's safety procedures when working on roofs, ladders, and other elevated locations.

Color Illustrations and Photographs

Full-color illustrations and photographs are used throughout each module to provide vivid detail. These figures highlight important concepts from the text and provide clarity for complex instructions. Each figure reference is denoted in the text in *italic type* for easy reference.

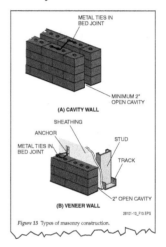

Figure 15 Types of masonry construction.

Notes, Cautions, and Warnings

Safety features are set off from the main text in highlighted boxes and are organized into three categories based on the potential danger of the issue being addressed. Notes simply provide additional information on the topic area. Cautions alert you of a danger that does not present potential injury but may cause damage to equipment. Warnings stress a potentially dangerous situation that may cause injury to you or a co-worker.

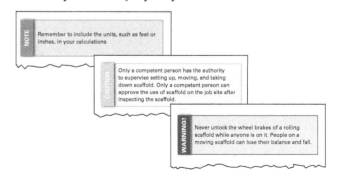

Going Green

Going Green looks at ways to preserve the environment, save energy, and make good choices regarding the health of the planet. Through the introduction of new construction practices and products, you will see how the greening of America has already taken root.

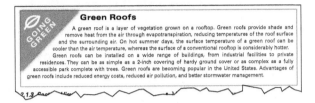

Did You Know?

The *Did You Know*? features offer hints, tips, and other helpful bits of information.

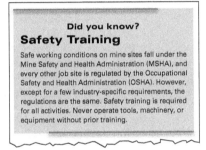

Step-by-Step Instructions

Step-by-step instructions are used throughout to guide you through technical procedures and tasks from start to finish. These steps show you not only how to perform a task but also how to do it safely and efficiently.

> The mason needs to determine whether the brick is too dry for a good bond with the mortar. The following test can be used to measure the absorption rate of brick:
>
> *Step 1* Draw a circle about the size of a quarter on the surface of the brick with a crayon or wax marker.
>
> *Step 2* With a medicine dropper, place 20 drops of water inside the circle.
>
> *Step 3* Using a watch with a second hand, note the time required for the water to be absorbed.
>
> If the time for absorption exceeds 1½ minutes,

Trade Terms

Each module presents a list of Trade Terms that are discussed within the text and defined in the Glossary at the end of the module. These terms are denoted in the text with bold, blue type upon their first occurrence. To make searches for key information easier, a comprehensive Glossary of Trade Terms from all modules is located at the back of this book.

> **M**asons are recognized as premier craftworkers at any construction site. Although masonry is one of the world's oldest crafts, masons also use 21st-century technology on the job. Masons build structures out of masonry units. The two main types of masonry units manufactured today are made of clay and concrete. Clay products are commonly known as brick and tile; concrete products are commonly known as concrete masonry units (CMUs) or block. Masonry units are also made from ashlar, glass, adobe, and other materials. In the most common forms of masonry, a mason assembles walls and other structures of clay brick or CMUs using mortar to bond the units together.

Review Questions

Review Questions are provided to reinforce the knowledge you have gained. This makes them a useful tool for measuring what you have learned.

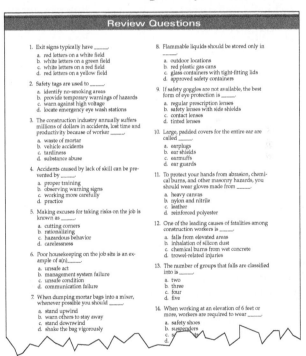

NCCER Standardized Curricula

NCCER's training programs comprise more than 80 construction, maintenance, pipeline, and utility areas and include skills assessments, safety training, and management education.

Boilermaking
Cabinetmaking
Carpentry
Concrete Finishing
Construction Craft Laborer
Construction Technology
Core Curriculum:
    Introductory Craft Skills
Drywall
Electrical
Electronic Systems Technician
Heating, Ventilating, and
    Air Conditioning
Heavy Equipment Operations
Highway/Heavy Construction
Hydroblasting
Industrial Coating and Lining
    Application Specialist
Industrial Maintenance Electrical
    and Instrumentation Technician
Industrial Maintenance
    Mechanic
Instrumentation
Insulating
Ironworking
Masonry
Millwright
Mobile Crane Operations
Painting
Painting, Industrial
Pipefitting
Pipelayer
Plumbing
Reinforcing Ironwork
Rigging
Scaffolding
Sheet Metal
Signal Person
Site Layout
Sprinkler Fitting
Tower Crane Operator
Welding

Maritime

Maritime Industry Fundamentals
Maritime Pipefitting
Maritime Structural Fitter

Green/Sustainable Construction

Building Auditor
Fundamentals of Weatherization
Introduction to Weatherization
Sustainable Construction
    Supervisor
Weatherization Crew Chief
Weatherization Technician
Your Role in the Green
    Environment

Energy

Alternative Energy
Introduction to the Power Industry
Introduction to Solar Photovoltaics
Introduction to Wind Energy
Power Industry Fundamentals
Power Generation Maintenance
    Electrician
Power Generation I&C
    Maintenance Technician
Power Generation Maintenance
    Mechanic
Power Line Worker
Power Line Worker: Distribution
Power Line Worker: Substation
Power Line Worker: Transmission
Solar Photovoltaic Systems
    Installer
Wind Turbine Maintenance
    Technician

Pipeline

Control Center Operations, Liquid
Corrosion Control
Electrical and Instrumentation
Field Operations, Liquid
Field Operations, Gas
Maintenance
Mechanical

Safety

Field Safety
Safety Orientation
Safety Technology

Supplemental Titles

Applied Construction Math
Careers in Construction
Tools for Success

Management

Fundamentals of Crew Leadership
Project Management
Project Supervision

Spanish Titles

Acabado de concreto: nivel uno,
    nivel dos
Aislamiento: nivel uno, nivel dos
Albañilería: nivel uno
Andamios
Aparejamiento básico
Aparajamiento intermedio
Aparajamiento avanzado
Carpintería:
    Introducción a la carpintería,
    nivel uno; Formas para
    carpintería, nivel tres
Currículo básico: habilidades
    introductorias del oficio
Electricidad: nivel uno, nivel dos,
    nivel tres, nivel cuatro
Encargado de señales
Especialista en aplicación de
    revestimientos industriales:
    nivel uno, nivel dos
Herrería: nivel uno, nivel dos,
    nivel tres
Herrería) de refuerzo: nivel uno
Instalación de rociadores: nivel
    uno
Instalación de tuberías: nivel uno,
    nivel dos, nivel tres, nivel cuatro
Instrumentación: nivel uno, nivel
    dos, nivel tres, nivel cuatro
Orientación de seguridad
Mecánico industrial: nivel uno,
    nivel dos, nivel tres, nivel
    cuatro, nivel cinco
Paneles de yeso: nivel uno
Seguridad de campo
Soldadura: nivel uno, nivel dos,
    nivel tres

Portuguese Titles

Currículo essencial: Habilidades
    básicas para o trabalho
Instalação de encanamento
    industrial: nível um, nível dois,
    nível três, nível quatro

Acknowledgments

This curriculum was revised as a result of the farsightedness and leadership of the following sponsors:

Arizona Masonry Contractors Association
Brick Industry Association
Central Cabarrus High School
Florida Masonry Apprentice & Educational
   Foundation, Inc.
Mason Contractors Association of America
Mortar Net Solutions

Mortenson Construction
Pyramid Masonry
Rhino Masonry, Inc.
Rocky Mountain Masonry Institute
Samuell High School
Skyline High School

This curriculum would not exist were it not for the dedication and unselfish energy of those volunteers who served on the Authoring Team. A sincere thanks is extended to the following:

Kenneth Cook
Steve Fechino

John Foley
Todd Hartsell

Moroni Mejia
Dennis Neal

NCCER Partners

American Fire Sprinkler Association
Associated Builders and Contractors, Inc.
Associated General Contractors of America
Association for Career and Technical Education
Association for Skilled and Technical Sciences
Carolinas AGC, Inc.
Carolinas Electrical Contractors Association
Center for the Improvement of Construction
Management and Processes
Construction Industry Institute
Construction Users Roundtable
Construction Workforce Development Center
Design Build Institute of America
GSSC – Gulf States Shipbuilders Consortium
Manufacturing Institute
Mason Contractors Association of America
Merit Contractors Association of Canada
NACE International
National Association of Minority Contractors
National Association of Women in Construction
National Insulation Association
National Ready Mixed Concrete Association
National Technical Honor Society
National Utility Contractors Association

NAWIC Education Foundation
North American Technician Excellence
Painting & Decorating Contractors of America
Portland Cement Association
SkillsUSA®
Steel Erectors Association of America
U.S. Army Corps of Engineers
University of Florida, M. E. Rinker School of
Building Construction
Women Construction Owners & Executives, USA

Contents

Module One
Elevated Masonry

Describes how to work safely and efficiently on elevated structures. Explains how to maintain a safe work environment, ensure protection from falls, how to brace walls from outside forces, and how to identify common types of elevated walls. Stresses safety around equipment such as cranes and hoists. (Module ID 28301-14; 15 Hours)

Module Two
Specialized Materials and Techniques

Introduces unique types of masonry situations that won't be encountered on every job, including sound-barrier walls, arches, and the use of acid brick, refractory brick, and glass block. Describes the handling and construction of these materials, and introduces the intricacies of each. (Module ID 28302-14; 60 hours)

Module Three
Repair and Restoration

Details techniques for identifying and repairing common masonry problems of weathering, settling, stain, and so on. Explains tuckpointing, the removal of efflorescence and stains, and crack repair. Includes sections on how to repair foundation walls, water intrusion and localized problems, and fireplace and chimney repair. (Module ID 28303-14; 20 hours)

Module Four
Commercial Drawings

Explains how to read and identify drawings for commercial structures, using previous experience from structural drawings as a baseline. Describes the requirements for these drawings, as well as how to interpret and create plans for architectural, structural, and shop drawings. (Module ID 28304-14; 25 hours)

Module Five
Estimating

Describes how to estimate building materials such as brick, block, grout, mortar, joint reinforcement, and masonry ties. Details multiple methods for estimating, as well as how to estimate for masonry elements such as openings and lintels. (Module ID 28305-14; 25 hours)

Module Six
Site Layout – Distance Measurement and Leveling

Covers the techniques needed to produce and read site plans and topographic maps. Describes the use of measuring devices such as tapes, range poles, plumb bobs, total stations, leveling instruments, and field notes. Also discusses the construction of batter boards and how to ensure correct measurements. (Module ID 28306-14; 20 hours)

Module Seven

Stone Masonry

Focuses on the application of natural stone in masonry construction. Describes types of stone and how stone is cut, finished, and stored. Discusses equipment and tools for handling stone. Details how to estimate and install stone using anchors and mortars, and explains how to install stone veneers. (Module ID 28308-14; 15 hours)

Module Eight

Fundamentals of Crew Leadership

Covers basic leadership skills and explains different leadership styles, communication, delegating, and problem solving. Job-site safety and the crew leader's role in safety are discussed, as well as project planning, scheduling, and estimating. Includes performance tasks to assist the learning process. (Module ID 46101-11; 20 hours)

Glossary

Index

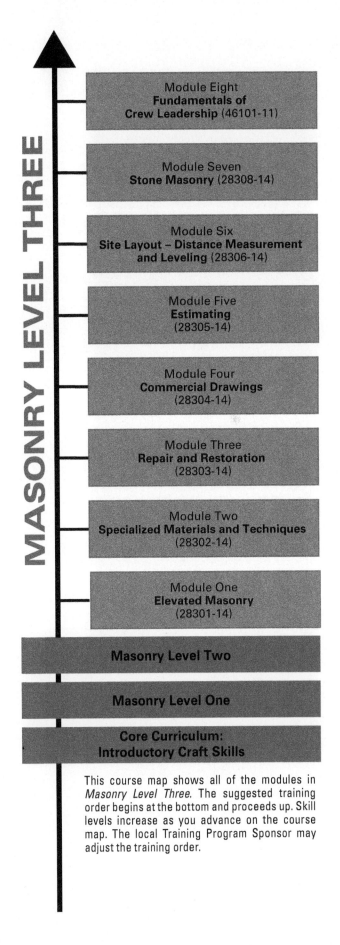

This course map shows all of the modules in *Masonry Level Three*. The suggested training order begins at the bottom and proceeds up. Skill levels increase as you advance on the course map. The local Training Program Sponsor may adjust the training order.

28301-14

Elevated Masonry

This module presents masonry techniques and safety principles for high-rise masonry construction. It also covers bracing and shoring. Advances in construction techniques, plus the work of OSHA, have resulted in a greater number of safety procedures at the elevated job site. The mason has the responsibility to be aware of these procedures and practice them. This module covers safety rules and appropriate personal protective equipment used in elevated masonry; wall bracing for wind and backfill; the design and construction of interior and exterior elevated masonry systems; and the proper moving, storage, use, and disposal of masonry materials in an elevated work environment.

Module One

Trainees with successful module completions may be eligible for credentialing through the NCCER Registry. To learn more, go to **www.nccer.org** or contact us at **1.888.622.3720**. Our website has information on the latest product releases and training, as well as online versions of our *Cornerstone* magazine and Pearson's product catalog.

Your feedback is welcome. You may email your comments to **curriculum@nccer.org**, send general comments and inquiries to **info@nccer.org**, or fill in the User Update form at the back of this module.

This information is general in nature and intended for training purposes only. Actual performance of activities described in this manual requires compliance with all applicable operating, service, maintenance, and safety procedures under the direction of qualified personnel. References in this manual to patented or proprietary devices do not constitute a recommendation of their use.

Objectives

When you have completed this module, you will be able to do the following:

1. Identify the proper personal protective equipment and safety precautions related to elevated masonry.
   a. Describe safety precautions related to an elevated work area.
   b. Discuss fall protection related to elevated work areas.
2. Describe how to properly brace a wall.
   a. Describe how to properly brace a concrete masonry wall for wind.
   b. Describe how to properly brace a wall for backfill.
3. Describe elevated masonry systems.
   a. List the construction sequence for elevated masonry systems.
   b. Describe how elevated masonry systems are designed.
   c. Identify common exterior walls used for elevated masonry systems.
   d. Identify common interior walls used for elevated masonry systems.
4. Describe how to properly handle materials at elevations.
   a. Explain safety precautions to be observed when working around cranes.
   b. Explain safety precautions to be observed when working around materials hoists.
   c. Explain safety precautions to be observed when moving and stocking materials.
   d. Explain safety precautions to be observed when working at elevated workstations.
   e. Explain how disposal chutes and waste bins are used when working from elevated workstations.

Performance Tasks

Under the supervision of your instructor, you should be able to do the following:

1. Properly brace a wall.
2. Demonstrate hand signals used for lifting materials.

Trade Terms

Controlled access zone
Cut
Guyed derrick

Lateral stress
Limited access zone
Reglet

Industry-Recognized Credentials

If you're training through an NCCER-accredited sponsor, you may be eligible for credentials from NCCER's Registry. The ID number for this module is 28301-14. Note that this module may have been used in other NCCER curricula and may apply to other level completions. Contact NCCER's Registry at 888.622.3720 or go to **www.nccer.org** for more information.

Code Note

Codes vary among jurisdictions. Because of the variations in code, consult the applicable code whenever regulations are in question. Referring to an incorrect set of codes can cause as much trouble as failing to reference codes altogether. Obtain, review, and familiarize yourself with your local adopted code.

Contents

Topics to be presented in this module include:

Figures and Tables

1.0.0 ELEVATED MASONRY SAFETY AND PERSONAL PROTECTIVE EQUIPMENT

Objective

Identify the proper personal protective equipment and safety precautions related to elevated masonry.

   a. Describe safety precautions related to an elevated work area.
   b. Discuss fall protection related to elevated work areas.

Trade Terms

Controlled access zone: A designated work area in which certain types of masonry work may take place without the use of conventional fall protection systems.

Cut: A common term for a scaffold level.

Limited access zone: A restricted area alongside a masonry wall that is under construction.

Working safely on a high-rise job means that you must work smart. Be aware of what you can do to keep your work area safe. Take precautions to prevent falls and protect yourself from falling objects. Always use personnel lifts correctly and safely.

This section provides a review of appropriate personal protective equipment (PPE) for masons, along with additional information that applies to working in elevation conditions.

Fall protection is required when workers are exposed to falls from work areas with elevations that are 6 feet or higher. The types of work areas that put the worker at risk include the following:

- Scaffolds
- Ladders
- Leading edges
- Ramps or runways
- Wall or floor openings
- Roofs
- Excavations, pits, and wells
- Concrete forms
- Unprotected sides and edges

Injuries from falls happen because fall protection systems are used inappropriately or not at all. They also occur due to carelessness. It is your responsibility to learn how to set up, inspect, use, and maintain your own fall protection equipment. Not only will this keep you alive and uninjured, it could also save the lives of your co-workers.

Falls are classified into two groups: falls from an elevation and falls on the same level. Falls from an elevation can happen when you are doing work from scaffolds, work platforms, decking, concrete forms, ladders, or excavations. Falls from elevations are almost always fatal. This is not to say that falls on the same level aren't also extremely dangerous. When a worker falls on the same level, usually from tripping or slipping, head injuries often occur. Sharp edges and pointed objects such as exposed rebar could cut or stab the worker.

The following safe practices can help prevent slips and falls:

- Wear strong work boots that are in good repair.
- Watch where you step. Be sure your footing is secure.
- Install cables, extension cords, and hoses so that they will not become tripping hazards.
- Do not allow yourself to get in an awkward position. Stay in control of your movements at all times.
- Maintain clean and smooth walking and working surfaces. Fill holes, ruts, and cracks. Clean up slippery material and litter.
- Do not run on scaffolds, work platforms, decking, roofs, or other elevated work areas.

It is vital to use fall protection equipment when working at elevations. The three most common types of fall protection equipment are guardrails, personal fall arrest systems, and safety nets.

1.1.0 Observing Work-Area Safety

Working in a high-rise construction site is more dangerous than working on a single level. Trip hazards caused by poor housekeeping can be deadly in an elevated work area. Many injuries can be prevented by keeping the work area safe. Work-area safety includes following procedures for fire prevention, identifying electrical hazards, keeping pathways clean, and using appropriate personal protective equipment properly.

1.1.1 Personal Protective Equipment

Proper clothing and safety apparel is an important part of working safely. All workers exposed to overhead hazards are required to wear protective headgear. Wear your hard hat at all times.

Shoes with safety toes are recommended for all construction workers. Heavy-duty work clothes and gloves give protection from bruises and cuts caused by sharp objects and falling material. Fall protection equipment must be worn when you are working 6 feet or more above a lower level.

Remember to take safety precautions when dressing for masonry work. The following guidelines are recommended:

- Confine long hair in a ponytail or in your hard hat. Flying hair can obscure your view or get caught in machinery.
- Wear appropriate clothing and personal protective equipment.
  - Always wear a hard hat.
  - Wear goggles when cutting or grinding.
  - Wear a high-visibility vest.
  - Wear close-fitting clothing, including long-sleeved shirts (minimum 4-inch sleeves) to give extra protection if skin is sensitive.
  - Wear gloves when working with wet mortar.
  - Wear pants over boots to avoid getting mortar on legs or feet.
  - Keep gloves and clothing as dry as possible.
- Wear face and eye protection as required, especially if there is a risk from flying particles, debris, or other hazards such as brick dust or proprietary cleaners.
- Wear hearing protection as required.
- Wear respiratory protection as required.
- Protect any exposed skin by applying skin cream, body lotion, or petroleum jelly.
- Wear sturdy work boots or work shoes with thick soles. Never show up for work dressed in sneakers, loafers, or sport shoes.
- Wear fall protection equipment as required.

1.1.2 Electrical Hazards

The most serious danger in using electrically powered tools is that of electrocution. Electricity can also cause burns, shocks, explosions, and fires. Electrical shocks can be minor and uncomfortable, or they can be severe, causing burns or death. Even a small amount of current can cause the heart to stop pumping in rhythm. If not corrected, this condition will result in death.

Electrical shock can also cause a loss of balance, muscle control, or consciousness. This could cause the victim to fall and/or drop a tool. A fall from a ladder or scaffold can be quite serious. To prevent electrical shock, tools must provide at least one of the following types of protection:

- *Double insulated* – Double insulation is more convenient than three-wire cords. The user and tools are protected in two ways: by normal insulation on the wires inside, and by a housing

that cannot conduct electricity to the operator in the event of a malfunction (*Figure 1*).
- *Powered by a low-voltage isolation transformer* – If your electrically powered tools do not have either a ground plug or double insulation, check with your supervisor to make sure that you are protected by a low-voltage isolation transformer.
- *Grounded with a three-wire cord* – Three-wire cords have two current-carrying conductors and one grounding conductor. Three-prong plugs are common on electrically powered tools (*Figure 2*). A three-prong plug should only be plugged into a three-prong, grounded receptacle. Any time an adapter is used to accommodate a two-hole receptacle, the adapter wire must be attached to a known ground. Never remove the third prong (grounding conductor) from a plug. If you are using a three-prong extension cord, make sure that it is properly grounded at its source.

Water can cause an electrical short. Prevent shock by keeping tools dry. When not in use, store power tools in a dry place. Do not use electrically powered tools in damp or wet places. Wear all appropriate personal protective equipment, such as gloves and safety footwear, when working with electrically powered tools.

The use of a ground fault circuit interrupter (GFCI) is one method used to overcome grounding and insulation problems. A GFCI is a fast-acting circuit breaker that senses small imbalances in the circuit caused by current leakage to ground. If an imbalance is detected, the GFCI interrupts the electric power within $\frac{1}{40}$ of a second. *Figure 3* shows a wall-mounted and portable GFCI.

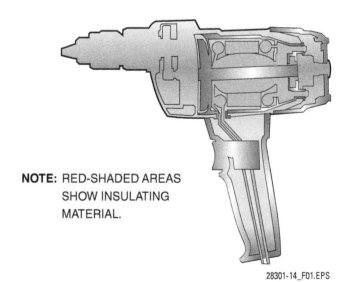

NOTE: RED-SHADED AREAS SHOW INSULATING MATERIAL.

28301-14_F01.EPS

Figure 1 Double insulation.

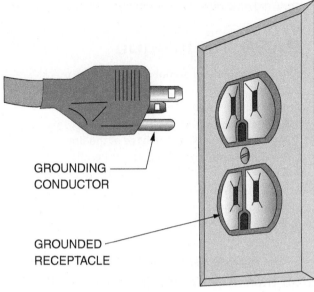

GROUNDING
CONDUCTOR

GROUNDED
RECEPTACLE

28301-14_F02.EPS

Figure 2 Grounded three-prong plug.

(A)

(B)

28301-14_F03.EPS

Figure 3 Wall-mounted and portable GFCIs.

A GFCI will not protect you from line-to-line contact hazards such as holding either two hot wires or a hot and a neutral wire in each hand. It does provide protection against the most common form of electrical shock, which is a ground fault. It also provides protection from fires, overheating, and wiring-insulation deterioration. GFCIs can be used successfully to reduce electrical hazards on construction sites. Inspect all equipment before using it. Never use electrical equipment with a damaged power cord. Use fiberglass ladders when working with electrical equipment.

Because malfunctioning electrical power tools can cause sparks, these tools can also cause fires and explosions. Make sure you know about any fire hazards in your work area, and avoid using electrically powered tools around flammable materials, fumes, and gases.

Finally, electrical cords and extension cords pose a tripping hazard. Extension cords should be brightly colored to make them more visible. Do not run cords and cables in walkways. Run them along a wall, rather than in the middle of a walkway or across a walkway. Avoid running cables and cords across elevated work areas and scaffolds. Occasionally, it may be necessary to run a cord or cable across a walkway. If this cannot be avoided, either tape the cord down and put a carpet over it, or place it in a cord runner designed to minimize the tripping hazard.

Don't Remove the Grounding Prong

An employee was climbing a metal ladder to hand an electric drill to the journeyman installer on a scaffold about 5 feet above him. When the victim reached the third rung from the bottom of the ladder, he received an electric shock that killed him. The investigation revealed that the extension cord had a missing grounding prong and that a conductor on the green grounding wire was making intermittent contact with the energized black wire, thereby energizing the entire length of the grounding wire and the drill's frame. The drill was not double insulated.

The Bottom Line: Do not disable any safety device on a power tool. A ground fault can be deadly.

Source: The Occupational Safety and Health Administration (OSHA)

1.1.3 Fire Prevention

Fire is a potential danger in all construction operations. It may present a greater hazard in high-rise construction because workers are in such vulnerable positions.

To reduce the risk of fire, protect all materials stored in the building or within 10 feet of the building with a noncombustible covering. Occasionally, combustible material must be stored within the structure. That part of the building must be fireproofed before this can happen. In either case, store combustible liquids away from other materials. Dispose of combustible waste in a secure container. Do not smoke at the work site or in the storeroom.

Make sure that at least one portable chemical fire extinguisher has been posted next to each storeroom, and that free access to fire hydrants on the street is maintained. Make sure all workers understand the fire procedures for the site. If you see that any of these items has not been taken care of, report it to your supervisor immediately.

1.1.4 Clean Work Areas

Some of the most important things that masons can do to protect themselves during elevated work are also the simplest. For example, clean work areas greatly reduce the possibility of injury. Remember to follow these guidelines:

- Keep all passageways free of materials, supplies, and other obstructions.
- Collect all scrap and waste material at the end of the day. Place it in containers or waste bins for regular removal.
- Make sure that all parts of the site are adequately lit. Replace any burnt-out bulbs.

1.2.0 Observing Fall Protection Requirements

According to OSHA, falls were the most common cause of construction worker fatalities in 2012, resulting in 278 deaths that year. That is more

---

PASS Technique

A fire extinguisher can put out small fires quickly, but it will not work unless you know how to use it properly. Remember the PASS technique:

P – **P**ull the pin from the handle.
A – **A**im the nozzle at the base of the fire.
S – **S**queeze the lever, button, or handle.
S – **S**weep the extinguisher from side to side as you spray.

Note that most fire extinguishers have to be recharged after each use.

---

than three times as many fatalities as the next most-common cause. To help reduce that number, OSHA has developed construction industry standards to prevent workers from falling off, onto, or through working levels, and to protect workers from being struck by falling objects.

High-rise construction typically requires the use of fall protection systems. Areas or activities where fall protection is needed include, but are not limited to, scaffolds, excavations, hoist areas, holes, ramps, runways, and other walkways. It is required for formwork and rebar work, leading-edge work, unprotected sides and edges, and other hazardous walking/working surfaces.

Contractors must protect their workers from fall hazards and falling objects whenever a worker is 6 feet or more above a lower level, or 4 feet above open machinery. Workers must be protected from falling into dangerous equipment. Work areas must also be protected from falling objects.

In high-rise construction, workers often find it necessary to work in places requiring wire rope or slings for access, in addition to personal lifelines. These positioning slings and ropes should be inspected daily by a qualified person. According to OSHA regulations, no wire rope should be used when more than 10 percent of the total wires are frayed or broken in any running foot.

No Training + No PPE = Death

A carpenter apprentice was killed when he was struck in the head by a nail that was fired from a powder-actuated tool in another room. The tool operator was attempting to anchor a plywood form in preparation for pouring a concrete wall. When he fired the gun, the nail passed through the hollow wall and traveled 27 feet before striking the victim. The tool operator had never received training in the proper use of the tool, and none of the employees in the area were wearing personal protective equipment.

The Bottom Line: You can be injured by the actions of others. Wear your PPE as a first line of defense against injury.

Fall protection is generally provided through the use of guardrails, safety nets, and personal fall arrest systems. All systems must be tested and approved before use.

1.2.1 Personal Fall Arrest Systems

Fall arrest equipment (*Figure 4*) catches a worker after the worker has fallen. Workers are required to use a fall arrest system when there is a risk of falling more than 6 feet. It should be used when working on scaffolds, high-rise buildings, roofs, and other elevated locations. It should also be used when working near a deep hole, near a large opening in a floor, or above protruding rebar. Use a full-body harness and lanyard for fall arrest protection where vertical free-fall hazards exist.

> **NOTE**
>
> In the past, body belts were often used instead of a full-body harness. However, as of January 1, 1998, they were banned from such use.
> This is because the body belt concentrates the arresting force in the abdominal area.

A personal fall arrest system uses specialized equipment, including a body harness, lanyards, deceleration devices, lifelines, anchoring devices, and equipment connectors. The body harness goes around the legs and shoulders, with strapping across the chest and back. A D-ring on the back is used to attach a lanyard. The lanyard is then attached to a lifeline, which is anchored to a point that is capable of holding more than 5,400 pounds without failure. The line should be long enough to allow work movements but short enough to limit a fall to 6 feet or less.

Lanyards are short, flexible lines with connectors on each end (*Figure 5*). They connect the body harness to the lifeline. There are many kinds of lanyards made for different situations. All must have a minimum breaking strength of 5,000 pounds. They come in both fixed and adjustable lengths and are made of steel, rope, or nylon webbing. Most, if not all, have a shock absorber that absorbs up to 80 percent of the arresting force when a fall is being stopped.

Workers on a suspended scaffold may also use harnesses with rope grabs and retractable lifelines, shown in *Figure 6*. The lifeline is secured to a point independent of the scaffold platform. The rope grab links the lifeline to the harness. The grab has a ratchet that locks in case of a fall.

Vertical lifelines (*Figure 7*) are suspended vertically from a fixed anchor point. A fall arrest device such as a rope grab is attached to the lifeline. Vertical lifelines must have a minimum breaking strength of 5,000 pounds.

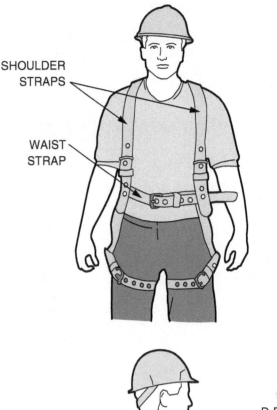

SHOULDER STRAPS

WAIST STRAP

D-RING

PELVIC STRAP

28301-14_F04.EPS

Figure 4 Personal fall arrest equipment.

28301-14_F05.EPS

Figure 5 Lanyard with shock absorber.

ROPE GRAB   RETRACTABLE LIFELINE

28301-14_F06.EPS

Figure 6 Rope grab and retractable lifeline.

Horizontal lifelines (*Figure 8*) are connected horizontally between two fixed anchor points. These lifelines must be designed, installed, and used under the supervision of a qualified, competent person. Horizontal lines must be able to support a minimum tensile load of 5,000 pounds per person attached to the line. Vertical and horizontal lifelines must not be used by more than one worker at the same time.

Anchor points, commonly called tie-off points, support the entire weight of the fall arrest system. The anchor point must be capable of supporting 5,000 pounds for each worker attached. Eyebolts (*Figure 9*) and overhead beams are considered anchor points to which fall arrest systems are attached.

The D-rings, buckles, snap hooks, and carabiners (*Figure 10*) that fasten and/or connect the parts of a personal fall arrest system are called connectors. There are regulations that specify how they are to be made. D-rings and snap hooks are required to have a minimum tensile strength of 5,000 pounds.

> **NOTE**
>
> Since January 1, 1998, only locking-type snap hooks are permitted for use in personal fall arrest systems.

28301-14_F07.EPS

Figure 7 Vertical lifeline.

Like all safety equipment, lanyards, lifelines, and safety harnesses should be carefully inspected before each use. Look for worn or frayed areas and check for metal fatigue. Do not use the equipment if you find any damage. Replace any cabling that has more than 10 percent of the total wires frayed or broken in any running foot. Only use equipment that meets or exceeds minimum OSHA standards.

1.2.2 Falling Objects

According to OSHA, falling objects killed 78 construction workers in 2012—10 percent of the total number of construction workers killed that year. In 2002, OSHA added two new rules to protect workers in high-rise construction from being struck by falling objects. First, all materials, equipment, and tools that are not being used must be secured in order to prevent them from being accidentally knocked or bumped off the platform. Second, the controlling contractor must bar other

28301-14_F08.EPS

Figure 8 Horizontal lifeline.

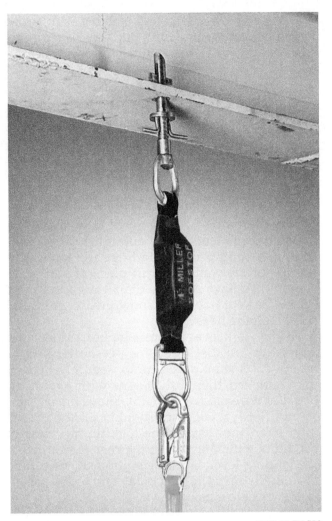

28301-14_F09.EPS

Figure 9 Push-through eyebolt and shock absorber.

construction processes below steel erection unless overhead protection is provided for the employees working below.

Debris netting, also called containment netting, is used on elevated construction sites to prevent dropped tools, materials, particulate waste such as dirt, and small items such as nails, fasteners, and screws from hitting people below. OSHA requires the installation of debris netting when a pile of materials or equipment on a scaffold is

(A) LOCKING SNAP HOOK

(B) CARABINER

28301-14_F10.EPS

Figure 10 Locking snap hook and carabiner.

higher than the top edge of a standard toeboard. In such cases, debris netting is installed from the top rail to the deck level so that it completely encases the work area like a protective curtain (*Figure 11*).

Typical debris netting is made from a fine mesh of heavy-duty, fire-retardant polyethylene plastic, and comes in rolls in a variety of widths and lengths depending on the application. OSHA requires that debris netting maintain flexibility in extreme temperatures. Debris netting is not designed to serve as fall protection.

Keep your work area safe. Prevent injury from falling objects by following these guidelines:

- Always wear a hard hat.
- Keep openings in floors covered, secured, and properly marked.

It Really Works

A mason was moving brick on a roof. The mason was standing on a metal roof panel when his right foot slipped. In an attempt to correct himself, the mason caught his left foot in a corrugation rib. The mason then fell forward to his knees on an underlying bottom insulation sheet and broke through the layer of ceiling drywall. After a short free fall, the clutch mechanism of his retractable lanyard engaged, gently slowing the worker's descent to a complete stop.

The Bottom Line: Personal fall protection equipment can save your life.

- Do not store materials other, than masonry and mortar, within 4 feet of the working edges of a guardrail system.
- Keep the working area clear by removing excess mortar, broken or scattered masonry units, and all other material and debris on a regular basis.
- Never work or walk under loads that are being hoisted by a crane.
- Erect toeboards or guardrail systems. When guardrail systems are used to prevent material from falling from one level to another, any openings must be small enough to prevent the passage of potential falling objects.
- Erect paneling or screening from the walking/working surface or toeboard to the upper edge of the top guardrail or midrail if tools, equipment, or materials are piled higher than the top edge of the toeboard.

28301-14_F11.EPS

Figure 11 Debris netting installed on scaffold during a renovation project.

- Raise and lower tools and material with a rope and bucket or other lifting device. Never throw tools or material to or from a raised surface.

1.2.3 Personnel Lifts

Many contractors switch to a temporary stairway above the fourth cut, or level, of scaffolding. Continue the temporary stairway upward as the work progresses. It must be maintained in serviceable condition until at least one permanent stairway has been completed.

Stairways must be adequately lighted. If temporary stairways are required, they should be adequately braced, wide enough for two persons, and equipped with railings and toeboards. Ramps or runways used in place of stairways should also have railings. Stairways and ramps should be kept free of ice, snow, grease, mud, and other slipping hazards.

As the work level becomes higher, ladders and stairways are discarded for personnel lifts or hoists. Lifts should be plumb, securely braced, and enclosed their full height with expanded metal or wire mesh (*Figure 12*). The doors or gates should be 6 feet tall. They should lock and unlock only from the inside. The locking mechanism should operate only when the cage is stopped at the landing level. Hoisting equipment should be thoroughly inspected each day and tested whenever subject to a new use.

Personnel lifts must be equipped with guardrails, toeboards, and hand controls that operate from the ground or the platform. Only persons riding the lift should be able to lock the doors. The lifts must also have safety brakes to prevent free fall if the cabling fails.

OSHA requires that personnel lifts be given a trial lift and proof testing before they are used to lift people. This is accomplished by loading the lift to 125 percent of capacity, operating it, and inspecting the cabling. The lift capacity must be posted so that it will not be overloaded by accident.

All personnel lift towers (*Figure 13*) that are outside the structure must be enclosed their full height on the side or sides used for entrance and exit to the structure. At the lowest landing, the enclosure on the sides not used for exit or entrance to the structure should reach a height of at least 10 feet. Other sides of the tower structure adjacent to floors or scaffold platforms should be enclosed to a height of 10 feet above the level of such floors or scaffolds. Towers inside structures should be enclosed on all four sides throughout their full height.

Figure 12 Personnel lift.

28301-14_F13.EPS

Figure 13 Personnel lift tower.

Typically, materials hoists do not have the safety features of personnel lifts. On some jobs, however, personnel hoists do double duty because they have enclosed cabs, internal controls, and safety brakes. Be sure that materials hoists are clearly marked so that everyone knows that they do not have the safety equipment required for carrying workers.

Although using cranes to lift people was common in the past, OSHA regulations, as spelled out in 29 *CFR* 1926.550, now discourage the practice. Using a crane to lift personnel is not specifically prohibited by OSHA, but the restrictions are such that it is only permitted in special situations where no other method is suitable. When it is allowed, certain controls must be in place:

- The rope design factor is doubled.
- The lift capacity is cut in half.
- Free falling is prohibited.
- Devices are required that provide warnings and prevent the lower load block (pulley) or hook from coming into contact with the upper load block, boom point, or boom-point machinery, likely causing the failure of the rope or release of the load or hook block.
- The platform must be specifically designed for lifting personnel.
- Before the lift is used, it must be tested with a comparable weight and then inspected.
- Every intended use must undergo a trial run with weights rather than people.

1.2.4 Access Zones

To prevent injury on any project where an elevated masonry wall is to be constructed, the contractor must establish controlled access zones and limited access zones before construction can begin. This section reviews the purpose of each of these access zones, and how to establish and maintain them.

A controlled access zone, or CAZ, is a designated work area in which certain types of masonry work may take place without the use of conventional fall protection systems such as guardrails, personal fall arrest systems, or safety nets. Controlled access zones are established prior to the beginning of construction. They are created and clearly marked to limit entrance to authorized workers only. OSHA guidelines specify that if there are no guardrails, masons are the only workers allowed in the CAZ. The height of the CAZ is equal to the height of the wall plus 4 feet. If the wall is higher than 8 feet, the CAZ must remain defined until the wall is braced.

The CAZ is marked on the side of the wall without the scaffold by a rope, wire, or tape control line that runs along the entire length of the unprotected edge, and is connected to each side of the guardrail or wall. The purpose of the control line is to restrict access to workers who are not authorized to work there. OSHA also requires that supporting stanchions must be clearly flagged or marked at maximum intervals of 6 feet. The stanchion flagging must be rigged to ensure that the lowest point is not less than 39 inches and not more than 45 inches from the working surface. The control line must be able to withstand at least 200 pounds of force applied to it.

For block and brick wall construction, the control line is erected a minimum of 6 feet and a maximum of 25 feet from the unprotected edge. For precast concrete wall construction, the control line should be erected a minimum of 6 feet and a maximum of 60 feet, or half the length of the precast member, whichever is less, from the edge. If overhand bricklaying techniques are used to construct the wall, the control line must be a minimum of 10 feet and a maximum of 15 feet from the edge.

On floors and roofs where guardrail systems are not in place before the start of overhand bricklaying operations, controlled access zones should be enlarged to enclose all points of access, materials handling areas, and storage areas. On floors and roofs where guardrail systems are in place but need to be removed to allow overhand bricklaying work or leading-edge work, remove only that portion of the guardrail necessary to accomplish that day's work.

A limited access zone (LAZ) marks a restricted area alongside a masonry wall that is under construction. Like the CAZ, the LAZ is established before construction begins. The LAZ is equal to the height of the wall plus 4 feet, for the entire length of the wall on the side without scaffold. If the maximum height of the wall is 8 feet, the LAZ must remain in place until the wall has been adequately supported. If the wall is higher than 8 feet, the wall must be braced until permanently supported.

Barricades are used to mark off the LAZ. Only employees who are working on the construction of the wall are permitted access to the LAZ.

Additional Resources

Fall Protection and Scaffolding Safety: An Illustrated Guide. 2000. Grace Drennan Ganget. Government Institutes.

"Online Safety Library: Scaffold Safety." Oklahoma State University. **www.ehs.okstate.edu**

"Scaffolding." OSHA. **www.osha.gov**

1.0.0 Section Review

1. To reduce the risk of fire in a building, use a noncombustible covering to protect materials stored in the building as well as those stored within _____.

   a. 40 feet of the building
   b. 30 feet of the building
   c. 20 feet of the building
   d. 10 feet of the building

2. OSHA requires that supporting stanchions in a controlled access zone be clearly flagged or marked at maximum intervals of _____.

   a. 3 feet
   b. 6 feet
   c. 9 feet
   d. 12 feet

SECTION TWO

2.0.0 CONCRETE MASONRY WALL BRACING

Objective

Describe how to properly brace a wall.
  a. Describe how to properly brace a concrete masonry wall for wind.
  b. Describe how to properly brace a wall for backfill.

Performance Task

Properly brace a wall.

All buildings are constructed from the ground level up. Concrete masonry walls are not often designed to be freestanding, self-supporting systems. In fact, both during construction and after the structure has been completed, the stability of the structure must be maintained. During construction, and immediately after a wall section has been laid up, you must consider the need for appropriate bracing while the mortar gains its final set. You must also maintain sufficient support for intersecting walls by applying the proper anchors or ties between walls, and plan the use of anchors to tie other building materials, such as joists, to the masonry. Otherwise, the wall will collapse, potentially causing injury and death as well as damage and delays.

Wall supports for masonry walls may be in the form of vertical columns of masonry units that are directly tied into the structure itself and located at various points along the length of the wall. These vertical columns, called pilasters, increase the base area of the structure and add bonding area along their vertical distance. After the design, the mason's skill is a major factor in determining the structure's strength, durability, safety, and appearance.

In addition to the static weight of the wall, which is called the dead load, wind speed must also be taken into consideration when building a wall, and bracing provided to ensure that the wall under construction is protected from being toppled by the wind. Once the wall has been attached to bracing, wind speeds in excess of 35 miles per hour require the evacuation of the immediate work area. Leave bracing in place until the wall's final lateral support has been provided in accordance with the design.

During construction, the maximum unbraced and braced heights of a masonry wall above grade are calculated in terms of the relationship between the nominal wall thickness and the density of the masonry unit. *Figure 14* shows how to visualize the maximum height above the base or highest line of lateral support that is allowed for walls of various thickness during the construction of the wall and prior to bracing. *Figure 15* shows how to visualize the typical vertical and horizontal brace-spacing requirements for an unsupported wall.

In the past, local building codes set limits for unsupported, nonreinforced concrete walls according to the thickness and the height of the unsupported wall. Often, these regulations specified various heights and thicknesses without considering the total design features or other available technical data. Newer building codes take a number of features into consideration, including overall design, flexural stresses, shear stresses, and wind loads. Therefore, more realistic estimates of unsupported wall heights or lengths are available to the designer and mason. *Table 1* provides guidelines for lateral support requirements for nonreinforced concrete masonry walls (refer to *Figure 14*).

Lateral support may be provided by vertical supports such as intersecting walls, pilasters, or columns. Horizontal support may be given by floors, roofs, or beams. The distance between lateral supports depends on the following:

- The type of construction (reinforced, partially reinforced, and nonreinforced)
- Wall thickness
- The degree of lateral and/or vertical loading expected from outside and inside the structure

Allowing for code restrictions for the placement of supports is the job of the architect or designer. However, you must be certain that each feature is laid up with the proper technique to ensure that the structure meets the code requirements.

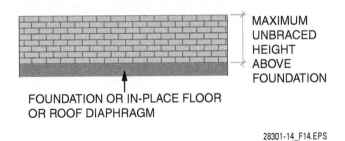

MAXIMUM UNBRACED HEIGHT ABOVE FOUNDATION

FOUNDATION OR IN-PLACE FLOOR OR ROOF DIAPHRAGM

28301-14_F14.EPS

Figure 14 Maximum height of unbraced wall.

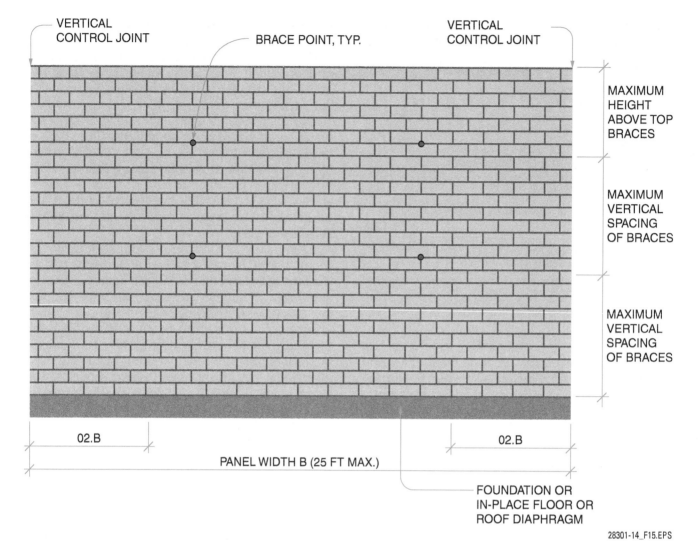

VERTICAL CONTROL JOINT   BRACE POINT, TYP.   VERTICAL CONTROL JOINT

MAXIMUM HEIGHT ABOVE TOP BRACES

MAXIMUM VERTICAL SPACING OF BRACES

MAXIMUM VERTICAL SPACING OF BRACES

02.B   02.B

PANEL WIDTH B (25 FT MAX.)

FOUNDATION OR IN-PLACE FLOOR OR ROOF DIAPHRAGM

28301-14_F15.EPS

Figure 15 Brace-spacing requirements for a wall.

2.1.0 Bracing for Wind

Most local building codes and OSHA requirements insist that temporary vertical bracing or shoring be used during construction to provide adequate support for uncured wall sections. Bracing is required primarily because of the force of wind.

Wind produces lateral forces on a wall that tend to overturn the wall. These forces produce a great deal of stress near the base of a wall and must be counteracted by bracing or shoring. A newly laid wall has very little internal strength and must be supported until the mortar cures and is self-supporting. To take care of this problem, temporary bracing similar to that shown in *Figure 16* should be provided.

Generally, masonry walls over 8 feet high should be braced. The supports should be no more than 20 feet apart. The OSHA regulations for wall bracing are covered in *29 CFR 1926.706(b)*, which states: "All masonry walls over eight feet in height shall be adequately braced to prevent overturning and to prevent collapse unless the wall is adequately supported so that it will not overturn or collapse. The bracing shall remain in place until permanent supporting elements of the structure are in place."

> **NOTE**
>
> In some areas, the foundation must withstand frost heaving or earthquakes. Check local codes for specific requirements for foundations.

Federal OSHA requirements do not define what constitutes adequate bracing. State or local building codes may have specific rules. Always check your local building codes.

The Masonry Contractors Association of America developed two standards for wall bracing to clarify the OSHA requirements. *Standard Practice for Bracing Masonry Walls under Construction* was developed to provide a detailed definition of adequate bracing for masonry walls. The *Masonry Wallbracing Design Handbook* contains over 200 diagrams of wall bracing for specific conditions.

2.2.0 Bracing for Backfill

In the case of masonry basement or foundation walls, the lateral pressure created by backfilled earth can also damage or overturn the structure. As with wind bracing, you can use bracing techniques and common trade skills to avoid such damage.

Beyond the use of temporary bracing of such basement walls, avoid backfilling until the first-floor construction is in place. You must also avoid backfilling with wet material or using water when compacting the backfill materials. During the process of backfilling, avoid the impact stresses created by backfill materials sliding down steep slopes directly against the structure.

Temporary bracing of wood planks or steel members should be placed diagonally against the wall at different horizontal distances according to the bracing system used and the expected lateral forces. This arrangement is shown in *Figure 17*.

The braced wall section should be allowed to cure for a period of three to seven days under satisfactory weather conditions. Even more important, bracing should be used during times when construction is postponed because of weather conditions. For instance, in areas where heavy rains occur, bracing may prevent wall failure as water, mud, and backfill exert hydrostatic pressure against the wall.

Table 1 Lateral Support Requirements for Non-Reinforced Concrete Masonry Walls

Standard Building Code Table 2105.1 Lateral Support (h/t)* Ratios For Exterior Bearing And Nonbearing Walls[1]				
	Design wind pressure (psf)[2]			
Wall construction[3]	15	20	25	30
Grouted, solid, or filled-cell masonry	26	22	20	18
Hollow masonry or masonry-bonded hollow walls	23	20	18	16
Cavity walls[4]	20	18	16	15

* h = clear height or length between lateral supports
t = nominal wall thickness

1 h/t ratios required for wind pressures greater than 30 psf must be determined by an engineering analysis in accordance with the Masonry Standards Joint Committee (MSJC) *Building Code Requirements for Masonry Structures*, ACI 530/ASCE 6/ TMS 402, as prescribed in Standard Building Code Section 2103.6.

2 All masonry units shall be laid in Type M, S, or N mortar unless otherwise required (see *Standard Building Code* Table 2102.9). Where Type N mortar is used and the wall spans in the vertical direction, the ratios shall be reduced by 10%.

3 These wind pressures include shape factors from *Standard Building Code* Section 1205.

4 In computing the h/t ratio for cavity walls, t shall be the sum of the nominal thicknesses of the inner and outer wythes.

28301-14_T01.EPS

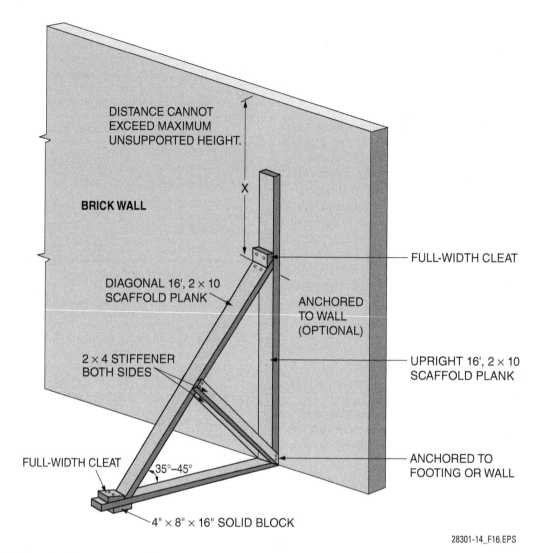

DISTANCE CANNOT EXCEED MAXIMUM UNSUPPORTED HEIGHT.

X

BRICK WALL

DIAGONAL 16', 2 × 10 SCAFFOLD PLANK

FULL-WIDTH CLEAT

ANCHORED TO WALL (OPTIONAL)

2 × 4 STIFFENER BOTH SIDES

UPRIGHT 16', 2 × 10 SCAFFOLD PLANK

FULL-WIDTH CLEAT

35°–45°

ANCHORED TO FOOTING OR WALL

4" × 8" × 16" SOLID BLOCK

28301-14_F16.EPS

Figure 16 Wind bracing for masonry wall.

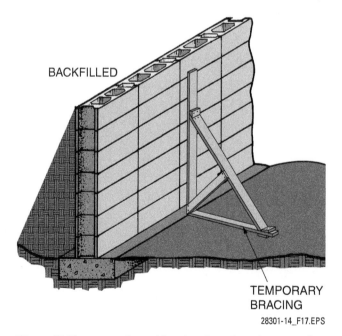

BACKFILLED

TEMPORARY BRACING

28301-14_F17.EPS

Figure 17 Temporary lateral bracing for a foundation wall.

Additional Resources

Bricklaying: Brick and Block Masonry. 1988. Brick Industry Association. Orlando, FL: Harcourt Brace & Company.

Concrete Masonry Handbook: for Architects, Engineers, Builders, Fifth Edition. 1991. W. C. Panerese, S. K. Kosmatka, and F. A. Randall, Jr. Skokie, IL: Portland Cement Association.

Masonry Wallbracing Design Handbook, Latest Edition. Rashod R. Johnson and Daniel S. Zechmeister. Algonquin, IL: Masonry Contractors Association of America.

Standard Practice for Bracing Masonry Walls under Construction. 2012. Algonquin, IL: Masonry Contractors Association of America.

2.0.0 Section Review

1. In general, masonry walls should be braced if their height exceeds _____.

   a. 8 feet
   b. 10 feet
   c. 12 feet
   d. 14 feet

2. Masonry basement and foundation walls should be braced to protect them against the lateral pressure created by _____.

   a. wind
   b. flooding
   c. adjacent construction
   d. backfilled earth

3.0.0 ELEVATED MASONRY SYSTEMS

Objective

Describe elevated masonry systems.

a. List the construction sequence for elevated masonry systems.
b. Describe how elevated masonry systems are designed.
c. Identify common exterior walls used for elevated masonry systems.
d. Identify common interior walls used for elevated masonry systems.

Trade Terms

Guyed derrick: An apparatus used for hoisting on high-rise buildings, consisting of a boom mounted on a column or mast that is held at the head by fixed-length supporting ropes or guys.

Lateral stress: Wind shear and other forces applying horizontal pressure to a wall or other structural unit.

Reglet: A narrow molding used to separate two structural elements, usually roof and wall, to divert water.

The growing popularity of high-rise masonry construction is a result of advances in masonry reinforcement, engineering, and materials. For the mason, techniques of high-rise construction require constant attention to detail and safety.

3.1.0 Identifying and Following Construction Sequences

All building construction requires substantial planning, but this is especially true with high-rise construction. The architect and engineer will plan and make design choices. The contractor will plan and be responsible for the method of construction.

There are several ways to construct a high-rise building. The most common method of high-rise construction is tier by tier. Each tier represents a vertical column height of two or three building floors. Another method involves erecting the structure in segments or bays. A bay is the distance between loadbearing supports such as beams or columns. This method is typically used on buildings whose height can be reached by ground-based lifting equipment.

Frequently, a combination of these methods is employed. The first set of horizontal or vertical segments is erected by ground-based mobile equipment. A tower crane is then erected to complete the lifting process. Most tall buildings require one or more tower cranes, as shown in *Figure 18*. The crane is raised as the tiers are finished.

For most tall buildings, the first step is setting the foundation. The second step is raising a skeleton frame of steel or concrete. Fill-in construction proceeds upward by tiers along the frame, typically in horizontal segments or bays around the building core. Masonry work may proceed by segments, panels, or floors.

The permanent structural floor does not have to be finished before construction proceeds upward. However, it must be covered over as soon as possible except for necessary openings. OSHA requires guardrails (*Figure 19*) wherever there is a danger of falling through an opening. The top guardrail is required to withstand a weight of 250 pounds and the middle rail must be able to withstand a weight of 200 pounds. The top rail should be between 38 inches and 42 inches from the bottom, with the middle rail positioned halfway between the top rail and the bottom surface. OSHA regulations also prohibit more than four uncovered floors above the highest permanently finished floor, to prevent injury from falling objects.

When construction has progressed to a height 60 feet above grade, OSHA requires that per-

28301-14_F18.EPS

Figure 18 High-rise building under construction.

manent ladders must be replaced by at least one temporary stairway (*Figure 20*). The temporary stairway must be continued upward as work progresses. It must be maintained in a serviceable condition until a permanent stairway has been completed.

An important factor affecting materials lifting, as well as the safety of all those on the job, is the loadbearing capacity of the unfinished structure. During construction, the partially completed structure must be able to support itself, any construction materials, and the added load of the lifting equipment. Structural engineers are responsible for determining the load capacity of the structure. Special bracing may be required during construction. This bracing will be removed after the structure is complete.

Call for Reinforcements

Temporary supports are often needed during high-rise construction. In this instance, shoring is in place for concrete floor slabs in a parking garage. The shoring will be removed when the concrete floor is set.

28301-14_SA01.EPS

28301-14_F19.EPS

Figure 19 Guardrails.

Construction personnel must not overload the structure and must not place too much material at any given point, to avoid floor failure. Working platforms are typically designed to support 50 pounds per square foot (psf). If the anticipated load exceeds 50 psf, the contractor must take appropriate steps to ensure that the deck will carry the load.

The erection abilities of cranes and derricks are sometimes restricted by the design of the structure. A tower crane or guyed derrick operates in a circular pattern. The total area it covers depends on its boom reach. If the structure has a square or rectangular footprint, only one piece of lifting equipment may be required. However, if the building is L-shaped, additional cranes may be required to reach all points on any floor.

Because of the need to move materials upward, planning for special erection aids is an important part of constructing tall buildings. Hitches or supports for attaching hoisting lines should be added to heavy elements to be lifted. These typically include columns, beams, or panels that are not easily adapted to the use of slings. Other erection aids include special plates or angle clips. In steel frame construction, these are added in the fabricating shop to the top and bottom of welded columns. When the column is erected, these aids give it stability until the floor framing is added.

28301-14_F20.EPS

Figure 20 Temporary stairway.

3.2.0 Understanding Building Design

High-rise masonry buildings usually have frames of reinforced concrete, steel, or a combination of these materials. *Figure 21* shows the construction of a steel frame for a high-rise building. These buildings can incorporate several different types of masonry walls.

The critical element in this type of masonry design is not loadbearing so much as resistance to lateral stress. Masonry walls carry little or no loads beyond their own weight. They act as load-transfer agents and displace wind loads onto the structural frame of the building. Lateral support is critical. It is provided for exterior walls by other structural elements of the building, angle irons, anchors, and other braces. The masonry can be supported laterally by intersecting walls and floors. The roof slab also gives lateral support to masonry walls.

The *International Building Code®* bases lateral support requirements on (wind) design pressure. This is one example of code requirements. Mini-mum required wall thickness varies among the different codes. Lateral support can be provided by cross walls, columns, pilasters, or by buttresses where the limiting span is measured vertically.

28301-14_F21.EPS

Figure 21 Steel frame for high-rise building.

Anchorage between walls and supports must be able to resist wind loads and other lateral forces acting either inward or outward. All lateral support members must have sufficient strength and stability to transfer these lateral forces to adjacent structural members or to the foundation. All of the codes contain provisions stating that specific limitations may be waived if engineering analysis is provided to justify additional height or width, or reduced thickness. Such waivers must be documented on drawings stamped by the engineer of record.

3.3.0 Constructing Exterior Walls

Solid masonry walls are typically used in foundations and residential buildings, while cavity and veneer walls are often found in high-rise construction. Two types of veneer walls are only used in high-rise construction: panel and curtain walls. Both are special veneer walls that do not support any load except their own weight. They are attached to a structural frame.

In a high-rise building, these frames are usually made of steel or concrete beams and columns. Panel and curtain walls provide weather protection, color, texture, and architectural interest to the structure (*Figure 22*). They require flashing, weepholes, and other moisture containment measures.

In addition to curtain and panel walls, parapet walls are also common in high-rise construction. The construction of these three types of structures is discussed in the following sections.

3.3.1 Panel Walls

Panel walls are exterior nonbearing walls wholly supported at each floor by a concrete slab or beam, or by steel shelf angles. Detail features of a panel-wall support using a steel shelf angle are shown in *Figure 23*. On multistory buildings, masonry veneer constructed between relieving angles and masonry infill between floors and columns of structural frames are also considered panel-wall sections.

Each panel must be made to resist lateral forces and transfer the load to adjacent structural members. Panel walls can be prefabricated masonry units, or they may be job-built. In addition to be-

28301-14_F22.EPS

Figure 22 Curtain wall under construction.

ing supported along the bottom edge, panels are also attached to the structure by anchors or clips, much like stone veneer panels. *Figure 24* shows how a stone panel is attached to the steel frame.

Since they are supported at each floor, panel walls require some type of pressure-relieving joint. This is usually located between the shelf angle and the top course of the panel below the shelf angle. The joint is usually filled with a neoprene material that allows for expansion and contraction of the panel while limiting moisture penetration.

All shelf angles must be installed with flexible joints to allow for different expansion and contraction rates of the panel wall and the support structure. The flexible anchors attached to the back of the panel permit the different expansion rates to occur without cracking the structure.

The Ingalls Building

The first reinforced-concrete high-rise office building is the 16-story Ingalls building. It was completed in 1903 in Cincinnati, Ohio. Before that the tallest concrete building was only six stories. The Ingalls building is designed to act as a monolithic unit. Each floor slab provides a rigid diaphragm to steady the building from the wind. The building stands 210 feet high and is still in use today.

Figure 23 Panel-wall mounting.

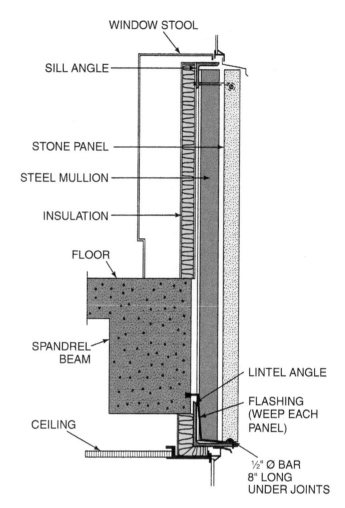

The flashing should be brought beyond the face of the wall and turned down to form a drip. A cavity filter or mortar net may also be installed to prevent mortar from clogging the drainage area. Traditionally, a soft sealant is placed below the flashing to prevent the water from re-entering the joint. This configuration is shown in *Figure 25*. The sealant can also be placed above the flashing. This forces water to drain through the weepholes and eliminates additional drainage below the brick.

3.3.2 Curtain Walls

Curtain walls are exterior nonbearing walls designed to span horizontally or vertically between lateral connections without intermediate support. Horizontal curtain walls span across column faces and intersecting interior walls, and are connected to them in order to transfer wind loads to the structure. Because the weight of multistory curtain walls is wholly supported by the foundation, they are built from the foundation to the roof. They

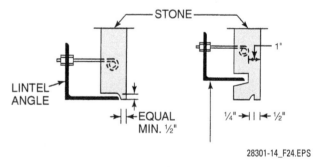

Figure 24 Stone-panel wall attachment.

are connected only at the floors and roof for lateral load transfer, without intermediate shelf angles.

Masonry curtain walls are designed by methods that rely on, or are derived from, observation or experiment. Such methods are called empirical methods. Curtain walls are also designed by engineering analysis. Regardless of the method used, masonry curtain walls are designed to span multiple structural bays. Curtain walls may be single-wythe or multiwythe design, and may incorporate reinforcing steel to increase their lateral load resistance or the distance between their lateral supports.

Curtain walls are tied to concrete or steel frames. Temperature changes will cause the masonry curtain to expand and contract at a different rate than the attached frame. Therefore, a curtain wall must be tied to the frame with flexible anchors made of galvanized steel or some other noncorrosive metal. *Figure 26* shows details of a block curtain wall anchored to a concrete frame and a steel frame. In *Figure 26A*, the anchor is embedded in the reinforced concrete beam. In *Figure 26B*, the anchor is welded to a steel beam. In both cases, the block core is grouted to hold the anchor to the block.

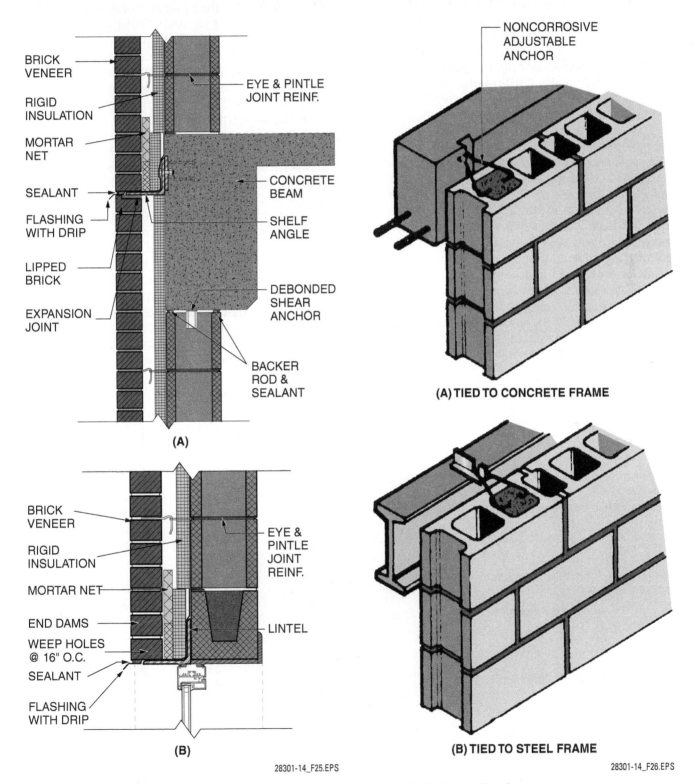

Figure 25 Flashing and sealant detail.

28301-14_F25.EPS

Figure 26 Curtain wall anchorage.

28301-14_F26.EPS

Curtain walls require flashing at the top and bottom of the walls, as well as above and below any wall openings.

3.3.3 Parapet Walls

The parapet is that part of a wall that extends above the roofline. A parapet may be added for architectural interest or it may be functional. For example, it may be used to support swing scaffolding or other equipment for window washing. Because the roof does not protect it, the parapet is exposed to the weather from both the front and the back (*Figure 27*). Wind creates increased lateral stress at the unsupported top of the parapet. Because of this exposure, it is subject to greater thermal movement.

Possible leak lines for parapets are at the coping joints and at the interface with the roofing. Metal copings are the most common because they can be made with the fewest joints. Metal copings should extend at least 2 inches below the top of the masonry. The coping should be sloped to drain to the roof of the building. Coping legs should turn out to form a drip shield and be caulked with a high-performance caulk. Flashing should be laid under the coping, and all items penetrating the flashing should be sealed with mastic. Flashing should stop inside the face shell. Flashing also provides a slip plane for differential movement.

Copings made from precast concrete, stone, and terra-cotta are commonly used on taller parapets designed to support window-washing equipment. These parapets (*Figure 28*) are structurally reinforced and anchored to the roof slab. The precast coping is flashed, sealed, and typically anchored to the wall.

Terra-cotta or other masonry coping must be carefully mortared to avoid leaking joints. The coping should overhang both sides of the wall and have integral drip notches. Because masonry coping joints are not impervious to water, through-wall flashing must be installed underneath, as shown in *Figure 29*. The coping joints should be raked out while the mortar is still plastic, then filled with elastomeric sealant. Even hairline cracks or separations at the top of the wall act as funnels for water to reach the interior of the wall.

The joint between the roof flashing and the parapet is the place on the parapet where the greatest leakage can occur. Roof flashing must be turned up onto the back face of the parapet wall and terminated above the level of the roof deck. A reglet is usually specified for this upper joint. To avoid disturbing the joint, the reglet is installed in two pieces. The mason installs the upper piece, and the roofer installs the lower piece.

In addition to the reglet over the upper edge, through-wall masonry flashing is also typically specified at the roof terminus. Where the ma-

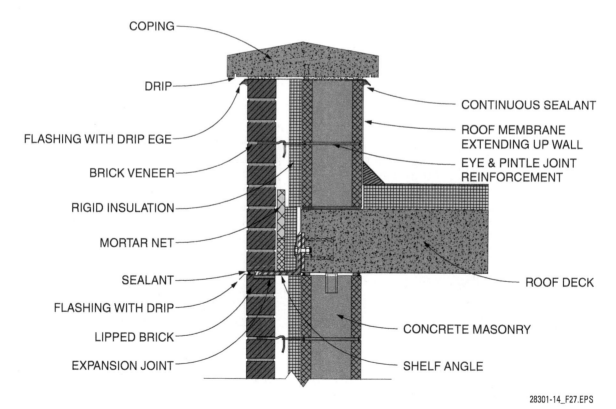

Figure 27 Parapet.

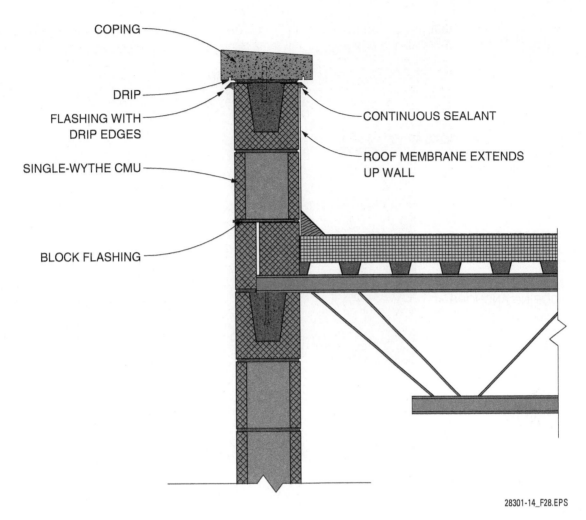

COPING

DRIP

FLASHING WITH
DRIP EDGES

SINGLE-WYTHE CMU

BLOCK FLASHING

CONTINUOUS SEALANT

ROOF MEMBRANE EXTENDS
UP WALL

28301-14_F28.EPS

Figure 28 Precast coping.

sonry and roof slab must move independently, the specifications should call for a flexible flashing connection or a roof-edge expansion joint. The flashing provides the slip plane for differential movement between the two elements.

Block parapet walls require full mortar joints even though the wall below the roofline may have face-shell bedded joints. The flashing must be fully embedded in mortar; copings also require a full mortar bedding. The roofing material is turned up to make a flashing behind the parapet.

3.4.0 Constructing Interior Walls

Interior walls are usually partition walls. They are nonbearing walls that support only their own weight. They are used as room separators, shaft enclosures, and barriers to fire, sound, or smoke. Partition walls can be single-wythe block, tile, or brick. These materials are commonly used for stairwell and elevator-shaft enclosures, as well as room dividers. Cavity walls are used for parti-

tions that must carry concentrated utility conduit and piping, as the cavity allows easy placement of the mechanical items.

Based on the *International Building Code®*, the ratio of height to thickness should be 36 to 1 for partition walls. *Tables 2* and *3* show the maximum wall height-to-thickness ratios. The partition must be securely anchored against lateral movement at the floor or ceiling. There is no requirement for intermediate pilasters or other lateral reinforcement inside these dimensions.

Single-wythe hollow brick, CMU, and vertical-cell tile partitions can be internally reinforced to stand over longer distances if there are no cross walls or projecting pilasters. Internal pilasters can provide the needed reinforcement to increase partition lengths. For block, a continuous vertical core can be reinforced with deformed steel bars and then grouted solid. Double-wythe cavity walls can be similarly reinforced without thickening the wall section.

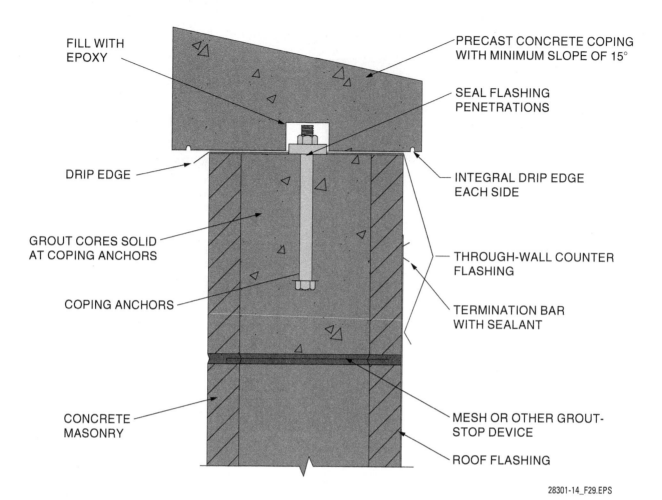

FILL WITH EPOXY

PRECAST CONCRETE COPING WITH MINIMUM SLOPE OF 15°

SEAL FLASHING PENETRATIONS

DRIP EDGE

INTEGRAL DRIP EDGE EACH SIDE

GROUT CORES SOLID AT COPING ANCHORS

THROUGH-WALL COUNTER FLASHING

COPING ANCHORS

TERMINATION BAR WITH SEALANT

CONCRETE MASONRY

MESH OR OTHER GROUT-STOP DEVICE

ROOF FLASHING

28301-14_F29.EPS

Figure 29 Through-wall flashing.

Table 2 Maximum Wall Length- or Height-to-Thickness Ratio

Construction	Maximum wall length-to thickness or height-to-thickness ratio*
Bearing walls	20
Solid or solid grouted	18
Nonbearing walls	
Exterior	18
Interior	36
Cantilever Walls**	
Solid	6
Hollow	4
Parapets (8" thick min.)**	3

* Ratios are determined using nominal dimensions. For multiwythe walls where wythes are bonded by masonry headers, the thickness is the nominal wall thickness. When multiwythe walls are bonded by metal wall ties, the thickness is taken as the sum of the wythe thickness.

** The ratios are maximum height-to-thickness and do not limit wall length.

28301-14_T02.EPS

Table 3 Maximum Wall Spans

Construction	Wall Thickness, (inches)			
	6	8	10	12
Bearing walls				
Solid or solid grouted	10 *	13.3	16.6	20
All other	9 *	12	15	18
Nonbearing walls				
Exterior	9	12	15	18
Interior	18	24	30	36
Cantilever Walls **				
Solid	3	4	5	6
Hollow	2	2.6	3.3	4
Parapets **	1.5	2	2.5	3

* 6-in. thick bearing walls are limited to one story in height.

** For these cases, spans are maximum wall heights.

28301-14_T03.EPS

Structural clay tile or glazed CMU is often used for partitioning in schools, hospitals, food-processing plants, sports facilities, airports, and public facilities. These provide a low-maintenance, high-durability surface. The partitions may be constructed of glazed materials. Double-wythe walls can provide different colors and finishes on each side. Lateral support spacing is governed by the same height-to-thickness ratio as for brick or block.

Additional Resources

"Materials Handling and Storage." OSHA. **www.osha.gov**

"Scaffolding." OSHA. **www.osha.gov**

WorkSAFE masonry safety resources. **www.worksafecenter.com**

3.0.0 Section Review

1. To prevent injury caused by falling objects, OSHA limits the number of uncovered floors above the highest permanently finished floor to _____.

   a. two
   b. four
   c. six
   d. eight

2. The critical element in high-rise masonry structural design is resistance to _____.

   a. lateral stress
   b. loadbearing stress
   c. torsional stress
   d. compressive stress

3. The part of a wall that extends above the roofline is called the _____.

   a. stud
   b. header
   c. parapet
   d. trimmer

4. The *International Building Code*® specifies the ratio of height to thickness for partition walls as _____.

   a. 24 to 1
   b. 36 to 1
   c. 48 to 1
   d. 60 to 1

SECTION FOUR

4.0.0 MATERIALS HANDLING

Objective

Describe how to properly handle materials at elevations.

a. Explain safety precautions to be observed when working around cranes.
b. Explain safety precautions to be observed when working around materials hoists.
c. Explain safety precautions to be observed when moving and stocking materials.
d. Explain safety precautions to be observed when working at elevated workstations.
e. Explain how disposal chutes and waste bins are used when working from elevated workstations.

Performance Task

Demonstrate hand signals used for lifting materials.

In high-rise construction, materials movement poses the greatest safety hazard to workers on the job. To make sure that the job site is as safe as possible, follow these guidelines:

- Check the amount of materials ordered and the places where it will be stored.
- Plan where and when materials will be moved to elevated workstations.
- Establish clear pathways for movement of all materials.
- In advance, determine individual responsibilities for each worker as materials are lifted and transported.
- Arrange a consistent system of signals for alerting workers to materials movement.
- Be prepared to move off scaffolding when supplies are loaded onto the platforms.
- Schedule materials deliveries and crane availability with the general contractor.

Timing Is Everything

Deliveries of materials may be regulated by local laws. For example, deliveries may be prohibited during rush hours. Some communities also prohibit deliveries at night or early in the morning. Check the local ordinances when ordering materials and scheduling deliveries.

4.1.0 Working around Cranes

Cranes and derricks use large, versatile boom-arm mechanisms to lift heavy loads. A derrick has a lift arm pivoted at its base, while a crane lift arm may be moveable or fixed on its vertical axis. Both derricks and cranes operate by motorized cables with hooks for raising and lowering heavy loads of materials. The boom arms can move loads both horizontally and vertically. Several types of cranes are used on large construction projects (*Figure 30*). The most common are as follows:

- Tower cranes stand alongside or in the middle of the building under construction. They are erected on their own foundation. They have one central tower with a boom at the top.
- Mobile cranes are mounted on crawler tracks or truck beds, and move about the job site. The boom is attached to the base and extends at an angle upward and outward.
- Conventional derrick cranes stand away from the building at a distance determined by the length of the boom arm. They are mounted on a fixed base.
- Traveling cranes are placed on scaffolding that is attached to the outside of the building face and moved upwards as the work progresses.

Always be alert to any materials-handling activity going on around you. When working with lifting equipment, remember to stay out of the path of the moving load. Establish and mark materials-movement pathways before hoisting starts. The supervisor should assign each worker responsibility for particular tasks during materials movement. Masons should leave the scaffold when masonry units are being placed on elevated workstations using lifting equipment.

> **WARNING!**
>
> Cranes are a necessary part of large-scale construction projects, but they pose many hazards. In high-rise construction, materials movement by crane poses the greatest safety hazard to workers on the job.

In addition to being alert at all times, follow these general safety rules when working around cranes:

- Wear a hard hat and safety shoes, and eye protection as needed.
- Keep out from under loads and away from the wheels or tracks of the equipment.
- Never stand between the crane cab and the materials truck.

SECTION FOUR

4.0.0 MATERIALS HANDLING

Objective

Describe how to properly handle materials at elevations.

a. Explain safety precautions to be observed when working around cranes.
b. Explain safety precautions to be observed when working around materials hoists.
c. Explain safety precautions to be observed when moving and stocking materials.
d. Explain safety precautions to be observed when working at elevated workstations.
e. Explain how disposal chutes and waste bins are used when working from elevated workstations.

Performance Task

Demonstrate hand signals used for lifting materials.

In high-rise construction, materials movement poses the greatest safety hazard to workers on the job. To make sure that the job site is as safe as possible, follow these guidelines:

- Check the amount of materials ordered and the places where it will be stored.
- Plan where and when materials will be moved to elevated workstations.
- Establish clear pathways for movement of all materials.
- In advance, determine individual responsibilities for each worker as materials are lifted and transported.
- Arrange a consistent system of signals for alerting workers to materials movement.
- Be prepared to move off scaffolding when supplies are loaded onto the platforms.
- Schedule materials deliveries and crane availability with the general contractor.

Timing Is Everything

Deliveries of materials may be regulated by local laws. For example, deliveries may be prohibited during rush hours. Some communities also prohibit deliveries at night or early in the morning. Check the local ordinances when ordering materials and scheduling deliveries.

4.1.0 Working around Cranes

Cranes and derricks use large, versatile boom-arm mechanisms to lift heavy loads. A derrick has a lift arm pivoted at its base, while a crane lift arm may be moveable or fixed on its vertical axis. Both derricks and cranes operate by motorized cables with hooks for raising and lowering heavy loads of materials. The boom arms can move loads both horizontally and vertically. Several types of cranes are used on large construction projects (*Figure 30*). The most common are as follows:

- Tower cranes stand alongside or in the middle of the building under construction. They are erected on their own foundation. They have one central tower with a boom at the top.
- Mobile cranes are mounted on crawler tracks or truck beds, and move about the job site. The boom is attached to the base and extends at an angle upward and outward.
- Conventional derrick cranes stand away from the building at a distance determined by the length of the boom arm. They are mounted on a fixed base.
- Traveling cranes are placed on scaffolding that is attached to the outside of the building face and moved upwards as the work progresses.

Always be alert to any materials-handling activity going on around you. When working with lifting equipment, remember to stay out of the path of the moving load. Establish and mark materials-movement pathways before hoisting starts. The supervisor should assign each worker responsibility for particular tasks during materials movement. Masons should leave the scaffold when masonry units are being placed on elevated workstations using lifting equipment.

> **WARNING!**
>
> Cranes are a necessary part of large-scale construction projects, but they pose many hazards. In high-rise construction, materials movement by crane poses the greatest safety hazard to workers on the job.

In addition to being alert at all times, follow these general safety rules when working around cranes:

- Wear a hard hat and safety shoes, and eye protection as needed.
- Keep out from under loads and away from the wheels or tracks of the equipment.
- Never stand between the crane cab and the materials truck.

26   NCCER – *Masonry Level Three* 28301-14

(A) TOWER

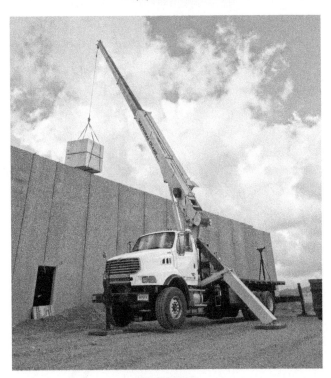

(B) MOBILE

28301-14_F30.EPS

Figure 30 Construction cranes.

- If you must guide a load down, use a guideline and do not get between the load and the crane.
- Before touching a load, hook, or cable, check to see that the boom is not touching or near a power line.
- If you guide a load, be alert for tipping or other movement of the crane, sling, or hook.

- When cranes pivot, so does the back of the crane. Be sure to mark off the appropriate space behind the crane to prevent workers from being hit by the back of the crane when it is pivoting.
- Unless you are assigned to such work, stay away from the crane and the materials-movement path.

Before cranes and derricks are used, they should be tested to ensure that they are capable of handling the required loads. Only certified operators should operate this equipment. The operators should recognize signals only from the designated person supervising the lift. Confused command channels can result in serious injury.

The methods and modes of communication vary widely in mobile crane operations. The method of communication refers to whether the communication is verbal (spoken) or nonverbal. The mode is what is used to facilitate the communication. This can include, for example, a bullhorn, a radio, hand signals, or flags.

4.1.1 *Verbal Modes of Communication*

Verbal modes of communication vary depending on the requirements of the situation. One of the most common modes of verbal communication used is a portable radio (walkie-talkie). Compact, low-power, inexpensive units enable the crane operator and signal person to communicate verbally. These units are rugged and dependable, and are widely used on construction sites and in industrial plants.

There are some disadvantages to using low-power and inexpensive equipment in an industrial setting. One disadvantage is interference. With low-power units, the frequency used to carry the signal may have many other users. The frequency can become crowded with signals from other units. Another disadvantage is high background noise. In attempting to send a signal in a high-noise area, the person sending the signal may transmit unintended noise, resulting in a garbled, unintelligible signal for the receiver. On the receiving end, the individual may not be able to hear the transmission due to a high level of background noise in the cab of the crane.

There are several solutions to the problems associated with radio use. To overcome the shortcomings associated with low-power units, more expensive units with the ability to program specific frequencies and transmit at a higher power level may be needed. Some of these more expensive units may require licensing. To overcome the background noise problem, the use of an ear-mounted noise-canceling microphone/headphone combination may be required (*Figure 31*).

Another solution is the use of an optional throat microphone (*Figure 32*). This device feeds the transmitted sound directly to the ear and picks up the voice communication from the jawbone at the ear junction. This prevents noise from entering the microphone and blocks out any background noise when listening. To avoid missed communication, the signal person's radio is usually locked in transmit so that the crane operator can tell if the unit is not transmitting. In any event, a feedback method should be established between the signal person and the crane operator so that the signal person knows the crane operator has received the signal.

Another mode of verbal communication is a hardwired system (*Figure 33*). These units overcome some of the disadvantages of radio use. When using this type of system, interference from another unit is unlikely because this system does not use a radio frequency to transmit information. As in a telephone system, occasional interference may be encountered if the wiring is not properly shielded from very strong radio transmissions. A hardwired unit is not very portable or practical when the crane is moved often. These units can also use an ear-mounted noise-canceling microphone/headphone combination to minimize the effects of background noise.

High-power handheld radios designed for use in industrial settings (*Figure 34*) are popular alternatives to communication systems that are worn on the head. They are designed to withstand dust, heat, shock, and immersion in water. They

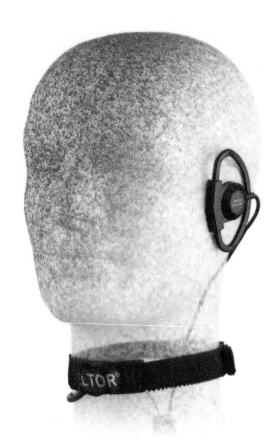

28301-14_F32.EPS

Figure 32 Throat microphone.

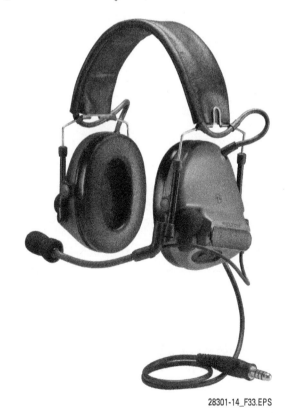

28301-14_F33.EPS

Figure 33 Hardwired communications system.

28301-14_F31.EPS

Figure 31 Noise-canceling microphone and headphones attached to a hard hat.

typically feature noise-suppression technology to reduce background sounds when transmitting. Their controls are designed to be used easily with gloves. Modern digital handheld radios feature displays and keypads that allow the user to send and receive text messages over the radio frequency.

4.1.2 Nonverbal Modes of Communication

Nonverbal modes of communication can vary tremendously. However, this is the most common type of communication used when performing crane operations. Several modes are available for use under this method. One mode is the use of signal flags, which may mean different-colored flags or a specific positioning of the flags to communicate the desired message. Another mode is the use of sirens, buzzers, and whistles in which the number of repetitions and duration of the sound convey the message. The disadvantage of these two modes is that there is no established meaning to any of the distinct signals unless they are pre-arranged between the sender and receiver. When sirens, buzzers, and whistles are used, background noise levels can be a problem.

The most commonly visual communication method is the set of hand signals established in American Society of Mechanical Engineers (ASME) consensus standard B30.5, *Mobile and Locomotive Cranes* (*Figures 35* and *36*). In accordance with *ASME B30.5*, crane operators are required to use standard hand signals when voice communication equipment is not used. The hand-signal chart must also be posted conspicuously at the job site. The signaler should also be versed in crane operations in order to understand and anticipate the crane's motions when signaled.

28301-14_F34.EPS

Figure 34 Handheld radio.

The advantage to using these standard hand signals is that they are well established and published in an industry-wide standard. This means that these hand signals are recognized by the industry as the standard hand signals to be used on all job sites. This helps ensure that there is a common core of knowledge and a universal meaning to the signals when lifting operations are being conducted. This may eliminate a significant barrier to effective communication.

Additions or modifications may be made for operations not covered by the illustrated hand signals, such as deployment of outriggers. The operator and signal person must agree upon these special signals before the crane is operated, and these signals should not be in conflict with any standard signal. If it is desired to give instructions verbally to the operator instead of by hand signals, all crane motions must be stopped before doing so.

Crane Operator Certification

OSHA now requires crane operators to be certified. NCCER's Mobile Crane Operator Certification program is recognized by OSHA and accredited by American National Standards Institute (ANSI). It offers 13 equipment-specific certifications, including capacity maximums for each. Many states have adopted, or are in the process of adopting, the NCCER certification. In those states, certified individuals are able to apply for an operator's license in the state without further testing.

Swing — Extend arm with closed fist, extend index finger. Use appropriate arm for desired direction.

Raise Boom and Lower Load — Extend arm, thumb up, open and close fingers.

Lower Boom and Raise Load — Extend arm, thumb down, open and close fingers.

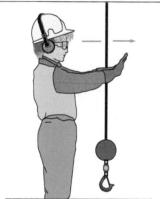

Travel — Extend arm, palm raised, and motion arm in the direction desired.

Extend Boom — Extend arms in front of body, palms up, fists closed, extend thumbs out to the sides.

Retract Boom — Extend arms in front of body, palms down, fists closed, extend thumbs inward.

Travel Both Tracks — With clenched fists, roll one fist over the other.

Travel One Track — Raise arm, fist clenched, to indicate lock track; roll other fist to travel. Raised hand indicates track to travel.

Stop — Extend arm, palm down, and hold. Move hand and forearm in a horizontal chopping motion.

28301-14_F35.EPS

Figure 35 Standard hand signals.

Emergency Stop — Same position as for Stop; extend and retract arms rapidly.

Dog Everything — Clasp hands, palm in palm, in front of the body.

Move Slowly — Placing the hand over any signal indicates a slow movement. "Hoist up" is used as an example.

Raise Load or Hoist Up — Fist up with pointer finger pointing straight up. Move hand in small horizontal circles.

Lower Load or Hoist Down — Fist down with pointer finger pointing straight down. Move hand in small horizontal circles.

Use Main Hoist — Rap on hard hat with closed fist.

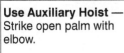

Use Auxiliary Hoist — Strike open palm with elbow.

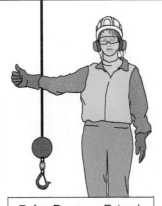

Raise Boom — Extend arm with closed fist, thumb extended up.

Lower Boom — Extend arm with closed fist, thumb extended down.

Figure 36 Standard hand signals (continued).

(A)

(B)

28301-14_F37.EPS

Figure 37 Scaffold mounted hoist.

When moving a mobile crane, audible travel signals must be given using the crane's horn:

- *Stop* – One audible signal
- *Forward* – Two audible signals
- *Reverse* – Three audible signals

There are certain requirements that mandate the presence of a signal person. When the operator of the crane cannot see the load, the landing area, or the path of motion, or cannot judge distance, a signal person is required. A signal per-

son is also required if the crane is operating near power lines or another crane is working in close proximity.

Signal persons must be qualified by experience, and be knowledgeable in all established communication methods. They must be stationed in full view of the operator, have a full view of the load path, and understand the load's intended path of travel in order to position themselves accordingly. In addition, they must wear high-visibility gloves and/or clothing, be responsible for keeping everyone out of the operating radius of the crane, and never direct the load over anyone.

Although personnel involved in lifting operations are expected to understand these signals when they are given, it is acceptable for a signal person to give a verbal or nonverbal signal that is not part of the *ASME B30.5* standard. In cases where such nonstandardized signals are given, both the operator and the signal person must have a complete understanding of the message that is being sent.

4.2.0 Working around Materials Hoists

A materials hoist can be mounted on a scaffold (*Figure 37*), a ladder, the ground (*Figure 38*), or a truck bed. The materials hoist includes a lift platform, lift cabling, and a gasoline, diesel, or electric motor. The movement is controlled by a series of pulleys and cables. On larger hoists, the lift platform may have a cage around it.

All hoists have load limitations. The maximum rated capacity must be marked on the hoist. A typical ladder hoist can lift 400 pounds to a height of up to 40 feet. Larger hoists can lift up to 5,000 pounds up to 300 feet. Larger hoists are usually attached to the side of the building under construction, to act as temporary elevators. Check the rated capacity before using a hoist.

> **WARNING!**
>
> Do not exceed a hoist's rated load. Hoist failure can cause serious injury and property damage.

Never use a hoist to transport personnel unless it has all the required safety devices. Personnel hoists have guardrails, doors, safety brakes, and hand controls in addition to the features of the materials hoist. Personnel hoists can be used for transporting materials, if needed.

When using a properly rated materials hoist to stock a workstation, follow these safety rules:

Figure 38 Large materials hoist.

28301-14_F38.EPS

- Make certain the hoist rigging is not worn, frayed, or off the pulley sheaves.
- Make certain that the load is balanced and in the middle of the hoist platform.
- Make certain that the hoist is enclosed on all sides and the gate is secured before it starts moving.
- If the hoist is operated from the ground level, make sure you understand the hand signals.
- Never ride on a materials hoist.

Oddly shaped or uncubed materials may need to be secured with safety straps or shrink wrapping to prevent them from shifting as the hoist rises. Be sure to balance the load before securing it with the straps. Check that any buckles, clamps, or ties are securely fastened before the hoist cage rises.

4.3.0 Moving and Stocking Materials

On any high-rise work site, materials handling requires special care. Masonry workstations are restocked as the job progresses, and materials for other craftworkers are moving through the job site as well. It is important to be aware of materials movement around you.

Stack masonry units carefully and safely. OSHA has guidelines regarding the stockpiling and handling of materials. Make sure you understand the following requirements before performing this work:

- Stack bagged material by stepping back the layers and cross keying the bags at least every 10 bags high.
- Do not store material on scaffolds or runways in excess of supplies needed for the immediate job.
- Do not stockpile palletized brick more than 7 feet in height. Taper back a loose brick stockpile when it reaches a height of 4 feet; it should be tapered back 2 inches for every foot of height above the 4-foot level.
- Taper back loose masonry units stockpiled higher than 6 feet; the stack should be tapered back one-half block per tier above the 6-foot level.
- All material stored in tiers should be stacked, racked, blocked, interlocked, or otherwise secured to prevent sliding, falling, or collapse.
- Maximum safe load limits of floors within buildings and structures should be posted in all storage areas. Do not exceed maximum safe loads.
- Keep aisles and passageways clear to provide for the free and safe movement of materials-handling equipment and workers.
- Do not place material stored inside buildings under construction within 6 feet of any hoist or inside floor opening, or within 10 feet of an exterior wall that does not extend above the top of the material stored.

Cubes of masonry and other building materials are loaded onto elevated workstations by pallet jack, forklift, crane, or derrick. *Figure 39* shows a typical reach-type forklift. Amounts and positions of elevated stockpiles must be carefully calculated so that the masons do not need to move more than the minimum amount of materials. Calculations require balancing weights across platforms and keeping within the rated loadbearing capacity of scaffold. Construction planning may call for closely set workstations; safety planning may call for stocking them with materials often instead of once daily, in order to avoid overloading the scaffold.

Masonry loads must be within the capacity of the lifting equipment and placed carefully so they do not block aisles. Any time workstations must be restocked, masons have to leave the scaffold. Lunchtime is often an appropriate time for restocking.

Even with the best planning, masons may still need to move masonry and other materials on elevated workstations. Remember to keep piles neat and vertically in line to avoid snagging clothes or

Figure 39 Reach-type forklift.

28301-14_F39.EPS

electrical cords. Keep the piles about 3 feet high so that it will not be hard to get at the brick or other material.

4.4.0 Using Elevated Workstations

Some work must be performed on an elevated workstation. Safety is the major consideration in organizing an elevated workstation, such as a personnel lift tower (refer to *Figure 13*).

Keeping materials neat and organized is particularly important on an elevated workstation, as space is tight. The danger and inconvenience from dropping items increases with height. Arrange materials and equipment with the following requirements in mind:

- Check the scaffold for proper assembly and rated loads before using it.

- Keep the work area clean and dry; water, mud, or dried mortar can create hazardous footing.
- Keep walkways and work areas free from stored materials, tools, rope, electrical cords, and trash. Store unused tools under the edges of the mortar pan.
- Stack masonry units far enough apart to leave room for mortar pans. Stacks should be no higher than 3 feet and no closer than 6 feet to any opening.
- Stack masonry directly over frame members, away from the edge of the scaffold platform, and never on the very ends of scaffold planks.

4.5.0 Using Disposal Chutes and Waste Bins

In 29 *CFR* 1926.852 of the OSHA regulations, there is a requirement for the use of waste chutes that have been set up at angles greater than 45 degrees: the chutes must be completely enclosed. On some jobs, separate chutes are used for designated materials. When dropping material down a waste chute, be careful to maintain your balance.

If internal drop holes are used inside a building framework, the area must be enclosed. The drop hole must be enclosed with a 4-foot-high barricade that surrounds the opening at a distance of 6 feet. When disposing of waste down the drop hole, do not lean over the edge.

Disposal chutes and drop holes should deliver their contents into waste bins or dumpsters. At the end of the day, fill the waste bins with dried mortar, broken masonry, and any other debris that can be removed from your workstation. Use the waste bin to keep your workstation clean.

Additional Resources

ASME B30.5, *Mobile and Locomotive Cranes*, Latest Edition. New York: ASME.

"Materials Handling and Storage." OSHA. **www.osha.gov**

"Scaffolding." OSHA. **www.osha.gov**

4.0.0 Section Review

1. The type of crane that is mounted on a fixed base situated away from the building at a distance that is determined by the length of the boom arm is called the _____.

   a. mobile crane
   b. tower crane
   c. derrick crane
   d. traveling crane

2. Materials hoists used to build a masonry wall can be mounted on each of the following *except* _____.

   a. a ladder
   b. the wall
   c. the scaffold
   d. a truck bed

3. When stacking bagged material, step the layers back and cross key the bags at least every _____.

   a. 5 bags high
   b. 10 bags high
   c. 15 bags high
   d. 20 bags high

4. When working in an elevated workstation, the maximum height of a masonry unit stack should be _____.

   a. 3 feet
   b. 4 feet
   c. 5 feet
   d. 6 feet

5. The height of the barricade that encloses an internal drop hole inside a building must be _____.

   a. 6 feet
   b. 5 feet
   c. 4 feet
   d. 3 feet

SUMMARY

High-rise construction often involves some type of masonry work. Many modern high-rise buildings have concrete or steel framework with masonry exterior walls. Exterior walls are constructed as veneer panel or curtain walls, depending on the design of the building. Masonry veneer walls do not carry any structural load, although they transmit lateral forces such as wind load to the main structure.

Personnel safety for high-rise construction requires, the practice of work-area safety rules, fire prevention, electrical hazard identification, the proper use of appropriate personal protective equipment, and maintaining a clear work area. Use of fall prevention and fall arrest equipment is an essential part of scaffold and leading-edge work.

Masonry panel walls can be supported at each floor level by shelf angles, while curtain walls are connected only at the foundation and the. Parapet walls are constructed with brick or concrete masonry units on many buildings. Panel walls, curtain walls, and parapets may require steel reinforcement and flashing.

Interior masonry walls for high-rise buildings include room partitions, stairwells, elevator shafts, and other structures. There are strict codes that specify the height-to-thickness ratios for partition walls.

Materials-handling equipment plays an important role in high-rise construction. Equipment, such as cranes and derricks, is used to lift materials and other equipment up to working floors from the ground. Materials hoists and personnel hoists are also used to move materials and personnel between levels of the building. Materials can be transported in personnel hoists.

Masons must be familiar with the basic hand signals used to direct crane and derrick operations. The crane operator should be given directions by only one designated signaler.

1. The most common types of fall protection equipment include guardrails, personal fall arrest systems, and _____.
   a. large air bags
   b. scaffold cages
   c. safety nets
   d. workstation enclosures

2. A ground fault circuit interrupter can break an electrical connection in as little as _____.
   a. $\frac{1}{40}$ of a second
   b. $\frac{1}{30}$ of a second
   c. $\frac{1}{20}$ of a second
   d. $\frac{1}{10}$ of a second

3. OSHA regulations require that a steel cable must be replaced if the quantity of frayed or broken wires in a running foot exceeds _____.
   a. 5 percent
   b. 10 percent
   c. 15 percent
   d. 20 percent

4. Debris netting must be installed when _____.
   a. scaffold reaches a third-story level
   b. an injury involving falling objects occurs
   c. material or equipment piles are higher than a standard toeboard
   d. masonry materials are being lifted onto scaffolds

5. Before it can be used, a personnel lift must be tested with a load equal to _____.
   a. 90 percent of its capacity
   b. 125 percent of its capacity
   c. 150 percent of its capacity
   d. 200 percent of its capacity

6. The width of a limited access zone adjacent to a wall being built is equal to _____.
   a. two-thirds the wall height
   b. the wall height
   c. the wall height plus 4 feet
   d. twice the wall height

7. Temporary bracing of a masonry wall must remain in place until _____.
   a. the final course of masonry is completed
   b. the building inspector approves its removal
   c. foundation backfilling is done
   d. permanent supporting elements of the structure are in place

8. The distance between loadbearing supports, such as beams or columns, is referred to as a _____.
   a. bay
   b. tier
   c. span
   d. panel

9. Working platforms are typically designed to support a per-square-foot load of _____.
   a. 100 pounds
   b. 75 pounds
   c. 50 pounds
   d. 25 pounds

10. Panel and curtain walls in high-rise construction require lateral bracing to _____.
    a. prevent sagging
    b. resist wind loads
    c. remain level
    d. prevent stress cracking

11. Pressure-relieving joints in masonry panel walls are usually filled with _____.
    a. butyl caulking
    b. a rubber gasket
    c. expanded foam sealant
    d. a flexible neoprene material

12. Because they have different rates of expansion, masonry curtain walls and steel or concrete building frames must be connected with _____.
    a. expansion joints
    b. flexible noncorrosive-metal anchors
    c. floating foundations
    d. sleeve connectors

13. Metal copings used on a parapet should extend below the top of the masonry by at least _____.

    a. 2 inches
    b. 2½ inches
    c. 3 inches
    d. 3½ inches

14. Interior nonbearing walls that support only their own weight are called _____.

    a. cubicle walls
    b. partition walls
    c. divider walls
    d. spacer walls

15. The partition lengths of internal walls can be increased if reinforcement is provided by _____.

    a. cross walls
    b. horizontal joint reinforcement
    c. internal pilasters
    d. projecting pilasters

16. A crane attached to scaffold that moves up the building face as work progresses is described as a _____.

    a. mobile crane
    b. tower crane
    c. derrick crane
    d. traveling crane

17. One of the most common modes of verbal communication between the crane operator and signal person is _____.

    a. cell phones
    b. a wired connection
    c. handheld radios
    d. bullhorns

18. Disadvantages of using radios for job-site communications include interference, crowded frequencies, and _____.

    a. equipment cost
    b. background noise
    c. battery life
    d. frequent breakdowns

19. A nonverbal means of communication sometimes used between the signal person and crane operator is _____.

    a. signs with printed commands
    b. sign language
    c. color-coded signal flags
    d. coded light signals

20. Material stored in a building under construction should be placed no closer to an inside floor opening than _____.

    a. 3 feet
    b. 6 feet
    c. 9 feet
    d. 12 feet

Trade Terms Quiz

Fill in the blank with the correct term that you learned from your study of this module.

1. A designated work area in which certain types of masonry work may take place without the use of conventional fall protection systems is called a(n) _____.

2. A(n) _____ is a narrow molding used to separate two structural elements, usually roof and wall, to divert water.

3. An apparatus used for hoisting on high-rise buildings, consisting of a boom mounted on a column or mast that is held at the head by fixed-length supporting ropes or guys, is called a(n) _____.

4. A(n) _____ is a restricted area alongside a masonry wall that is under construction.

5. Wind shear and other forces applying horizontal pressure to a wall or other structural unit causes _____.

6. _____ is a common term for a scaffold level.

Trade Terms

Controlled access zone
Cut
Guyed derrick

Lateral stress
Limited access zone
Reglet

Trade Terms Introduced in This Module

Controlled access zone: A designated work area in which certain types of masonry work may take place without the use of conventional fall protection systems.

Cut: A common term for a scaffold level.

Guyed derrick: An apparatus used for hoisting on high-rise buildings, consisting of a boom mounted on a column or mast that is held at the head by fixed-length supporting ropes or guys.

Lateral stress: Wind shear and other forces applying horizontal pressure to a wall or other structural unit.

Limited access zone: A restricted area alongside a masonry wall that is under construction.

Reglet: A narrow molding used to separate two structural elements, usually roof and wall, to divert water.

Additional Resources

This module presents thorough resources for task training. The following resource material is suggested for further study.

ASME B30.5, Mobile and Locomotive Cranes, Latest Edition. New York: ASME.

Bricklaying: Brick and Block Masonry. 1988. Brick Industry Association. Orlando, FL: Harcourt Brace & Company.

Concrete Masonry Handbook: for Architects, Engineers, Builders, Fifth Edition. 1991. W. C. Panerese, S. K. Kosmatka, and F. A. Randall, Jr. Skokie, IL: Portland Cement Association.

Fall Protection and Scaffolding Safety: An Illustrated Guide. 2000. Grace Drennan Ganget. Government Institutes.

Masonry Wallbracing Design Handbook, Latest edition. Rashod R. Johnson and Daniel S. Zechmeister. Algonquin, IL: Masonry Contractors Association of America.

"Materials Handling and Storage." OSHA. **www.osha.gov**

"Online Safety Library: Scaffold Safety." Oklahoma State University. **www.ehs.okstate.edu**

"Scaffolding." OSHA. **www.osha.gov**

Standard Practice for Bracing Masonry Walls under Construction. 2012. Algonquin, IL: Masonry Contractors Association of America.

WorkSAFE masonry safety resources. **www.worksafecenter.com**

Figure Credits

Courtesy of Dennis Neal, FMA&EF, CO01, Figure 3b, Figures 12–13, Figures 18–20, Figure 22, Figure 30a, Figure 37b, Figure 38

DBI/SALA & Protecta, Figure 5

Courtesy of Honeywell Safety Products, Figures 6–9, Figure 10a

Courtesy of PERI Formwork Systems, Inc., Figure 11, SA01

Courtesy Council for Masonry Wall Bracing, Figures 14–15

Courtesy of Associated Builders and Contractors, T01

Topaz Publications, Inc., Figure 21

Steven Fechino, Figure 23

National Concrete Masonry Association, Figure 25, Figures 27–29, T02–03

Manitowoc Cranes (The Manitowoc Company, Inc.), Figure 30b

3M Company, Figures 31–33

Product image (IC-F3001/F4001) courtesy of Icom America-Inc., Figure 34

Courtesy of Beta Max Hoist, Figure 37a

Courtesy of Skyjack, Figure 39

Section Review Answers

Answer	Section Reference	Objective
Section One		
1. d	1.1.3	1a
2. b	1.2.4	1b
Section Two		
1. a	2.1.0	2a
2. d	2.2.0	2b
Section Three		
1. b	3.1.0	3a
2. a	3.2.0	3b
3. c	3.3.3	3c
4. b	3.4.0	3d
Section Four		
1. c	4.1.0	4a
2. b	4.2.0	4b
3. b	4.3.0	4c
4. a	4.4.0	4d
5. c	4.5.0	4e

NCCER CURRICULA — USER UPDATE

NCCER makes every effort to keep its textbooks up-to-date and free of technical errors. We appreciate your help in this process. If you find an error, a typographical mistake, or an inaccuracy in NCCER's curricula, please fill out this form (or a photocopy), or complete the online form at **www.nccer.org/olf**. Be sure to include the exact module ID number, page number, a detailed description, and your recommended correction. Your input will be brought to the attention of the Authoring Team. Thank you for your assistance.

Instructors – If you have an idea for improving this textbook, or have found that additional materials were necessary to teach this module effectively, please let us know so that we may present your suggestions to the Authoring Team.

NCCER Product Development and Revision
13614 Progress Blvd., Alachua, FL 32615

Email: curriculum@nccer.org
Online: www.nccer.org/olf

❏ Trainee Guide   ❏ Lesson Plans   ❏ Exam   ❏ PowerPoints   Other _____

Craft / Level: _____   Copyright Date: _____

Module ID Number / Title: _____

Section Number(s): _____

Description: _____

_____

_____

_____

Recommended Correction: _____

_____

_____

_____

Your Name: _____

Address: _____

_____

Email: _____   Phone: _____

28302-14

Specialized Materials and Techniques

Common specialized types of masonry construction include sound-barrier walls, arches, acid and refractory brick, glazed masonry units, and glass block. This module provides a general introduction to specialized masonry materials and construction techniques

Module Two

Trainees with successful module completions may be eligible for credentialing through the NCCER Registry. To learn more, go to **www.nccer.org** or contact us at **1.888.622.3720**. Our website has information on the latest product releases and training, as well as online versions of our *Cornerstone* magazine and Pearson's product catalog.

Your feedback is welcome. You may email your comments to **curriculum@nccer.org**, send general comments and inquiries to **info@nccer.org**, or fill in the User Update form at the back of this module.

This information is general in nature and intended for training purposes only. Actual performance of activities described in this manual requires compliance with all applicable operating, service, maintenance, and safety procedures under the direction of qualified personnel. References in this manual to patented or proprietary devices do not constitute a recommendation of their use.

Objectives

When you have completed this module, you will be able to do the following:

1. Describe applications for sound-barrier walls.
   a. Identify pier-and-panel barrier walls.
   b. Identify continuous walls.
2. Describe the use of masonry arches.
   a. Identify common types of arches.
   b. Explain how to calculate and lay out common arches.
   c. Explain how to construct formed arches.
   d. Explain how to construct a jack arch.
3. Describe applications for acid brick.
   a. Explain uses of acid brick.
   b. Identify acid brick materials.
   c. Describe how to lay an acid brick floor.
4. Describe applications for refractories.
   a. Identify refractory brick shapes and sizes.
   b. Explain how to lay refractory brick.
   c. Describe the curing and heat-up process for refractory brick.
5. Identify the types and uses of glazed masonry units.
   a. Describe structural glazed tile and its applications.
   b. Describe glazed block and its applications.
6. Identify the types and uses of glass block.
   a. Describe applications and uses of glass block.
   b. Identify variations in glass block.
   c. Explain how to install glass block.

Performance Tasks

Under the supervision of your instructor, you should be able to do the following:

1. Lay out each of the following arches:
   - Semicircular arch
   - Segmental arch
   - Gothic arch
   - Multicentered arch
2. Build one of the following arches:
   - Semicircular arch
   - Segmental arch
   - Gothic arch
   - Multicentered arch
3. Construct a 4-foot × 4-foot wall from glazed masonry units.
4. Construct a 4-foot × 4-foot wall from glass block.

Trade Terms

Acid brick
Arch
Glazed block
Grog
Mullion

Plasticity
Refractory
Structural glazed tile

Industry-Recognized Credentials

If you're training through an NCCER-accredited sponsor, you may be eligible for credentials from NCCER's Registry. The ID number for this module is 28302-14. Note that this module may have been used in other NCCER curricula and may apply to other level completions. Contact NCCER's Registry at 888.622.3720 or go to **www.nccer.org** for more information.

Code Note

Codes vary among jurisdictions. Because of the variations in code, consult the applicable code whenever regulations are in question. Referring to an incorrect set of codes can cause as much trouble as failing to reference codes altogether. Obtain, review, and familiarize yourself with your local adopted code.

Contents

Topics to be presented in this module include:

Contents (*continued*)

Figures and Tables

SECTION ONE

1.0.0 SOUND-BARRIER WALLS

Objective

Describe applications for sound-barrier walls.
a. Identify pier-and-panel barrier walls.
b. Identify continuous walls.

Sound-barrier walls are widely used to reduce traffic noise in communities located near highways. They are often built using masonry units. Sound-barrier walls are becoming an important part of the landscape, but they do not have to be merely functional. As sound-barrier walls become more common, there will be greater interest in adding design features to them.

In the case of a new highway, the sound barrier is considered a part of the total highway design. It is not enough to simply construct a freestanding wall of some monolithic material. This type of structure may be effective in reducing noise levels, but it will likely be considered an obtrusive part of the landscape. *Figure 1* is an example of a well-designed pier-and-panel brick wall that complements the surrounding landscape.

Structurally sound masonry barrier walls can be designed in various ways. The most popular designs are the pier-and-panel wall and the continuous wall. Construction techniques for sound-barrier structures conform to conventional masonry construction for long walls.

1.1.0 Installing Pier-and-Panel Barrier Walls

Pier-and-panel wall systems are best suited for quick site erection. Panels can be fabricated on or off site, or laid in place. The choice will often depend on factors such as distance from the highway, site accessibility, and characteristics of the surrounding area. Also, pier caissons are typically constructed faster and require less concrete than strip footings. Strip footings under the panel are not required, as the panel can span from pier to pier.

The pier-and-panel wall is composed of a series of single-wythe panels, usually 4 inches thick. These panels are braced at intervals by piers. The layout of a section of a pier-and-panel wall

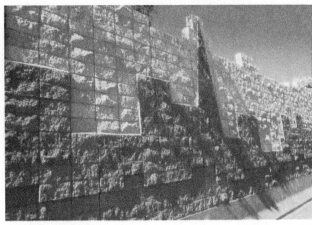

28302-14_F01.EPS

Figure 1 Sound-barrier wall.

is shown in *Figure 2*. This type of wall is relatively easy to build and is economical. It is easily adapted to varying terrain and is acoustically adequate for a highway sound barrier. Because the pier and panel are not bonded together, any horizontal movement can be absorbed at the end of the panel.

The pier-and-panel wall can also be built with returns of varying angles. However, the most easily constructed and economical return is one that is perpendicular to an adjacent panel. The panels, usually built from 8 to 20 feet long, are placed between piers of reinforced masonry, concrete, or steel. Any space left between the bottom of the wall and the ground must be adequately backfilled to prevent noise penetration.

The panels, supported on piles or clip angles attached to piers, act as thin, supported beams. Horizontal joint reinforcement will be required if the flexural stresses exceed the allowable stresses specified by the local building code. If horizontal reinforcement is required, it must be distributed the full height of the panel.

If vertical reinforcement is required, hollow brick units can be used to facilitate the reinforcement and grouting process. However, it is recommended that the piers be stiff enough so vertical reinforcement in the panels is not necessary.

The piers act as vertical cantilevers and must be designed to resist all lateral loads transferred from the panels. The piers are usually anchored to, or embedded in, reinforced concrete piles, which vary in depth according to local soil conditions. The piles must be designed to resist all shear and axial loads and potential overturning due to wind and seismic forces.

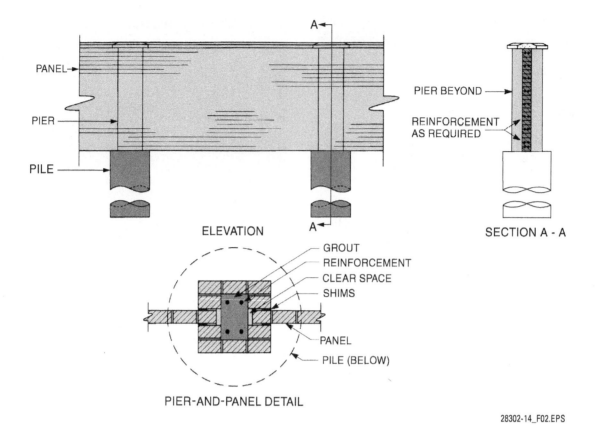

ELEVATION

SECTION A - A

PIER-AND-PANEL DETAIL

28302-14_F02.EPS

Figure 2 Pier-and-panel assembly.

1.2.0 Installing Continuous Walls

A continuous wall is supported by a reinforced footing that runs the length of the wall (*Figure 3*). Unlike the panel walls, this type of wall is subjected primarily to out-of-plane bending. Continuous walls can be built of hollow concrete block or reinforced, grouted hollow or multiwythe brick. These last two types of continuous walls are shown in *Figure 3*.

To function properly, this wall must be supported on a continuous foundation, usually made of reinforced concrete. The foundation must be sufficient to support the weight of the wall, and the continuous reinforcement must be strong enough to resist rotation caused by loads placed perpendicular to the wall.

The reinforced masonry wall is anchored to the foundation by steel reinforcement placed in the cells of hollow masonry or between the wythes in a multiwythe wall. The steel reinforcement should be designed to resist the flexural tension developed in the wall. It should be fully seated in both the foundation and grouted masonry.

Expansion joints should be placed at a maximum of 30 feet on center and detailed in a staggered fashion, as shown in *Figure 4*, for multiwythe construction.

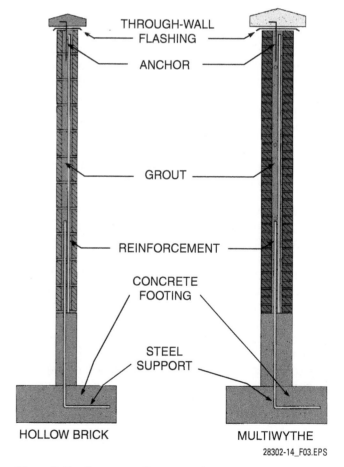

HOLLOW BRICK

MULTIWYTHE

28302-14_F03.EPS

Figure 3 Continuous-wall cross sections.

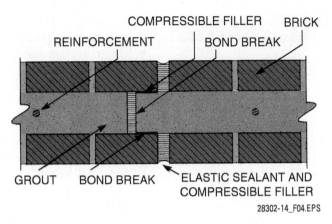

COMPRESSIBLE FILLER   BRICK

REINFORCEMENT   BOND BREAK

GROUT   BOND BREAK   ELASTIC SEALANT AND COMPRESSIBLE FILLER

28302-14_F04.EPS

Figure 4 Staggered expansion joint.

Additional Resources

Technical Note TN45, *Brick Masonry Noise Barrier Walls—Introduction*. 2001. Reston, VA: The Brick Industry Association. **www.gobrick.com**

Technical Note TN45A, *Brick Masonry Noise Barrier Walls—Structural Design*. 1992. Reston, VA: The Brick Industry Association. **www.gobrick.com**

1.0.0 Section Review

1. The most easily constructed and economical return to use with a pier-and-panel wall is _____.

   a. the same thickness of an adjacent panel
   b. the same width as an adjacent panel
   c. parallel to an adjacent panel
   d. perpendicular to an adjacent panel

2. To function properly, a continuous wall must be supported on _____.

   a. a series of pilasters
   b. a series of columns
   c. a continuous foundation
   d. a grouted, reinforced foundation

SECTION TWO

2.0.0 ARCHES

Objective

Describe the use of masonry arches.
   a. Identify common types of arches.
   b. Explain how to calculate and lay out common arches.
   c. Explain how to construct formed arches.
   d. Explain how to construct a jack arch.

Performance Tasks

Lay out each of the following arches:
- Semicircular arch
- Segmental arch
- Gothic arch
- Multicentered arch

Build one of the following arches:
- Semicircular arch
- Segmental arch
- Gothic arch
- Multicentered arch

Trade Term

Arch: A form of construction in which a number of units span an opening by transferring vertical loads laterally to adjacent units and thus to the supports.

The appeal of an arch lies in its diversity of forms. It may be used to express balance, unity, proportion, scale, rhythm, sequence, and character. The arch is not only a decorative structure; it is also an important structural member and can be used to support large loads while allowing an opening through the structure.

The structural efficiency of the arch is due to the fact that most of the arch will be in compression under normal loading conditions. This is important because many materials have a greater resistance to compression than tension. For example, brick masonry may be 20 times stronger in compression than in tension. Brick arches can therefore be useful in both a structural sense and a decorative sense.

Arches are classified into two groups, depending on span and loading:

- Minor arches have a maximum span of 6 feet and a rise-to-span ratio less-than or equal-to 0.15. These arches are most often used in building walls and typically function as lintels over openings. The maximum equivalent uniform load does not exceed 1,000 pounds per foot.
- Major arches have spans or loadings that exceed the maximum for minor arches. That is, spans are greater than 6 feet or loads are greater than 1,000 pounds per foot, with a rise-to-span ratio greater than 0.15.

Many arch styles have been developed over the years. Some have names such as the Gothic, Tudor, and Venetian arches. Others are known by their shapes, such as jack, semicircular, segmental, and multicentered arches. The jack arch (*Figure 5*) is a flat arch. The semicircular arch (*Figure 6*), as its name suggests, has a curve that is a semicircle (one-half of a circle). The segmental arch (*Figure 7*) has a curve that is flatter than a semicircle. The multicentered arch (*Figure 8*) is a longer curve that is made up of several arcs. Generally, the jack, segmental, and multicentered arch styles are used for minor arches. The semicircular arch is often used for major arches since it is the most structurally efficient. In addition, there are

28302-14_F05.EPS

Figure 5 Jack arch.

28302-14_F06.EPS

Figure 6 Semicircular arch.

28302-14_F07.EPS

Figure 7 Segmental arch.

28302-14_F08.EPS

Figure 8 Multicentered arch.

other types of arches that have specific names. The Gothic arch (*Figure 9*), also called the pointed arch, has a relatively high rise whose sides consist of arcs or circles. The bull's-eye arch (*Figure 10*) is a complete circle.

2.1.0 Understanding Arch Design

Due to the nature of its construction, the arch exerts not only a downward pressure, but also a strong sideways or outward pressure at its base. This tendency is called the thrust of an arch. In order for the arch to remain stable, this thrust must be resisted by abutments, buttresses, or the strength of the adjoining wall. How the arch is designed is directly related to the amount of thrust it must resist.

All masonry arches are fixed arches, and three conditions must be maintained to ensure pure arch action:

- The span length must remain constant.
- The elevation of the ends must remain unchanged.
- The angle of the skewback must be fixed.

If any of these conditions are violated during the arch layout or construction, critical stresses may develop, resulting in failure of the structure.

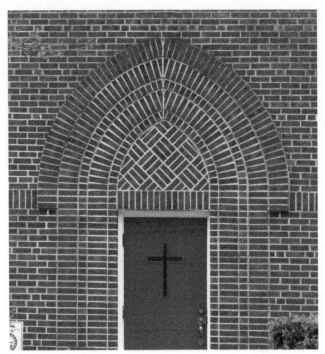

28302-14_F09.EPS

Figure 9 Gothic arch.

28302-14_F10.EPS

Figure 10 Bull's-eye arch.

To understand how arches are constructed, the various terms associated with them must first be understood first. *Figure 11* shows the various parts of an arch:

- *Abutment* – The structure that supports the arch. There is an abutment on either side of the arch.
- *Arch axis* – The median line of the arch ring.
- *Camber* – The relatively small rise of a jack arch (not shown).

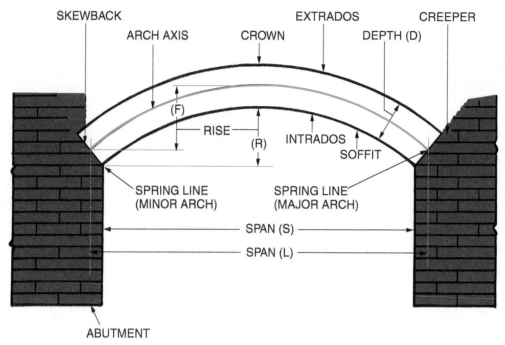

SKEWBACK   EXTRADOS   CREEPER

ARCH AXIS   CROWN   DEPTH (D)

(F)

RISE

(R)

INTRADOS

SOFFIT

SPRING LINE (MINOR ARCH)   SPRING LINE (MAJOR ARCH)

SPAN (S)

SPAN (L)

ABUTMENT

28302-14_F11.EPS

Figure 11 Parts of an arch.

- *Creepers* – The brick in the wall adjacent to the arch that are cut to conform to the curvature of the extrados.
- *Crown* – The top of the arch ring; in symmetrical arches the crown is at midspan.
- *Depth* – The minimum dimension (D) across an arch. This is perpendicular to the tangent of the arch axis.
- *Extrados* – The curve that bounds the upper edge of the arch.
- *Intrados* – The curve that bounds the lower edge of the arch.
- *Keystone* – The center unit placed at the crown that wedges the arch in place (not shown).
- *Rise* – The rise (R) of a minor arch is the maximum height of the arch soffit above the level of its spring line. The rise (F) of a major parabolic arch is the maximum height of the arch axis above its spring line.
- *Skewback* – The inclined surface on which the arch joins the supporting wall or abutment.

- *Soffit* – The undersurface of the arch.
- *Span* – The horizontal dimension between abutments. For a minor arch, the span is taken as the clear span (S) of the opening used. For a major parabolic arch, the span (L) is the distance between the ends of the arch axis at the skewback.
- *Spring line* – For minor arches, the line where the skewback cuts the soffit. For major parabolic arches, the term refers to the intersection of the arch axis with the skewback.

2.2.0 Laying Out an Arch

This section covers the recommended procedures for laying out four of the most common arches: semicircular, segmented, Gothic, and multicentered. Other types of arches can be constructed using similar procedures.

A Brief History of Arches

Arches are among the oldest decorative structures known. Discoveries dating as far back as 1400 BC show that craftspersons were able to design and build arches that not only withstood the required loads but were attractive as well. The earliest arches were built by the Babylonians, the Chinese, and the Egyptians, but it was the Romans who developed and expanded their use. The semicircular arch, still known as the Roman arch, was often used in Roman architecture.

2.2.1 Semicircular Arches

In order to lay out the curvature of a semicircular arch, the arch's span and rise must be known, and the radius must be found in order to mark the curve on the wood form.

In a semicircular arch, the rise equals one-half of the span (*Figure 12*). That is because the rise is the same as the radius, while the span is the same as the diameter. Therefore, to find the rise of a semicircular arch if the span is known, simply divide the span by one-half. To find the span of a semicircular arch if the rise is known, multiply the rise by 2.

2.2.2 Segmental Arches

In order to lay out the curvature of a segmental arch, the arch's span and rise must be known, and the radius must be found in order to mark the curve on the wood form (*Figure 13*). The calculations required for determining the radius are given later in this section.

Multiplying one-half of span A by itself and dividing by the rise B will give the distance C. To obtain the radius, add B and C and divide by 2. An example problem will help to illustrate these calculations.

Find the radius of an arch with a span of 4 feet (48 inches) and a rise of 12 inches. Use the following steps to find the answer:

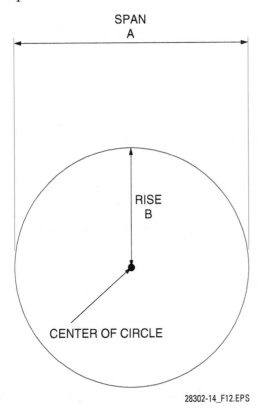

SPAN
A

RISE
B

CENTER OF CIRCLE

28302-14_F12.EPS

Figure 12 Determining the radius of a semicircular arch.

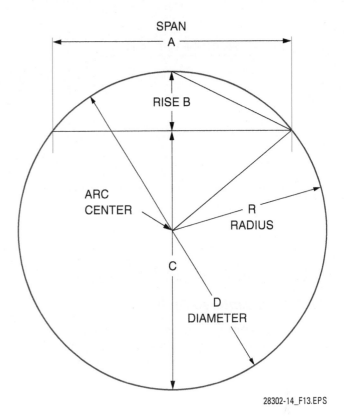

SPAN
A

RISE B

ARC
CENTER

R
RADIUS

C

D
DIAMETER

28302-14_F13.EPS

Figure 13 Determining the radius of a segmental arch.

Step 1 Take one-half of the span:

$$48 \times \frac{1}{2} = 24 \text{ inches}$$

Step 2 Multiply one-half the span by itself:

$$24 \times 24 = 576 \text{ square inches}$$

Step 3 Divide by the rise:

$$576 \div 12 = 48 \text{ inches}$$

Step 4 Determine the radius:

$$(48 + \text{rise}) \div 2 =$$
$$(48 + 12) \div 2 = 30 \text{ inches}$$

The radius of an arch is half the spring line. Some skilled masons may choose to correct the optical illusion of flatness in semicircular arches by raising the radial center point, or the exact center of the circle, 1 or 2 inches above the spring line.

2.2.3 Gothic Arches

To lay out the curvature of an equilateral Gothic arch, determine the spring line of the arch (see *Figure 14A*). The layout will use the entire length of the spring line as the left radius and the right radius points. Trace the left and right radius lines until they intersect (see *Figure 14B*). These lines represent the shape of the arch.

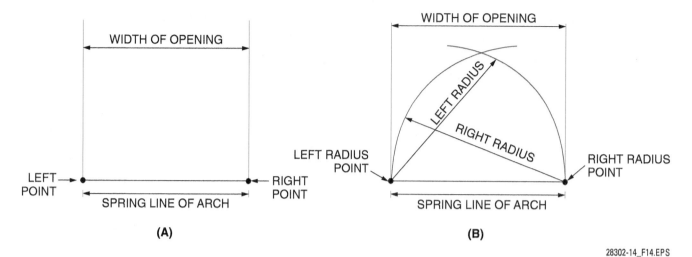

WIDTH OF OPENING

LEFT POINT → ← RIGHT POINT

SPRING LINE OF ARCH

(A)

WIDTH OF OPENING

LEFT RADIUS POINT

LEFT RADIUS

RIGHT RADIUS

RIGHT RADIUS POINT

SPRING LINE OF ARCH

(B)

28302-14_F14.EPS

Figure 14 Determining the radius of a Gothic arch.

2.2.4 *Multicentered Arches*

Layout of a multicentered arch is based on the amount of rise available within the masonry opening and the desired left and right radius points at the ends of the spring line. To lay out a multicentered arch, see *Figure 15* and follow these steps:

Step 1 Determine the radius of the center arch. The radius will be determined from below the spring line and will typically extend beyond the left and right radius points.

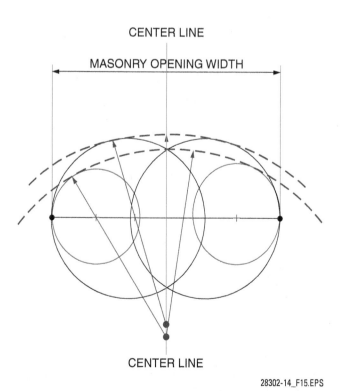

CENTER LINE

MASONRY OPENING WIDTH

CENTER LINE

28302-14_F15.EPS

Figure 15 Determining the radius of a multicentered arch.

Step 2 Starting from the arc center of the center arch, draw a line that represents the point where the change in arc radius is desired. This point should not require the end radius to extend above the center arc.

Step 3 The intersecting point will become the focus of the end arc that will terminate at the left and right parts of the spring line.

2.3.0 Constructing a Formed Arch

Masonry arches are fixed arches. To function correctly, the span length must remain constant, the elevation of the ends must remain unchanged, and the angle of the skewback must be fixed. Otherwise, critical stresses may develop and cause the structure to fail.

The semicircular arch is the strongest of all masonry arches because the loads are evenly distributed over the entire area of the arch. Although arches can support a load, they cannot, at least in the initial stages of construction, support themselves. For this reason, a curved wood form must be built to support the arch until it has cured. This form is called the arch template.

2.3.1 *Constructing the Arch Template*

For arches up to a span of 6 feet, a good construction grade of ¾-inch plywood is sufficient for the front and back of the arch template. Two pieces of plywood are cut to the specified curve. Then, 2 × 4 wood pieces are cut and spaced between them as spreaders that serve to hold the form together. Note that the horizontal base of the template is precisely the same as the span of the arch.

A piece of ¼-inch plywood is cut and fit over the top, or over the part of the arch template that the brick will lie on. The ¼-inch plywood bends easily over the curved form without breaking. A sufficient number of 2 × 4 pieces should be spaced between the front and back to prevent the ¼-inch plywood from sagging.

The purpose of the arch template is to support the dead load of the arch and the wall above the arch until the masonry has cured and is able to carry the weight. The required curing time depends on the size of the arch, the weight it must support, and the curing conditions. For minor arches, three days is usually enough time if the temperature is not freezing and the mortar is setting correctly. Major arches, which are larger, should not be disturbed or have the forms removed before 10 days have passed, if conditions are the same. These allowances are not, however, absolute rules. The most important considerations are the prevailing weather conditions and the setting time of the mortar.

After the arch template is constructed and brick piers are built to the height where the arch is to begin, the arch template is set in place. The arch template is supported by a pair of legs made from 2-inch lumber and placed against the abutment (*Figure 16*). The legs should be plumb and level and cut about ½ inch to 1 inch shorter than the actual height of the opening.

Four wood wedges are laid on the top of the legs, two on each, to aid in the final leveling of the arch. The arch template is set on top of the wedges and adjusted into the proper position by tightening up on the wedges. Care must be taken during this process so that when finished, the arch template is at the correct height, level and plumb with the face of the brickwork. Appropriate braces

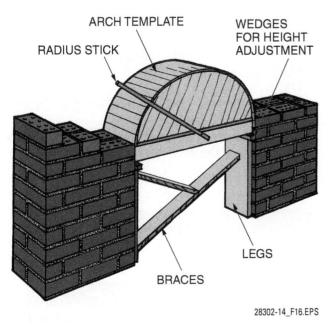

ARCH TEMPLATE

RADIUS STICK

WEDGES FOR HEIGHT ADJUSTMENT

LEGS

BRACES

28302-14_F16.EPS

Figure 16 Wooden arch template set in place.

can be fastened to the form to hold the arch center securely in place.

CAUTION

Leave the arch template in place until the mortar has set. If the support is removed before the mortar is set, the masonry units may fall.

2.3.2 Marking Arch Brick on the Template Face

Specifications may require that specially shaped arch brick be ordered from a manufacturer, or they may require that the brick be cut. If ordered, the brick will be sent separately and carefully marked. If cut, standard brick of uniform shape is used, and joint thickness is varied to obtain the desired curve.

The brick spacing is marked off with a mason's rule. This process is completed by bending the rule or modular steel tape to fit the curve of the bottom of the arch template and marking on the face of the form with a sharp pencil. The course-counter rule should be used because it allows any differences in the joint thickness to be adjusted evenly. By marking on the arch template, the placement of each brick is established.

The height of each arch brick course is difficult to determine because the use of a mason's line is not possible. To aid in this step, a radius stick (*Figure 16*) is attached at the arch center point with a finishing nail. As the arch is laid, swing the radius stick and it will indicate the exact position of the top of each arch brick, eliminating any guesswork. To be of value, the radius stick must extend to or past the top of the arch.

2.3.3 Laying the Arch Brick

Construction of an arch always begins at the two ends, or piers, and the brick are laid up to the center, or key, of the arch. Select the best edge of the brick (the one that is straightest and unchipped) to lie on the form in order to obtain a neat appearance. The mortar should be mixed to a consistency that permits the brick to be laid easily without movement as other brick are placed. Since the bottom of the arch rests completely on the wood form, the mortar joints must be cut out and refinished to obtain a finished appearance. Care must be taken in brick placement to avoid knocking the previously laid brick out of position and weakening the bond. The brick are always laid so that the masonry units are centered at the top of the arch.

Arch brick can be laid in several positions and patterns. The number of course rings should be determined before beginning, as this affects each course layout. Determine if any half brick are required to center the masonry units at the top of the arch. If half brick are required, these should be the first units placed at the end of the arch.

One of the most popular bonding patterns used for jack arches consists of soldier courses, either stacked or bonded. The minimum depth of a soldier-course arch is often 12 inches, or a brick and a half.

The most popular bond arrangement for semicircular brick arches is either rowlocks or header course rings. Two or three course rings provide greater strength than a single course, and this method is therefore recommended. The shorter height of the rowlock or header brick allows them to be turned on a sharper radius. The advantage is a smoother curve with more uniform mortar joints.

The mortar joints should all be the same size and as small as possible to give a neat, attractive appearance. Also, a small joint such as a ¼-inch to ⅜-inch joint is less likely to shrink and absorb water. The mortar joints at the intrados and the mortar joints at the extrados should be about the same size. This uniformity is not difficult to achieve with the rowlock or header arch. With a soldier arch, the mortar joint will be noticeably larger at the outer edge of the arch unless specially tapered brick are used. The specifications should be consulted for specific job requirements.

The mortar joints should be tooled as soon as they are thumbprint-hard and still pliable, and the arch should be brushed when work is completed. It is a good practice to have a line attached on the bottom and center of the arch as a guide to make sure the arch does not bulge out of position. A straightedge such as a plumb rule can be used on a small arch to keep the center from bulging out. This check should be performed as the arch brick are installed. Adjustment will not be possible once the arch is complete.

Parging the back of the arch with mortar helps to prevent water from leaking through from the face of the wall. It also helps to strengthen the

arch and make it more secure. If an arch does not equal the full width of the masonry wall, it should be tied or bonded into the backing work with metal wall ties. The wall ties should be embedded, every three or four courses, at least half the thickness of the width of the brick in the facing of the arch and the backing materials.

2.3.4 Removing the Arch Template

After the arch has cured, the arch template should be carefully removed. Removal is accomplished by gently driving out the wedges holding the form in place. When doing this, take care not to chip the bottom of the arch brick.

If excess mortar falls against the template, use a special chisel, called a plugging chisel, to cut out the joints. The plugging chisel (*Figure 17*) is made with an angled blade on the end so it does not chip the edge of the brick, but cuts out only the mortar joint. The special tapered blade cleans mortar from the joints easily, without binding in the joint.

If needed, the mortar joints should be cut out to a depth of ½ inch. Then, with a brush and bucket of potable water, dampen the mortar joints and repoint with fresh mortar. Dampening the joints prevents them from drying too rapidly and curing improperly. The same mortar mix that was originally used to build the arch must be used for repointing.

The following are important points to remember when building semicircular arches:

- Select brick that are approximately the same size before laying any of them on the form.
- Choose brick free of chips or cracks.
- Select brick that matches well with the color of the structure.
- Use dry brick because they are easier to hold in place on the arch template.
- Use well-filled mortar joints.
- Mark all spacings on the arch template before laying any of the brick.
- Work from each edge at the abutments up to the crown of the arch.

Neatness Counts

Before laying brick on the arch template, clean off any chips or nails protruding from the construction of the carpentry work. Brush off the form with a fine brush to remove all chips.

28302-14_F17.EPS

Figure 17 Plugging chisels.

- If the arch has a keystone, mark this off first. Then the arch brick must be laid out between the keystone and the bottom of the arch.
- Always try to make the top of the arch work with the stretcher course of brick in the wall. It is considered better workmanship to have the stretcher course of brickwork over the top of an arch without having any splits of brick.
- Use a range line across the bottom and center of the arch to make sure the arch has not bulged out of position.
- To avoid excessive mortar in the base brick, put a cotton wick or bond-breaking materials in the joints along the template.
- Provide enough wall ties to tie the arch and backing work together.
- Do not remove the arch template until the arch has cured enough to carry its own weight without sagging.
- If needed, clean out the mortar joints at the bottom of the arch and repoint with mortar.

Building arches requires superior workmanship in addition to an understanding of the principles of arch construction. Following the practices and techniques outlined here will help you successfully build semicircular arches.

> **WARNING!**
> Observe good safety practices by wearing eye protection when mixing mortar or cutting masonry units for arches. Mortar dust or flying masonry chips can cause severe damage to eyes.

2.4.0 Constructing a Jack Arch

A jack arch is a flat arch often used over small openings such as windows or doors. Because it has no curvature, it is the weakest of all arches. Normally, if the opening is more than 2 feet, the arch will have to be supported by steel angle irons. Unlike other types of arches discussed in this module, the jack arch does not require the use of an arch template.

The top and bottom of the arch brick must be in line with two things: the angle iron at the bottom, and a line at the top. Since the bricks are laid on a slant, they must be cut with a masonry saw to make them level with the iron and line. Specially molded brick with a slight taper are recommended so that a uniform mortar-joint thickness is possible.

If a long jack arch is laid perfectly horizontal, it will appear to sag in the middle. Sagging is only an optical illusion and can be corrected by having a slight camber (upwards curvature) in the angle iron. Of course, the longer the jack arch, the more evident the sag and the more need for increased camber.

Figure 18 shows a common and a bonded jack arch. A common jack arch is one full brick in height, laid in a soldier position. The arch is commonly laid with cut pieces (rather than full-length pieces) so mortar joints can be staggered. It is the simplest type, easiest to install, and is mostly decorative.

A bonded jack arch is laid out and built the same way as a common jack arch except that it is either a brick and a half or two brick in height. To make the arch stronger, the brick is bonded by staggering the mortar joints, similar to a brick wall. Unlike the wall, however, the arch brick is laid in a vertical position and careful attention must be given to joint alignment. Since the height of the bonded arch is greater, the brick skewback on each side is larger, usually a full four courses.

(A) COMMON

(B) BONDED WITH KEYSTONE

28302-14_F18.EPS

Figure 18 Common jack arch and bonded jack arch with a keystone.

Bonded jack arches can also have a keystone in the center for architectural effect and design. If the arch has a keystone, the stone should be centered in the arch and marked on the frame. Allow for mortar joints on each side and then lay the brick to that point.

2.4.1 Laying Out a Jack Arch from the Radial Center Point

Before laying a jack arch on the angle iron, determine the inclination of the skewbacks. The inclination may be determined by locating the radial center of the arch or the center at the bottom of the arch opening. Once this point is located, a small nail is driven at the center point and a line attached as shown in *Figure 19*.

The line is then held from the radial center to the edge of the opening at the point where the jack arch starts on the angle iron. The line can be attached to a wood board, and the brick skewbacks can be cut with either a sharp brick chisel or a brick saw to the correct angle.

2.4.2 Marking Off Spacing of the Arch Brick

When specified, arches are built with full brick. The units to the sides of the arch are cut, rather than the arch brick. It is also important that the brick is evenly spaced in the arch. Measure and mark the arch before the brick is laid. Mark the spacing of each brick on the bottom of the angle iron and the top of the arch.

One method for making these marks is to lay a wood board (2 × 4 or 2 × 6) on top of the skewbacks so that it spans the opening, as drawn in *Fig-*

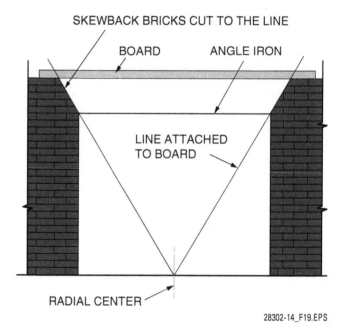

28302-14_F19.EPS

Figure 19 Locating the radial circle of a jack arch.

The George Washington Masonic National Memorial

The George Washington Masonic National Memorial is located in Alexandria, Virginia. Construction began in 1922 and was completed a decade later, in time for the bicentennial of George Washington's birth. The giant masonry structure, designed to resemble the ancient lighthouse of Alexandria in Egypt, is dedicated to the nation's first president, who was also a member of a Masonic lodge. It features a 17-foot-tall bronze statue of Washington and murals depicting events from Washington's life. The memorial is a popular tourist destination, and features a museum, research center, and a library. Its facilities include a performing arts center, concert hall, and banquet facility.

28302-14_SA01.EPS

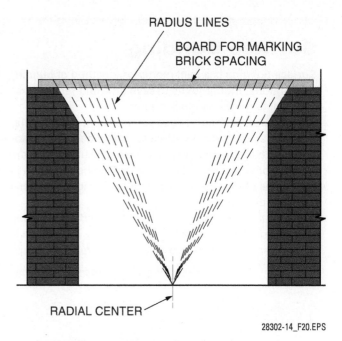

RADIUS LINES

BOARD FOR MARKING BRICK SPACING

RADIAL CENTER

28302-14_F20.EPS

Figure 20 Determining brick spacing for a jack arch.

What Does the Term *Jack Arch* Mean?

"Jack" is an old slang term that means "fake" or "false." A jack arch is called that because, while the brick is laid out to suggest a curvature like other types of arches, the construction is actually flat.

ure 20. Use this board to mark the top of the arch brick. Attach a line to the radial center of the arch. Use this line to mark the location of each brick on the angle iron and the top of the board. Remember that the center of the arch must be the center of the keystone. Preplan the work so that the correct markings are made from each side to the center of the arch. It is very important to mark the top and bottom of the arch to maintain the correct angle.

2.4.3 Laying the Jack Arch in Mortar

After the jack arch has been laid out on the board and angle iron, three different lines will be used to build the arch. Once the wall is built and the skewbacks installed, one line is attached at the bottom of the arch and across the face. Another line is used at the top edge to keep the top of the arch in line and as a guide for making the angle cut for the top of the brick. Finally, a range line is used across the center of the arch to make sure the arch does not bulge out of position.

After the lines are attached, the mortar should be mixed so that it is stiff enough to squeeze out when laid, but without too much pressure on the brick. Only solid mortar joints should be used. It is best to keep the mortar joints relatively small. Small joints are more attractive and allow the work to set up quickly.

The arch brick should be laid from the ends to the center of the arch. The spacing marks on the

board and on the angle iron must be checked often or the brick alignment may be lost. It is also a good practice to check the radius line at intervals as the work progresses. Check all work frequently to ensure that consistency is maintained.

When the last brick is laid in the center of the arch, both ends of the brick should be buttered in place. Both ends of the keystone should also be buttered to make sure the mortar joint is well filled and a complete bond established. The joints should be tooled as soon as they are thumbprint-hard, and then the work should be brushed. When the work is finished, any holes under the bottom of the arch should be pointed up neatly with mortar; a slicker does this job well.

Though jack arches are the easiest of all arches to construct, workmanship is as essential here as it is with any masonry system.

Surface Bonding

Surface bonding is a fast, mortarless method of building with concrete block. Walls may be erected without conventional mortar joints to save time and reduce cost. Surface bonding may also be used as a refinishing coat over existing brick or block masonry.

In a typical surface-bonded construction, only the first course of masonry units is embedded in mortar. Sometimes the top course is also embedded. The masonry units are dry-laid. A surface-bonding cement is applied either by hand trowel or spray. Once the coating has dried, the walls have a structural strength equal to that of conventionally mortared walls. A surface-bonded wall is seamless and water resistant. The coating is applied to both sides of the wall, which provides complete weather protection both inside and out. Normally, no other wall finish is required.

The wall surface can be smooth, textured, or patterned, and colored to suit architectural requirements. Surface-bonded mortar can be finished in a variety of textures by using sponges, paint rollers, brushes, or various types of floats. Test panels should be made when developing special finish textures.

Additional Resources

Bricklaying: Brick and Block Masonry. 1988. Brick Industry Association. Orlando, FL: Harcourt Brace & Company.

Concrete Masonry Handbook: for Architects, Engineers, Builders, Fifth Edition. 1991. W. C. Panerese, S. K. Kosmatka, and F. A. Randall, Jr. Skokie, IL: Portland Cement Association.

2.0.0 Section Review

1. The strong sideways or outward pressure exerted by an arch at its base is called the arch's _____.

   a. thrust
   b. force
   c. buttress
   d. spring line

2. The type of arch in which the rise equals one-half the span is the _____.

   a. segmental arch
   b. semicircular arch
   c. multicentered arch
   d. Gothic arch

3. When using stacked or bonded soldier courses on a jack arch, the minimum depth of the arch is often _____.

   a. 12 inches
   b. 16 inches
   c. 20 inches
   d. 24 inches

4. In a jack arch, brick should be laid _____.

   a. from the inner course to the outer course of the arch
   b. from the outer course to the inner course of the arch
   c. from the ends to the center of the arch
   d. from the center to the ends of the arch

SECTION THREE

3.0.0 ACID BRICK

Objective

Describe applications for acid brick.
  a. Explain uses of acid brick.
  b. Identify acid brick materials.
  c. Describe how to lay an acid brick floor.

Trade Term

Acid brick: Masonry units that are manufactured with special properties to withstand harsh chemical environments without disintegrating.

A technical but more accurate name for acid brick is chemical-resistant masonry. Chemical-resistant masonry units include block and stone as well as brick. However, brick is by far the most common. Acid brick are masonry units that are able to resist chemical exposure.

Acid brick construction is a highly specialized part of masonry. All work is done to close tolerances, and the workmanship must be very precise. The products are expensive and require special construction techniques. In addition, the work requires special mortars and mastics that must be handled with extreme care and attention to safety precautions.

3.1.0 Using Acid Brick

Acid brick is used to serve one of two purposes: to create a loadbearing structure that is capable of withstanding a chemical environment or to provide a lining that helps protect another structure by limiting chemical attack. Acid brick structures do not resist tension well, but are capable of carrying heavy compressive loads. Examples of loadbearing acid brick structures are freestanding brick linings of chimneys at refineries and ammunition manufacturing facilities.

Acid brick linings are more common than loadbearing acid brick structures. When used as a lining, acid brick protects the structure behind it from blows or abrasion. It reduces the temperature or limits chemical penetration to a level the backup surface can accept. For example, rubber linings are used in pickling tanks in steel mills. The pickling solution used in these tanks often exceeds a temperature of 210°F. The rubber linings, however, cannot withstand temperatures

above approximately 160°F. The installation of a 4- to 8-inch-thick acid brick lining over the rubber can lower the surface temperature of the rubber to about 140°F. In this application, the acid brick protects the rubber membrane from thermal exposure. It also protects it from bumps and scrapes caused by steel objects placed in the tank.

Special chemical-resistant mortars are required for acid brick structures to perform satisfactorily. Even then, the structures cannot completely eliminate chemical penetration. While they may effectively limit chemical penetration, they should never be expected to resist it completely.

Chemical-resistant masonry structures are excellent in compression and may be employed in loadbearing design. They are weak in tension because they cannot be reinforced. Therefore, if tensile forces are anticipated, support the acid brick with another material. Generally, this will be a steel or concrete substrate. This provides a type of reinforcing, but it is substantially different than the reinforcing required of concrete structures. The acid brick structure becomes a lining, and the substrate is actually a backing wall.

3.2.0 Identifying Acid Brick Materials

Acid brick structures have four parts: the chemical-resistant masonry units, membrane materials that make the system liquid-tight, mortars to bond the masonry units, and the external support structure of concrete or steel. Each of these four parts must be properly designed and constructed in order to achieve the desired performance. *Figure 21* shows a cross section of a floor requiring acid brick construction. If any one of the four parts fails, the entire system fails and must be rebuilt.

3.2.1 Masonry Materials

The techniques of acid brick manufacturing are somewhat similar to those for common brick. Common brick is a molded or extruded clay-and-water mixture which is then fired or baked at sufficient temperatures to drive out the water and harden the brick. All common brick has a high water-absorption rate, typically from 8 to 15 percent, and may be destroyed by strong acid or alkali.

Acid brick is also made from a clay-and-water mixture, but one that has few acid-soluble components. Some acid brick is manufactured using nonclay components such as alumina, silica, zircon, and other minerals. These are fired at higher temperatures than common brick, and for a longer period of time. This produces an exception-

 28302-14 Specialized Materials and Techniques   Module Two   15

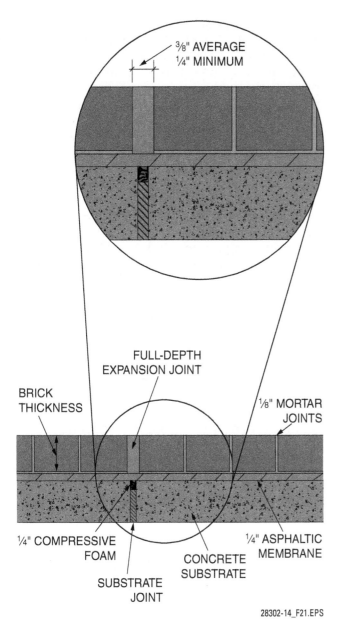

Figure 21 Acid brick floor.

ally hard brick that will not shrink and has a low water-absorption rate, often as low as 1 percent. The most common types of acid brick are the following:

- *Red-shale brick* – Red-shale brick is available in the same modular sizes as common brick (*Figure 22*). It is most often used in chemical-resistant systems. Red-shale brick has the lowest absorption rate, around 1 percent. It is also exceptionally brittle. Therefore, use it in systems that require low absorption rates, such as process vessels. Do not use it where it will be subjected to thermal shock.
- *Fireclay* – Fireclay brick usually has a higher absorption rate than red shale, about 5 percent. It is less brittle. Use it for outdoor structures

where rapid thermal changes are likely. The standard size of fireclay brick is $9 \times 4\frac{1}{2} \times 2\frac{1}{2}$ inches or $9 \times 4\frac{1}{2} \times 3$ inches. Other sizes and special shapes are also available. This brick is more dimensionally true than red shale, making uniformity of joints possible.

- *Carbon brick* – Carbon brick is not used as often as red shale or fireclay. It is used in areas exposed to strong alkali and hydrofluoric acid, especially when continuous wetting occurs. Carbon brick has higher absorption rates than red shale or fireclay. It is more shock resistant. Use it where pressure changes are rapid. A disadvantage of carbon brick is its high rate of heat transfer. If used as a lining for thermal insulation, its thickness must be greater than that of red shale or fireclay.

3.2.2 Membranes

Membranes act as barriers between the chemical-resistant masonry units and the supporting substrate. Acid brick linings are not, in themselves, liquid-tight. Therefore, they require a membrane to fully protect the concrete or steel substrate.

Sometimes a substrate acts as its own membrane. Steel often functions in this manner. The acid brick do not completely protect the steel from corrosive chemicals, but they do limit the exposure to a level that should not damage the steel. More often, however, even steel requires some sort of membrane. Membranes fall into two classes: true membranes and semimembranes.

A true membrane is a total barrier that allows no penetration to the substrate. A semimembrane provides some penetration, but the amount is limited to what the substrate can handle.

Natural rubber or neoprene sheets are commonly used for steel substrates. Hot asphalt or mastic membranes are the most popular for concrete. Mastics do not bond well to impervious surfaces and should not be used over a steel substrate. Other membrane materials are also available. Whatever the material, follow the manufacturer's instructions exactly when applying membranes.

3.2.3 Mortars

Effective acid brick structures require chemical-resistant mortars. These mortars will not completely resist chemical penetration, but will retard it to acceptable levels. Chemical-resistant mortars have been around since the early part of the twentieth century. However, most were not especially effective until the development of synthetic materials.

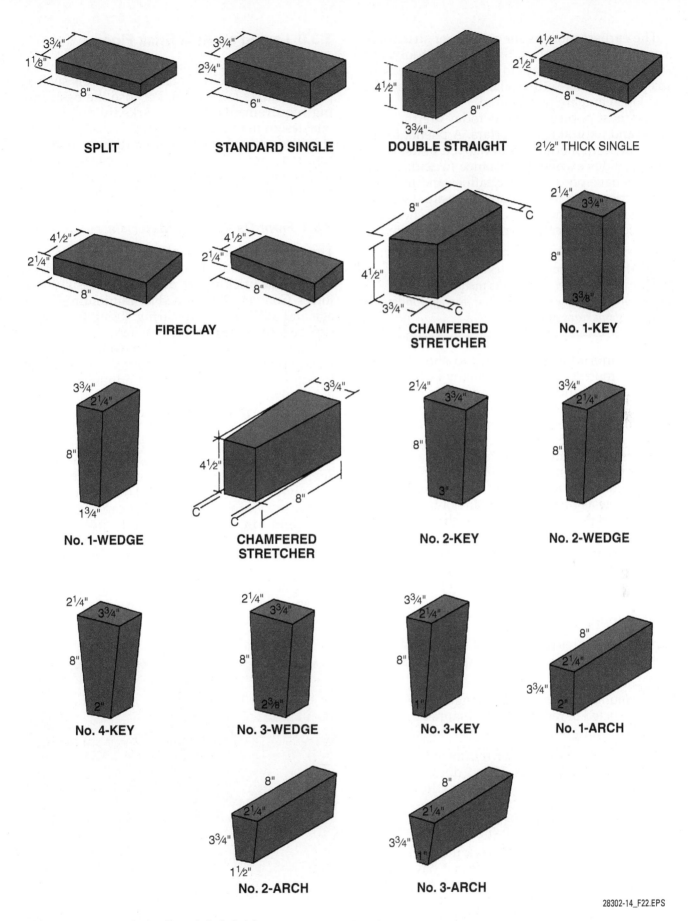

SPLIT   **STANDARD SINGLE**   **DOUBLE STRAIGHT**   2½" THICK SINGLE

FIRECLAY   **CHAMFERED STRETCHER**   **No. 1-KEY**

No. 1-WEDGE   **CHAMFERED STRETCHER**   **No. 2-KEY**   **No. 2-WEDGE**

No. 4-KEY   **No. 3-WEDGE**   **No. 3-KEY**   **No. 1-ARCH**

No. 2-ARCH   **No. 3-ARCH**

28302-14_F22.EPS

Figure 22 Common shapes for red-shale brick.

The earliest mortars for acid-proof structures were silicates. However, this material took anywhere from 30 to 90 days to set, which proved unacceptable. Faster-setting silicates have since been developed and are widely used.

The most popular mortars today are the synthetic and natural resin mortars. A phenol-formaldehyde resin mortar was developed in the 1930s and provides excellent resistance to acid. These mortars have poor storage qualities and must be kept refrigerated. More troublesome are the allergic reactions that they may cause. The reactions are rarely serious, but can be irritating on a job site.

> **WARNING!**
>
> These resin mortars contain hazardous chemicals. Always read the manufacturer's safety data sheets (SDSs) and take appropriate personal precautions.

Furan mortars were developed shortly after phenolic mortars and exhibit many of the same chemical resistances. They offer a long shelf life, which makes storage easy. Normally they do not cause allergic reactions. Furan mortars are often preferred for these reasons.

Polyester-resin and epoxy mortars are also used for acid brick construction. Polyesters provide excellent protection in most acid conditions, but are not as stable as furans, so they do not have a long storage life. Epoxy mortars offer much better bonding characteristics than other mortars, but are usually expensive. They were originally developed as toppings for concrete, but their great strength made them attractive for use with acid brick.

3.2.4 Support Structures

Support structures for acid-proof masonry should meet the following three basic requirements:

- Liquid-tight
- Continuous and formed in a single piece
- Constructed or reinforced in a manner that prevents bending

For these reasons, steel or reinforced concrete make the best substrates. Wood does not work well. Acid brick structures are weak in tension, so wood flooring, which bends and flexes under even light loads, will normally fail. Further, wood swells when wet and shrinks when dry. This causes warping that would crack the brickwork. Concrete and steel, however, do not have these problems. They may be expected to remain rigid and strong, and are easily formed in a continuous piece. Reinforced fiberglass vessels can also be used in certain applications.

3.3.0 Laying an Acid Brick Floor

Acid brick linings are used for structures such as floors, trenches, and containment vessels such as process tanks (*Figure 23*). Because the basic laying requirements do not markedly differ from one design to the next, this section discusses construction only. A vessel with a dish-shaped bottom may not require expansion joints, but the basic components of masonry units, membrane, and mortar will almost always be necessary.

3.3.1 Floor Design and Materials

The preferred floor substrate is reinforced concrete. A steel-plate floor can be expected to vibrate. Acid brick is not designed to resist these forces. Concrete, if designed for minimum shrinkage and sufficient loadbearing strength, may be expected to provide a stable surface.

Ideally, the concrete substrate would be continuously poured and sloped about ½ inch per foot to provide positive drainage. Continuous pouring is not always possible, but a watertight surface can be achieved if the floor joints are designed with water stops.

Before applying the membrane and masonry, make sure that the concrete surface is uniform. Screeding and float-finishing the surface will usually achieve uniformity. After the concrete hardens, a 72-inch straightedge can reveal high spots and low spots. Grind high spots and fill spots lower than ¹⁄₁₆ inch below the straightedge.

As a final check, pour potable water over the floor slab and allow it to drain. Standing puddles are not acceptable and should be corrected before laying proceeds. The floor surface must be absolutely dry before the membrane is applied. If the

28302-14_F23.EPS

Figure 23 Laying acid brick in a process tank.

NCCER – *Masonry Level Three* 28302-14

surface is flooded, allow sufficient drying time or use other methods to eliminate the moisture.

3.3.2 Membrane Selection and Application

The importance of the membrane in acid-proof structures has already been stressed. The membrane, not the masonry, is the final line of resistance to chemical penetration.

The most common membrane in floor construction is a three-part asphaltic membrane ¼ to ⅜ inch thick. Cover the substrate with a primer, which is the first part of the membrane. Follow that with an oxidized asphalt that contains no fillers. This has a softening point in the 219°F to 230°F range. It is applied hot, around 350°F, to the substrate.

Use a wood or metal squeegee to apply it. Make several applications until a thickness of ⅛ inch is reached. Then, overlay it with a layer of fiberglass fabric. Roll the fabric smoothly into place with edges overlapping 2 inches. Spread another application of asphalt over the fiberglass fabric to a thickness of ⅛ inch.

Apply the hot asphalt only with squeegees. Do not use a mop because the application will not be as smooth as that of a squeegee. Also, mop strings are often left in the membrane and will act as wicks that can conduct acid all the way through the membrane.

> **WARNING!**
> Always wear the appropriate personal protective equipment when working with hot materials. This includes coveralls, tight-fitting gloves, goggles, and a face shield.

3.3.3 Brick Selection

Shale or fireclay is normally used for acid-brick floor construction. Other types of brick have been tried, but they have disadvantages with regard to cost and absorbency. Shale and fireclay brick are quite economical and also have very low absorption rates, about 1 and 5 percent, respectively.

The choice between shale and fireclay will often depend on such factors as cost, availability, and module size. However, in outdoor applications where freeze-thaw effects or other extremes of weather are expected, fireclay brick will perform better because it is less brittle. If the absorption rate is a factor, as in a meatpacking house that needs a nonskid surface, then shale will be the standard selection. As with all material selection, follow the manufacturer's recommendations.

3.3.4 Mortar Selection

Special mortars are used with acid brick. Always make sure that the brick and mortar are compatible. Each type of mortar has a special chemical composition that is best suited for a particular type of brick. These mortars are chemical resistant in varying degrees. Some can also withstand temperatures up to 3,500°F. Check the manufacturer's SDS to determine the properties of the brick and the recommended mortar.

All mortars should possess a fine grain size and smooth workability. This is a necessity in good brick construction requiring thin joints. Mortars come dry-mixed in 100-pound bags and only need water. They are also available in wet-mixed form in 50-, 100-, or 200-pound pails or drums. Wet-mixed mortars are ready to use but may have a limited shelf life.

3.3.5 Expansion Joints

There are basically three types of expansion joint materials used with acid brick linings:

- Elastic but incompressible materials such as rubber are used where a sliding, rather than compression/expansion, movement is expected.
- Deformable but incompressible materials such as flexible epoxy can change shape without exerting pressure to return to its original shape. Joints of these materials are popular in acid brick floor design. All acid brick expands because of absorption and temperature change. A flexible epoxy can accommodate movement without pushing back at the brick. An excess of the joint material will appear above the brick, and this may be trimmed without damaging the joint.
- Compressible and recompressible materials such as foam rubber and vinyl sponge are preferred when a protruding joint is not permitted.

Floor designs usually require a deformable but incompressible material ⅜ inch in width. The joints should have straight vertical sides and should be completely filled from the top of the membrane to flush with the top of the brick. Normally, acid brick construction requires expansion-joint placement at the following points:

- At a maximum distance of 20 feet
- At points of movement in the substrate
- At the periphery of all floors, one brick in from the wall or curb
- Around all fixed objects and penetrations through the floor

Expansion-joint construction is probably the most important part of acid brick construction. If the joint is improperly constructed, the entire job may have to be rebuilt.

3.3.6 Laying the Brick

With a few minor differences, acid brick are laid much like conventional brick. Mortar should not be soupy, but of a firm consistency. Follow the manufacturer's instructions exactly. Fill the joints without voids, 1/16 to 1/8 inch thick.

Place most acid brick by buttering each brick on two sides and firmly pressing it into position. Sulfur mortar is occasionally used and requires a completely different procedure. The mortar is heated to between 260°F and 300°F, then poured around previously placed and spaced brick units. Manufacturer's directions must be followed carefully.

Cut acid brick with a masonry saw just as conventional brick, but do not cut it to less than one-half its dimension. Where a completed course would require less than one-half brick to finish, cut the brick next to it so that a larger piece can be used between them. This ensures that all brick meet minimum size requirements.

The following steps are recommended for laying acid brick:

Step 1 Examine the membrane. The fabric should be completely covered, with no surface cracks evident. Powdered silica may be dusted over the surface to keep it from sticking to workers' shoes. All silica must be swept from the surface before laying brick.

Step 2 Prewax the top surface of all brick with a water-soluble wax. Do not get the wax on the brick sides, as it will prevent the mortar from bonding properly to the brick.

Step 3 Mix the mortar in the exact proportions and manner indicated by the manufacturer. Do not add any water, aggregate, portland cement, or other foreign matter to the mix unless specifically authorized by the manufacturer. When the mix becomes too stiff to use, discard it. Do not attempt to reclaim or temper the mix for any reason.

Step 4 Spread the bed-joint mortar for two or three masonry units on the membrane. Flatten the mortar to a thickness of 1/8 inch and provide full coverage of all units.

Step 5 Completely butter two edges of a brick and set it into the bed-joint mortar. Tap and press the brick tightly against the adjoining brick, adjusting the level to provide a smooth surface and to maintain a 1/8-inch joint between adjacent brick. All joints should be completely full. The excess mortar should be squeezed out of the top of the joint and cut off with the trowel, leaving the floor surface smooth and clean.

Step 6 Keep the floor free from traffic and water until initial set has occurred. The temperature is normally maintained between 60°F and 90°F, but follow the manufacturer's recommendations as some mortars require different temperatures.

Step 7 After the floor is completed, remove the wax from the floor surface with hot potable water, leaving the exposed surface clean. If additional cleaning is required, use an appropriate cleaning solution per the manufacturer's recommendations.

Additional Resources

The Complete Guide to Masonry & Stonework. 2010. Minneapolis, MN: Creative Publishing International (video).

Corrosion and Chemical Resistant Masonry Materials Handbook. 1987. Walter Lee Sheppard. Norwich, NY: William Andrew.

3.0.0 Section Review

1. Acid brick structures do not effectively resist _____.

    a. distortion
    b. torsion
    c. compression
    d. tension

2. The two classes of membrane used with acid brick are _____.

    a. true membranes and semimembranes
    b. full membranes and partial membranes
    c. impermeable and semipermeable membranes
    d. steel and plastic membranes

3. A type of expansion joint material that can be used with acid brick linings when a protruding joint is not permitted is the _____.

    a. deformable joint
    b. elastic joint
    c. compressible joint
    d. incompressible joint

SECTION FOUR

4.0.0 REFRACTORIES

Objective

Describe applications for refractories.
  a. Identify refractory brick shapes and sizes.
  b. Explain how to lay refractory brick.
  c. Describe the curing and heat-up process for refractory brick.

Trade Terms

Grog: Burned, pulverized refractory material such as broken pottery or firebrick, utilized in the preparation of refractory bodies.

Plasticity: A complex property of a material involving a combination of qualities of mobility and magnitude of yield value; a material's ability to be easily molded into various shapes.

Refractory: A specialized masonry unit that can withstand high temperatures; used to form an insulating layer where extreme heat would damage other components of the structure. Refractories require special mortars that can also withstand high temperatures.

28302-14_F24.EPS

Figure 24 Refractory brick.

The term *refractory* describes any nonmetallic material or object that can withstand high temperatures without becoming soft. Refractories, also called refractory brick, are special masonry units that are used to line blast furnaces, crucibles for melting metals, and other places where resistance to temperature and corrosion is required. *Figure 24* shows various types of refractory brick. Over the years, the most widely used refractory material has been firebrick. This type of brick contains aluminum silicates and minor amounts of titanium and iron oxides. Other commonly used refractory substances are silica, magnesite, and graphite.

Manufacturers often speak of their product's refractoriness, or its ability to withstand high temperatures. This is measured using a numerical value. Pyrometric ceramic cones are used to determine this value in an American Society for Testing and Materials (ASTM) test. A series of cones, each with a different chemical composition, is numbered from 22 to 42. Each cone will soften and bend at a specified temperature, providing a convenient method of measuring the effects of time and temperature in the burning of ceramic products. It also provides

a quick means of judging the performance capabilities of refractory products. For example, in the standard ASTM test, if a refractory product softens and behaves in the same manner as Cone #3 (which has a value of 33), the product is said to have a pyrometric cone equivalent (PCE) of 33.

Metal industries have long made extensive use of refractory products to reduce iron ore to iron and to convert iron to steel. Most of their operations require structures that can last a long time while subjected to extremely high temperatures. Refractory products are used to line these structures. They are the principal means of withstanding the heat and corrosion these operations produce. Refractory products are also used for many other applications. The American space program, for example, makes extensive use of refractory material in its launchpads.

The information provided here is of a general and basic nature. New refractory products and installation procedures are constantly being developed. Therefore, it is very important to review the manufacturer's material information sheet before installing new materials and follow the suggested construction and curing procedures.

4.1.0 Understanding Refractory Brick

Refractory brick are the backbone of the refractory industry. Medium- and low-duty brick are used in barbecue grills (*Figure 25*) and fireplaces. High-duty brick are found in furnaces and even rocket launchpads.

Of the more than 100 elements in the Earth's crust, only a few have the ability to form stable refractory compounds. Silicon, aluminum, magnesium, calcium, chromium, zirconium, and carbon form the oxides that make up most refractory products.

The clays used for making firebrick are flint fireclays, semiflint fireclays, plastic fireclays, and kaolin. In low- or medium-duty brick, plastic fireclays with a PCE rating of 29 to 33 are used. In high- or super-duty brick, a higher percentage of flint and semiflint fireclays with PCEs of 33 to 35 are used. They increase the temperature rating and lower the shrinkage, but also reduce the plasticity of the brick.

The properties of firebrick vary depending on the clays and the nonplastics used, the density of forming, and the firing temperature. The properties of most concern are fusion point, creep under compression, spalling resistance, slag resistance, stability against gases and vapors, and abrasion resistance. Each refractory application requires consideration of these properties. Project engineers usually provide detailed performance specifications.

Pogo Sticks

Acid brick and firebrick are often used in tanks where brick must be laid overhead. Specialty tools are available to support brick in these applications. This adjustable support is called a pogo stick. It is positioned to support the brick in order to maintain the tight tolerances needed when laying acid brick and firebrick.

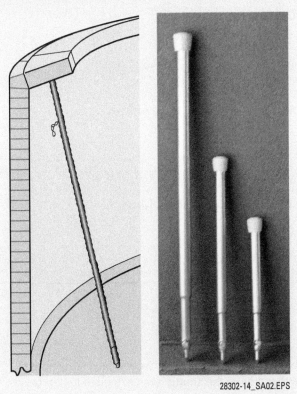

28302-14_SA02.EPS

28302-14_F25.EPS

Figure 25 Barbecue grill.

4.1.1 Brick Shapes and Sizes

Refractory bricklaying often requires forming cylindrical linings, arches, and other complicated constructions. All refractory brick manufacturers offer standard and special brick shapes.

A standard shape is a refractory brick or tile having dimensions that are used by all refractory manufacturers. This includes straight units as well as curved units. The basic size in the United States is the 9 × 4½ × 2½-inch size, sometimes called a 9-inch straight or square. The 9-inch series shown in *Figure 26* is the most commonly used and is recommended for most work. Also available and commonly used is a 9 × 4½ × 3-inch series. Similar series are available in other basic sizes, including a modified 9-inch series, a large 9-inch series, a 12-inch series, and a 13½-inch series.

Also available as standard refractory units are curved units for lining circular vessels. Several examples of these shapes are shown in *Figure 27*. Included are circle units, cupola units, and rotary-kiln units. These can be used to form a circular lining of practically any dimension. Other thicknesses are also available.

All manufacturers publish handbooks and product sheets illustrating in detail the various shapes they produce. When planning a refractory project, always check these references for the exact dimensions available.

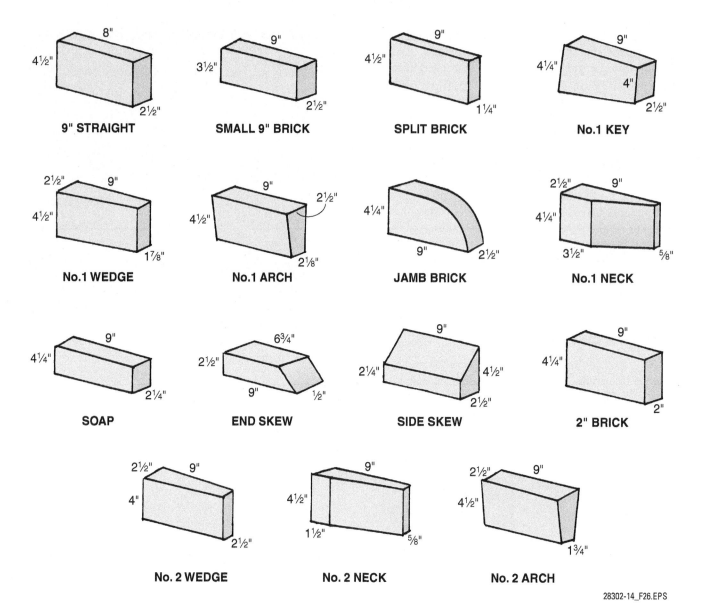

Figure 26 Standard shapes, 9-inch refractory.

Special shapes are of two varieties: those with existing molds for regular production and those for which molds must be made. The first group is often shown in product specification sheets and may include some or all of the following shapes:

- Malleable furnace
- Blast-furnace linings
- Boiler-setting tile
- Kiln car
- Tap-out and slag-hole block
- Burner block
- Suspended-arch tile
- Coke oven

28302-14_F26.EPS

Rocket Launching

The space shuttle's three main engines were fueled by liquid oxygen and hydrogen that were mixed and ignited in a controlled explosion. The two solid-fuel rocket motors on either side of the shuttle system's giant fuel tank used powdered aluminum and aluminum perchlorate for fuel. To protect the shuttle and the launchpad structures from the high-temperature exhaust, the flames were directed into two large trenches under the launchpad. These trenches were built of refractory brick and concrete. The flame trench was 490 feet long, 58 feet wide, and 40 feet high. The refractory lining was eventually replaced by newer technology late in the space shuttle program.

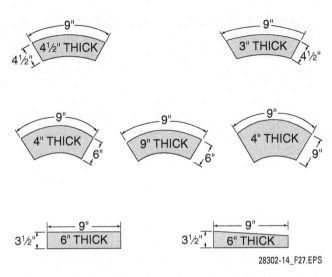

Figure 27 Circle, cupola, and rotary-kiln units.

Special shapes will have to be made to order when new refractory structures are designed, as new molds must be made for them. Small quantities will be quite expensive because of the mold cost.

4.2.0 Laying Refractory Brick

Although it requires a completely different process, the mason who has mastered laying traditional units will adapt easily to laying refractories. The units are different sizes, the mortar contains different materials, but the principles are the same—particularly the principle of sound workmanship. Refractory materials must be laid to exacting specifications. A failed furnace or launchpad is, at best, extremely costly; at worst, it is extremely dangerous.

4.2.1 Refractory Mortars

The refractory mortar is as important as the brick. If the mortar fails, the whole structure fails. These mortars are very fine-grained to permit the full, tight joints required in refractory work. They are designed to fill the void when the brick are set tightly together, forming the thinnest possible joint. They must also be able to withstand very high temperatures without shrinking.

Refractory mortar is always premixed (*Figure 28*). The material is available dry in 100-pound bags, or ready to use in 15-, 25-, and 100-pound pails or 200-pound drums. Many of these mortars have a limited shelf life and should not be stocked for long periods.

Refractory mortars serve three basic purposes:

- Bonding the brickwork into a solid unit that is capable of resisting shocks and stresses

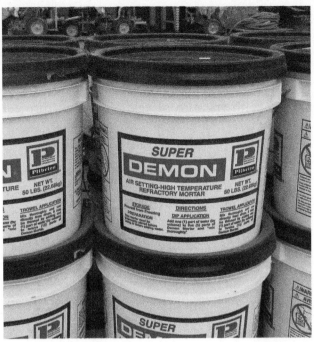

Figure 28 Refractory mortar.

- Providing a firm bearing surface between courses of brick
- Making a wall gastight and preventing penetration of slag into the joints

Sometimes mortars are thinned with water and used as coatings. As such, they provide additional insurance against wear from the destructive environment of a furnace.

There is generally some trade-off between plasticity and ability to withstand heat in the mortar, as well as the brick. The best mortars are a combination of plastic clay and a volume-constant grog of refractory materials used to reduce shrinkage. Raw fireclay is used as a mortar only for low-temperature applications.

Refractory mortars come in two basic types: heat-setting and air-setting. Heat-setting mortars are generally composed of grog and a bond clay. The nonplastic grog makes up as much as 60 percent of the total mix. This reduces shrinkage to a very low level. Choose the clays carefully so that the mortar is sufficiently workable and able to withstand the heat. As the name implies, heat-setting mortars reach full strength as high temperatures fuse the bond.

Air-setting mortars are composed of a plastic fireclay and either a precalcined fireclay or raw flint clay crushed to pass through a 35-mesh screen. A sodium silicate solution is added to give the mortar its air-setting qualities. Adjust the ingredients to meet individual service requirements. Adjust the water content of an air-setting mortar until it has the consistency of a thick batter. Air drying will set these mortars. They form a strong, almost monolithic structure with the brickwork. The mortar joint is often stronger than the surrounding brick.

Mortar may be applied in several ways, but always with the aim of achieving the thinnest possible joint. One method uses a conventional buttering technique. Mix the mortar to a batter consistency. Use a trowel to spread a thin bed joint to the top of the exposed course of brickwork. Butter the brick on the bottom and one end. Then tap or push it into place.

Another method is known as the dip method. Thin the mortar with potable water to a more fluid consistency. Dip the bottom and one end of the brick into the thin batter. Then push it into place. The thickness of the joint can be controlled by the consistency of the batter. The thicker the batter, the thicker the mortar joint. A combination method consists of pouring a batter onto the exposed top course and then dipping the brick to be laid on them. This results in more completely filled joints than the straight dipping method.

4.2.2 Floors and Hearths

Foundations of furnaces are usually made of reinforced concrete. Concrete loses considerable strength, however, when temperatures increase above 750°F. There is usually an insulating layer between the foundation and the furnace floor. Usually, a layer of hollow tile placed on the concrete foundation is sufficient. In smaller furnaces, two or three courses of insulating brick may be laid on the concrete and then heavy brick is placed on top of them. If the insulation is inadequate, a great deal of heat is lost through the foundation.

Figure 29 Building a furnace floor.

Most hearth and floor structures are laid with standard refractory brick (*Figure 29*). In some cases, larger block is used in the hearth to minimize the number of joints. Hearth block 18 × 9 × 4½ inches is generally used in blast furnaces.

Place floor units using the dip method. Flood a small area of the base with a thin coating of mortar, or dip the individual brick into a creamy mortar and set it into place. After all units are placed, cover the entire surface with mortar. Work the mortar into the joints to completely fill all voids.

4.2.3 Walls

The primary purpose of furnace walls is to retain the heat in the working chamber. The thickness of the refractory lining depends on the service conditions, and may consist of a single wythe of masonry 2½ or 4½ inches thick, or multiple wythes that have a total thickness of up to 18 inches. Under extreme conditions, even thicker walls may be required. Multiple-wythe walls must be tied together with header courses every fifth course.

Wall units are placed much like conventional masonry, with the exception of the very thin mortar joints. The brick can be placed using the buttering or dip method. With either method, be sure to completely fill all joints.

As with all masonry construction, refractory walls require expansion joints. In fact, all but the smallest refractory structures must have allowances for expansion due to heating. Expansion joints in a wall are either straight vertical joints or broken joints. In high-temperature furnaces, place joints no more than 10 or 15 feet apart. Typically, place them at the corners first, and then inward as required.

4.2.4 Rotary-Kiln Linings

Rotary kilns are used to make portland cement. They are shaped like large pipes and run some 300 feet long and 10 to 15 feet high. Flame jets at one end heat materials up to 3,000°F. When lining a rotary kiln, start at the discharge end of the kiln and move uphill toward the feed end. This way, the slope of the kiln makes it easier to lay each ring of masonry tightly against those laid previously. This helps maintain tight joints.

Take care in the layout process to ensure that each ring of masonry is placed perpendicular to the axis of the kiln. Strike a chalkline the length of the kiln parallel to the kiln axis. Use this as a baseline, and strike other axial lines by measuring offsets at 3- to 6-foot intervals. Establish lines perpendicular to these axial lines, marking the circumference of the kiln. Follow these lines and the masonry will remain perpendicular to the kiln's axis.

Place the masonry units dry and wash-coated, or use the dip method. The dip method provides for maximum joint protection and lining stability. Many masons, however, prefer the dry method. Place the masonry units dry and cover them with a thin mortar. The thin mortar will run down into the joints. Brush it with a broom to ensure that all joints are filled.

Lay kiln liners bonded or unbonded. Bonded construction is similar to the running bond in standard masonry construction. Unbonded construction, much like stack-bond construction, has the joints lined up along the axis of the kiln. The best lining, however, is achieved by using a one-third or one-fourth bond, as shown in *Figure 30*. This reduces the bending moment exerted on adjacent brick, which is a major cause of cracking in bonded construction. This also reduces spalling and grooving. This occurs in unbonded construction when the straight joints shift with respect to one another.

There are several methods to install the kiln linings. The two basic methods are the kiln jack method and the arch-form method. Properly laid, either method will result in a well-laid lining. The kiln jack method is becoming less common, even for small- and medium-sized kilns. The arch-form method, or some variation, is now generally preferred.

4.2.5 Kiln Jack Method

Start the kiln jack method of installation by laying liner units on the bottom side of the kiln. Stop about one unit above the center line. *Figure 31* shows the procedure in the three drawings.

Almost any number of courses may be laid in the bottom of the kiln before placing the supporting jacks. The number of courses is determined by the ease with which material can be brought to the mason. Some masons will lay a 3-foot section, while others will lay as much as 30 or 40 feet. A section 6 to 10 feet long is usually most convenient.

After laying a section of the kiln bottom, place longitudinal timbers measuring 3 × 6 inches or 3 × 8 inches along the last masonry unit laid on each side, and support in place with jacks. Space the jacks 1½ to 3 feet apart. They hold the lining in place when the kiln is rotated.

After the lining has been tightly jacked into place, rotate the kiln about 100 degrees. Continue each ring of masonry up the side. Rotate the kiln again, about 90 degrees. Lay the last section of masonry. The last masonry unit placed in each ring is called the key. It must be slipped in from the side of the ring. Secure the key and place it tightly in order for the ring to remain stable.

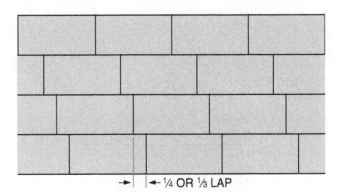

¼- OR ⅓-POINT BONDING

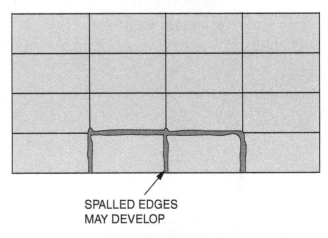

SPALLED EDGES
MAY DEVELOP

UNBONDED

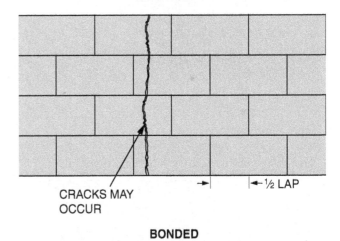

CRACKS MAY
OCCUR

BONDED

28302-14_F30.EPS

Figure 30 Kiln-lining bond patterns.

4.2.6 Arch-Form Method

The arch-form method is preferred in many instances because the kiln does not have to be rotated. Begin the laying process the same as in the kiln jack method. Lay the masonry units on the bottom half of the kiln. After the bottom is placed, install a wood form to support the top units as they are placed. The construction and location of the wood arch form is shown in *Figure 32*.

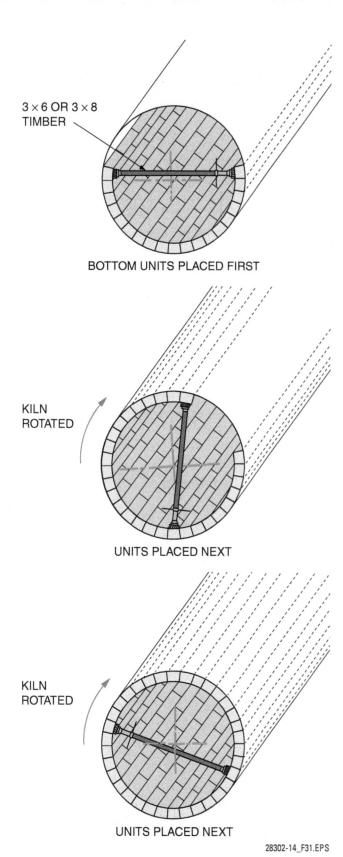

BOTTOM UNITS PLACED FIRST

3 × 6 OR 3 × 8 TIMBER

KILN ROTATED

UNITS PLACED NEXT

KILN ROTATED

UNITS PLACED NEXT

28302-14_F31.EPS

Figure 31 Kiln jack installation.

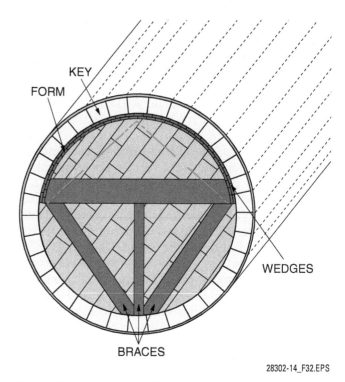

KEY

FORM

WEDGES

BRACES

28302-14_F32.EPS

Figure 32 Arch-form installation.

The wooden form is typically long enough to support only one ring of masonry. Brace it into position, leaving a small clearance between the liner and the masonry units. This clearance allows the mason to slide each masonry unit into place and secure it with a wood wedge. After the key unit is placed, remove the form braces. Move the arch form for the next ring of brick.

4.3.0 Curing and Heating Up Refractory Brick

The best temperature range for laying and curing refractory work is 60°F to 90°F. Do not allow the mortar to dry during the first 24 hours after laying. Spray the installation frequently with a fine mist of potable water to cure. Prevent the rapid loss of water by spraying with a curing compound, covering with a moisture barrier, or covering all openings of the vessel with a moisture barrier.

After curing, begin a slow drying procedure with a low draft in the system. The draft prevents the sudden buildup of gases, vapor, or heat during drying. Increase the temperature to about 250°F. Use the main or auxiliary burner of the kiln. Maintain this temperature for about four hours. Check if any steam is evident. Maintain the temperature until all steaming ceases.

Use the main burners in the system to heat up the kiln. Do this slowly the first time. A rapid initial buildup of heat decreases the life of the masonry work. Increase the temperature in the system to about 1,000°F and maintain it for eight hours. Increase the temperature 100 degrees per hour until a temperature of 1,500°F is reached. Maintain this temperature for 30 minutes per inch of thickness of the lining. Gradually reduce the temperature to shut the system down, or maintain it if the kiln is ready for use.

The Right Tools

Specialty tools help masons lay brick in rotary kilns. An aluminum form can be purchased instead of using a wooden one. The Quickset Bricking System includes adjustable pogo sticks, a 180-degree aluminum arch, and an expandable mobile support platform. The system can be adjusted to fit different types of furnace conditions and designs. Masons can install 6 feet of overhead brick before repositioning the machine.

28302-14_SA03.EPS

Additional Resources

Dekker Mechanical Engineering Series, Book 95, *Refractory Linings: Thermomechanical Design and Applications*. 1995. Boca Raton, FL: CRC Press.

Kiln Construction: A Brick by Brick Approach. 2006. Joe Finch. Philadelphia: University of Pennsylvania Press.

4.0.0 Section Review

1. Each of the following is a type of clay used to make firebrick *except* _____.

   a. plastic fireclay
   b. flint fireclay
   c. resinous fireclay
   d. kaolin

2. Concrete begins to lose much of its strength when exposed to temperatures above _____.

   a. 750°F
   b. 850°F
   c. 950°F
   d. 1,050°F

3. Cured refractory brick should be dried using a low draft for four hours at a temperature of

   _____.

   a. 200°F
   b. 250°F
   c. 300°F
   d. 350°F

SECTION FIVE

5.0.0 GLAZED MASONRY UNITS

Objective

Identify the types and uses of glazed masonry units.

   a. Describe structural glazed tile and its applications.
   b. Describe glazed block and its applications.

Performance Task

Construct a 4-foot × 4-foot wall from glazed masonry units.

Trade Terms

Glazed block: Glazed masonry units made from concrete.

Structural glazed tile: Glazed masonry units made from burned clay or shale.

28302-14_F33.EPS

Figure 33 Structural glazed tile in hospital corridor.

Hollow masonry units with integral ceramic-glaze finishes are called glazed masonry units. The ceramic glaze gives the face an enduring beauty and hardness. Glazed masonry units are used for a wide variety of load-bearing and nonbearing installations, including structural and partition walls, multiwythe walls, and veneers. Glazed masonry units made from burned clay or shale are called structural glazed tile. Glazed masonry units made from concrete are called glazed block. This section covers common applications of both types of glazed masonry units, as well as their bonding, coursing, and laying requirements.

5.1.0 Using Structural Glazed Tile

Structural glazed tile is used in the construction of interior and exterior walls, sewage- and water-treatment plants, bakeries, hospitals (*Figure 33*), food establishments, and other structures where a hard, durable, stain-resistant surface is required. Structural glazed tile walls are suitable for construction above and below grade, and may be either loadbearing or nonbearing. Nonbearing partition walls are frequently made of structural glazed tile. These walls are easily built, light in weight, and have good heat- and sound-insulating properties.

5.1.1 Properties

Structural glazed tile is manufactured by mixing finely ground clay with water in a pug mill. The clay is worked through a die and cut into units. The tile is then glazed, predried, and fired in a kiln. The firing process hardens the clay and fuses the glaze to the body of the tile. Each tile is then wrapped separately and transported to the job site. Although the tile is sold through distributors, it is shipped directly to the job to reduce handling. This reduces chipping and breakage.

Structural glazed tile has the following popular characteristics:

- A dense, hard surface that does not absorb stains
- A minimum amount of maintenance required
- Permanent color that does not fade
- Zero flame-spread that does not support combustion, char, or give off toxic fumes
- Structural strength that eliminates the need for backup support walls
- Highly sanitary finish that can be kept cleaner than other masonry units because of its hardness

Physical requirements for structural glazed tile are outlined in ASTM C126, *Standard Specification for Ceramic Glazed Structural Clay Facing Tile, Facing Brick, and Solid Masonry Units*, which also governs glazed brick. ASTM C126 covers compressive strength, absorption rate, number of cells, shell and web thickness, dimensional tolerances, and properties of the glazed finish. For exposed exterior applications, the tile body should also meet the durability requirements for ASTM C652, *Standard Specification for Hollow Brick (Hollow Masonry Units Made from Clay or Shale)*, Grade SW hollow brick units.

NCCER – *Masonry Level Three* 28302-14

The color of structural glazed tile varies somewhat between manufacturers. The most popular finishes are whites, creams, and speckles. Decorator colors are generally used in panels, or for special effect. They are rarely laid in large walls. Matte finishes have become common because of the popularity of earthy, rustic tones. Matte finishes, like gloss finishes, are easily cleaned because glazed tile is nonabsorbent. Glazed tile is also available with a clear glaze that reveals the natural color of the clay.

Grade and type classifications are the same for both glazed tile and glazed brick. ASTM C126 covers two grades and two types of units:

- *Grade S (select)* – For use with comparatively narrow mortar joints
- *Grade SS (ground edge)* – For use where variations of face dimensions must be very small
- *Type I (single-faced units)* – For general use where only one finished face will be exposed
- *Type II (two-faced units)* – For use where two opposite finished faces will be exposed

Both grades can be produced in Type I or Type II. Grade S tile can be used for most applications. If very strict dimensional requirements are necessary, use Grade SS. Type I units are used for the vast majority of structural tile work. Type II units are difficult to lay with a consistent appearance on both faces and are rarely used.

Plastic clay is extruded through dies that set the shapes of all structural glazed tile units. Various designs are produced with relative ease. There is a large assortment of sizes and patterns. This has been reduced to only the most economical and useful units through standardization.

Use ¼-, ⅜-, or ½-inch mortar joints with structural glazed tile. Use ¼-inch joints for facing tile. Nominal dimensions include the joint thickness and are multiples of the 4-inch modular system.

Various sizes of structural glazed tile are shown in *Figure 34*. Two common sizes of structural glazed tile are the 6T and 8W series. The 6T series is available with nominal face dimensions of 5 ½ × 12 inches. It is often referred to as a 6 × 12 face. Common nominal thicknesses are 2, 4, 6, and 8 inches. The 8W series has nominal face dimensions of 8 × 16 inches and is also available in 2-, 4-, 6-, or 8-inch thicknesses.

The 8W series is generally sold with ground ends to provide a more uniform size when laying in a stack bond. The 8W series also has the same face size as a standard concrete block when laid in mortar using the specified joint. Both the 6T and 8W series fit the modular grid and lay up to a height of 16 inches for bonding to its backing.

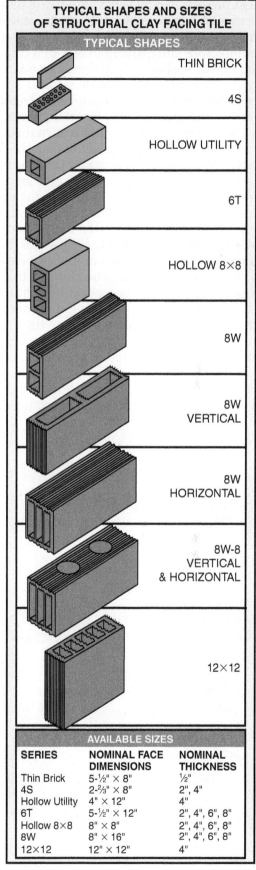

TYPICAL SHAPES AND SIZES OF STRUCTURAL CLAY FACING TILE

AVAILABLE SIZES		
SERIES	NOMINAL FACE DIMENSIONS	NOMINAL THICKNESS
Thin Brick	5-½" × 8"	½"
4S	2-⅔" × 8"	2", 4"
Hollow Utility	4" × 12"	4"
6T	5-½" × 12"	2", 4", 6", 8"
Hollow 8×8	8" × 8"	2", 4", 6", 8"
8W	8" × 16"	2", 4", 6", 8"
12×12	12" × 12"	4"

28302-14_F34.EPS

Figure 34 Sizes of structural glazed tile.

Most projects can be completed using only basic shapes. When possible, use only basic shapes, because special or unique shapes are more costly to manufacture and difficult to use. In a complete layout, however, special shapes may be necessary.

Figure 35 shows the use of several special shapes such as sills, caps, cove-base stretchers, and coved internal corners. Special shapes provide a wall with no internal or external corners that will catch dirt or be subject to chipping. All-purpose combination corners and jamb units are also available. With these units, the interior coring is arranged to permit ready cutting to the proper lengths needed to work the bond in the wall.

One of the most frequently used shapes in glazed structural tile is the bull nose. Its characteristic roundness on the corner makes it easier to keep clean and less likely to be chipped than tile with square corners. It is especially popular for hallways and corridors because it does not present the hard, sharp corner of a square tile.

The openings in a structural tile are called cells or cores, as they are in block and brick. Unlike standard concrete masonry units (CMUs), these cores may be either vertical or horizontal. The term *coring* is used to refer to the direction of the cores. The standard method of tile construction is horizontal coring, called nice construction. Vertical coring is required if the structural tile is to be reinforced or grout-filled, known as end construction.

Structural tile is available with a scored or smooth back. Scoring on the back of the tile allows the mortar to readily stick to the tile and makes the unit easy to pick up. Smooth-backed tile is not often used for general construction.

5.1.2 Bonding and Coursing

A structural glazed tile wall can be built in several bond patterns. The conventional half-lap bond and stack bond patterns shown in *Figure 36* are most commonly used. The one-third and one-quarter lap bond patterns can be used, but are rare, because special cutting of all starter units would be required.

The stack bond is also called block bond or plumb bond. In the stack bond pattern, each tile is laid directly over the one underneath, with all the head joints lining up in a plumb position. When the tile is laid in all stretchers, the bond is called a stretcher stack bond. This particular bond is the pattern most frequently used.

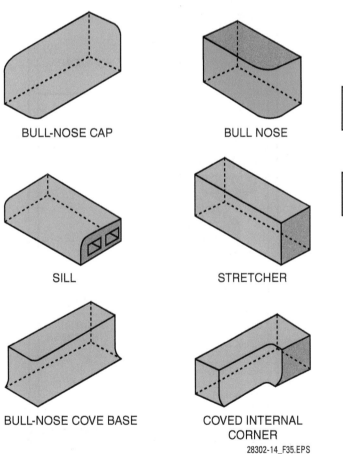

BULL-NOSE CAP   BULL NOSE

SILL   STRETCHER

BULL-NOSE COVE BASE   COVED INTERNAL CORNER

28302-14_F35.EPS

Figure 35 Special tile shapes.

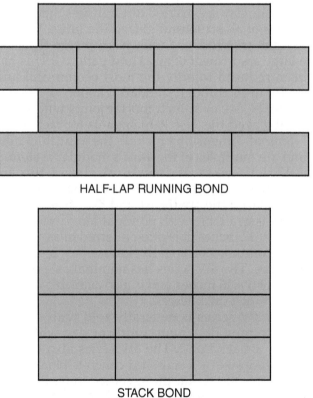

HALF-LAP RUNNING BOND

STACK BOND

28302-14_F36.EPS

Figure 36 Structural glazed tile bonding patterns.

When tile is laid in the stack bond, they all must be the same height and length. Continuous head and bed joints show even the slightest deviations; therefore, Grade SS glazed tile should be specified. Horizontal joint reinforcement is also used with structural tile and is especially recommended when a stack bond is used. The joint reinforcement is normally placed 16 inches on center.

When determining the vertical or horizontal coursing for structural tile, remember that the sizes differ from standard masonry units. The 6T series has actual face dimensions of $5\frac{1}{16} \times 11\frac{3}{4}$ inches, which courses to $5\frac{5}{16}$ inches vertically and 12 inches horizontally with the standard $\frac{1}{4}$-inch mortar joint. The 6T series will course one unit per foot horizontally and three units per 16 inches vertically.

Coursing of the 8W series is the same as for standard concrete masonry units, requiring the use of a $\frac{1}{4}$-inch mortar joint. The actual face dimensions of an 8W unit are $7\frac{3}{4} \times 15\frac{3}{4}$ inches.

Mortar for structural glazed tile must meet the compressive strength requirements of the job and match the color of the tile on the job specifications. The color consistency of the mortar is much more critical than with most masonry work. White and off-white mortars are most frequently used. These are commercially available mixed, or they can be job-mixed using portland cement mortar, hydrated lime, and white sand.

For general-purpose mortar, Type N is recommended in the proportions of 1-part portland cement, 1-part hydrated lime, and 6-parts fine sand. Regardless of the mortar used, the joints should be well filled and free from any material that can cause staining.

Structural glazed tile does not absorb as much moisture from the mortar as brick or block and therefore does not set as quickly. Use a mortar that is a little stiffer and richer than normal. Using a fine sand in the mix gives the best results. Do not mix the mortar more than two hours before using it or it will lose its strength. Also, tempering too often affects the strength of bond and elasticity of the mortar. Only temper the mortar once, and then only with an amount of water equal to that lost to evaporation.

A special epoxy mortar is commercially available for pointing mortar joints for a denser finish. This mortar extends the life of new glazed-tile walls and is easy to keep clean. In structures such as dairies, laboratories, bakeries, and chemical plants, epoxy mortar provides a joint that is non-toxic, unaffected by bacteria, safe for use around food, low in absorption, and stain resistant.

Epoxy mortar consists of several components such as resin, hardener, and powder. It has a working life of approximately 45 minutes after being mixed at 75°F and hardens in 16 hours. It is used only for pointing mortar joints for structural glazed tile and should never be used in the laying process.

Lay glazed tile in mortar with a joint width of $\frac{1}{4}$ inch. As soon as the mortar can hold the weight of the tile, rake it out to a depth of $\frac{1}{4}$ inch. Wipe off mortar stains on the face of the tile with a cloth or burlap. Allow the wall to cure at least 24 hours before pointing with epoxy mortar.

Mix epoxy mortar according to the manufacturer's instructions. If possible, use a mechanical mixer. Apply the mortar to the face of the wall with a rubber-faced trowel and tool. Remove all mortar from the face of the tile at the end of the pointing process.

Begin cleaning immediately after pointing. Wash the tile with warm potable water and a sponge. Use a circular motion with the sponge. Never allow epoxy mortar to remain on the face of the wall for more than 45 minutes, or it will become permanently attached.

5.1.3 Laying Structural Glazed Tile

Glazed tile is made very accurate in length. Use a rule or steel tape to mark the bond on the floor over a distance. Mark checkpoints on the base about every three tile units to ensure uniform joint thickness. Remember that the 6T unit has a 12-inch nominal length, while the 8W series has a 16-inch nominal length.

Construct the corner or lead using the same procedures as in brick or block construction. Build up the corner ahead so that the tile units do not pull loose when the line is attached and tightened. The exception occurs when a door, window jamb, or concrete column is against the end of the wall. Then the line can be attached to one of these objects and pulled tight. Lay a tile against the object to hold the line, or use a trig.

When the lead is built, level the tile with care so the face is not chipped. Put the level in the center of the tile and settle the tile into position gently with a hammer or trowel. After the lead is built, tool the joints and wipe the tile gently with a clean cloth or burlap.

When there is a corner return on one or both ends of the wall, use a starter piece at every course on the corner. The glazed end of the tile is not wide enough to lay out over the tile below for a starter piece.

Structural Glazed Tile Nomenclature

Most structural glazed tile is numbered, such as 4S or 6C. The first number represents the tile's thickness or bed depth. This is followed by a letter code, such as S for stretcher or G for cove base. This is followed by letters designating horizontal and vertical axis conditions, return and reveal, back-surface finish, and right- or left-handed unit.

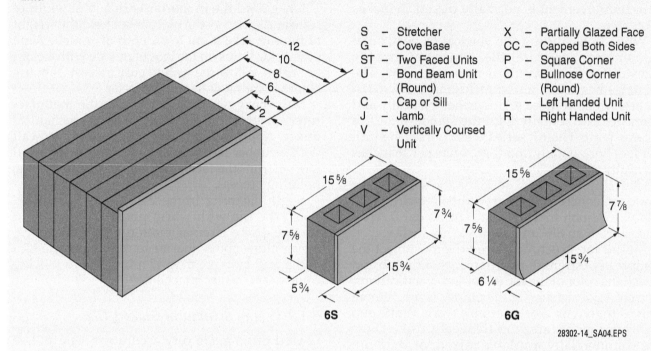

S – Stretcher	X – Partially Glazed Face
G – Cove Base	CC – Capped Both Sides
ST – Two Faced Units	P – Square Corner
U – Bond Beam Unit (Round)	O – Bullnose Corner (Round)
C – Cap or Sill	L – Left Handed Unit
J – Jamb	R – Right Handed Unit
V – Vertically Coursed Unit	

28302-14_SA04.EPS

For a 6T-series tile in running bond, the corner piece is cut to 9¾ inches, as shown in *Figure 37*. This length is necessary because of the 4-inch corner-return tile used as a starter piece for the second course. If the tile ended at a jamb or intersecting wall, use a half tile for a starter piece.

For an 8W-series tile in a similar situation, use a starter piece 11¾ inches long. This allows for the 4-inch corner return in the second course of the running bond.

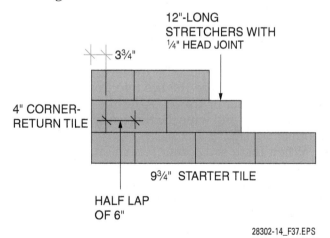

28302-14_F37.EPS

Figure 37 6T-series corner construction.

If there must be a piece of tile cut, locate it at a corner or against a jamb. The cut piece is not obvious at these points. Center cuts of glazed tile show poor planning and poor workmanship. Avoid them by using good judgment in planning the initial layout of the tile.

Only cut structural glazed tile with a masonry saw (*Figure 38*). A diamond-tipped blade will yield the best results. If a worn or inferior blade is used, the surface of the tile has a tendency to chip. As always, extreme care should be used when cutting masonry units with a power saw.

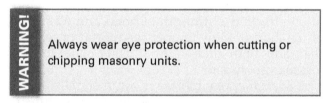

WARNING!

Always wear eye protection when cutting or chipping masonry units.

After the corner or leads are built, attach a string line and lay up the stretchers. Lay glazed tile the same as other masonry units, but take special care not to chip or damage the edges. Never attempt to tap a tile unit into place by using the blade of the trowel.

Figure 38 Masonry saw.

If cove-base tile is laid on the first course, take care that the lip of the cove at the base of the tile is properly aligned. If it is not, the floor tile will not fit against it correctly. Alignment can be checked by straightedging the lip of the cove-base tile with the plumb rule. The bottom lip of the cove tile may be easily broken as the wall is being built. Covering the lip with sand or other protective covering will help to prevent breaking. If the lip is broken, the tile must be cut out and replaced.

Tool the joints after the tile has been placed and the mortar is thumbprint-hard. Tool the bed and head joints with a concave joint. Use a plastic or glass tool for tooling to prevent any black marks in the joints.

After the wall has cured, clean it with warm potable water and soap. Use a fiber brush to scrub the wall. Rinse it thoroughly to remove all the soap. Do not use acid compounds or metal scrapers to clean tile walls because they can damage the finished faces.

5.2.0 Using Glazed Block

As with structural glazed tile, glazed block is used to construct interior and exterior walls in a wide variety of installations where a durable, stain- and water-resistant surface is required. They may be used in above- and below-grade construction, and may be either loadbearing or nonbearing. They are also available with sound-absorbing capabilities. In grouted applications, a 4-hour fire rating can be achieved using 8-inch-wide glazed block.

5.2.1 Properties

Glazed block are made from a mixture of hydraulic cement, water, mineral aggregates, and in some cases other materials such as water-repellent admixtures, in conformance with ASTM C90, *Standard Specification for Loadbearing Concrete Masonry Units*. A nontoxic glazed surface is heat-treated and cast on the block following the requirements laid out in ASTM C744, *Standard Specification for Prefaced Concrete and Calcium Silicate Masonry Units*. Because it is molded to the underlying concrete block, the finished face of a glazed block cannot pop off, as can happen with some types of structural glazed tile made from burned clay.

The facing dimensions of a standard glazed block are 7¾ inches × 15¾ inches. This represents the modular dimensions of the block (7⅝ inches × 15⅝ inches) plus a ¹⁄₁₆-inch lip around the edges of the block. Glazed block are also available in nominal 2-, 4-, 6-, 8-, 10- and 12-inch standard thicknesses, as well as other standard and special shapes. Basic units typically include stretchers, jambs, caps, and cove bases (*Figure 39*). Other commonly manufactured face dimensions include nominal 16 × 16 inch, 12 × 12 inch, 8 × 18 inch, 4 × 16 inch, and 8 inch × 8 inch.

A standard coding system is used to indicate variations from the nominal 8-inch or 16-inch lengths. The codes identify the nominal thickness, the type of unit, the type of cap or jamb used if required, the length of the unit if it varies from the nominal length, whether the return unit is right- or left-handed, and the scoring pattern if required.

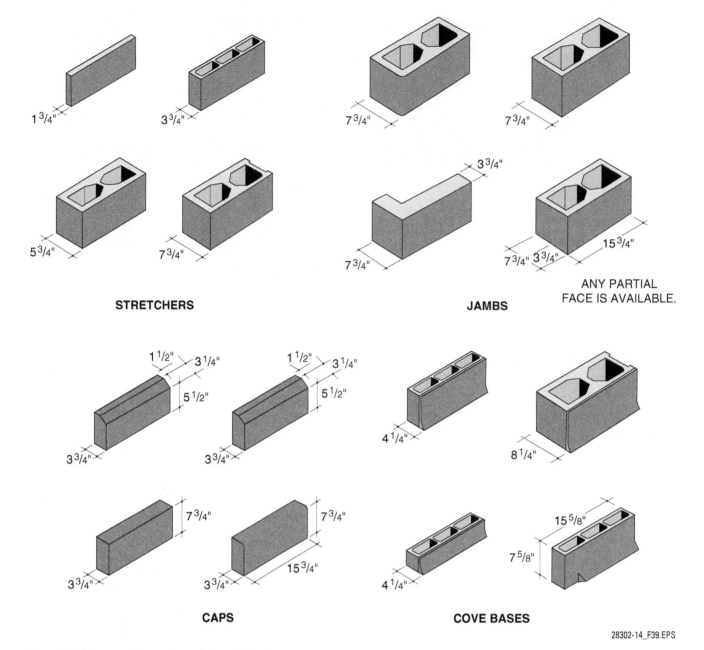

STRETCHERS

JAMBS

ANY PARTIAL FACE IS AVAILABLE.

CAPS

COVE BASES

28302-14_F39.EPS

Figure 39 Common dimensions of glazed block.

The first element in the coding system is a number that denotes the nominal thickness in inches. The second element is a letter or combination of letters that identifies the type of unit:

- S = Stretcher
- C = Cap or sill
- G = Cove base
- J = Jamb
- T = Two-faced
- V = Vertically coursed unit
- U = Bond-beam unit
  (specify open or solid bottom)
- X = Partially unglazed face
- CC = Capped both sides

The next element is a letter identifying the type of cap or jamb unit:

- P = Square
- O = Bull nose

That element is followed by a number indicating the unit's length as it varies from the nominal 16-inch length. For example, a block labeled 8SP8 is an 8-inch thick square stretcher that is 8 inches long. This element is followed by an R or an L, denoting whether the unit is right- or left-handed.

Other special indicators are used to indicate a unit height that varies from the nominal 8 inches, as well as the scoring pattern.

5.2.2 Bonding and Coursing

Glazed block walls can be built using the same conventional half-lap bond and stack bond patterns as shown in *Figure 36*. When the appearance of a stack bond is preferred, use stack bond construction rather than attempting to simulate a stack bond using scored units in a running bond pattern. The glazed face dimensions allow the use of a uniform ¼-inch joint size using modular coursing (*Figure 40*). Use Types M, S, or N mortar to lay glazed block. For exterior installations, use mortar that includes manufacturer-approved water-repellent additives.

The ideal application for exterior walls is the cavity-wall construction method, incorporating flashing, venting, and weepholes per conventional block construction. However, single-wythe loadbearing construction is easily accommodated using glazed CMU, and can be properly flashed using BlockFlash® single-wythe through-wall flashing. Glazed block can be laid using the techniques you have already learned for laying other types of concrete masonry units. Faces should be

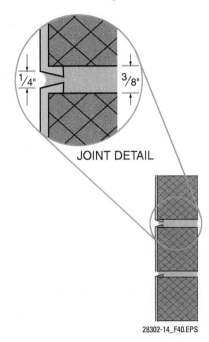

JOINT DETAIL

28302-14_F40.EPS

Figure 40 Joint dimensions for glazed block.

level, plumb, and true to the line. Discard damaged or cracked units. Ensure that the face-joint dimensions are uniformly ¼ inch horizontally and vertically, and tool the joints when they are thumbprint-hard. Cut pieces should be sized and placed so as to ensure consistency and bond pattern.

5.2.3 Laying Glazed Block

Begin by aligning the base course on the floor slab. If vinyl floor tile will be installed, ensure that the cove base is kept tight to the slab. Thicker flooring requires the cove base to be raised to the desired height. Cut glazed block with a powered masonry saw fitted with an abrasive or wet diamond blade. When estimating block and mortar for a project, remember that glazed units are wider on one or more sides than standard block.

For maximum aesthetic effect, lay glazed block so that the glazed face protrudes slightly beyond the block above it. Alternately, use a 2-inch glazed block with a standard-weight block backup. This approach provides for a consistent depth of material while also ensuring moisture retention in the substrate of the glazed block.

Glazed block should be laid with full mortar coverage on head and bed joints. Take precautions to prevent blocking of cores that will be grouted or filled with masonry insulation later in the construction process. Mortar joints should be tooled with a concave pattern. Use manufacturer-approved techniques to remove mortar from unit faces before it sets. Be careful not to damage the glazed surface of the block when cleaning. Never use paint remover, lacquer or epoxy thinners, methylene chloride, acetone, or muriatic acid to clean the glazed facing of a block.

Tuckpoint the joints of scored units with a manufacturer-approved water-resistant grout to match the block's appearance and to prevent water penetration. When using manufacturer-approved epoxy to seal joints against moisture, rake back the joint as required. If epoxy is not used, do not use a rake joint. Never float grout across the glazed facing of a block.

Additional Resources

ASTM C90, *Standard Specification for Loadbearing Concrete Masonry Units*, Latest Edition. West Conshohocken, PA: ASTM International.

ASTM C126, *Standard Specification for Ceramic Glazed Structural Clay Facing Tile, Facing Brick, and Solid Masonry Units*, Latest Edition. West Conshohocken, PA: ASTM International.

ASTM C652, *Standard Specification for Hollow Brick (Hollow Masonry Units Made from Clay or Shale)*, Latest Edition. West Conshohocken, PA: ASTM International.

ASTM C744, *Standard Specification for Prefaced Concrete and Calcium Silicate Masonry Units*, Latest Edition. West Conshohocken, PA: ASTM International.

5.0.0 Section Review

1. Series 8W structural glazed tile has a nominal length of _____.

   a. 20 inches
   b. 16 inches
   c. 12 inches
   d. 8 inches

2. The facing dimensions of a standard glazed block include a lip around the block edges of _____.

   a. ½ inch
   b. ¼ inch
   c. ⅛ inch
   d. ¹⁄₁₆ inch

SECTION SIX

6.0.0 GLASS BLOCK

Objective

Identify the types and uses of glass block.
 a. Describe applications and uses of glass block.
 b. Identify variations in glass block.
 c. Explain how to install glass block.

Performance Task

Construct a 4-foot × 4-foot wall from glass block.

Trade Term

Mullion: A thin, vertical bar that divides lights in a window or panels in a door.

G lass block is a hollow, partially evacuated unit of clear, pressed glass. Partially evacuated means that the air is partly removed from inside the block to provide a dead airspace that acts as insulation. A typical square glass block is shown in *Figure 41*. Manufactured shapes, in addition to square, include rectangular, radial, and other special shapes. Glass block cannot be trimmed or cut. It is available in a variety of sizes to accommodate a wide range of designs.

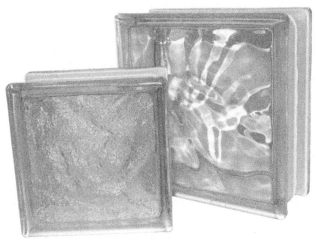

28302-14_F41.EPS

Figure 41 Typical glass block.

6.1.0 Understanding Applications and Uses of Glass Block

Glass block walls, partitions, and windows have distinctive appearances and can be serpentine, curved, or straight. The chief aesthetic advantage that glass block offers is a limit on visibility without severely reducing light. A standard transparent block admits 84 percent of daylight, but this can be reduced according to the wishes of the designer. Daylight may be used to illuminate interior spaces as shown in *Figure 42*. It can also be directed by the glass block to a particular area or focal point.

Interior lighting shining through glass block can produce a stunning visual effect (*Figure 43*). When used in interior structures, glass block overcomes the barriers that define space with-

28302-14_F42.EPS

Figure 42 Glass block windows.

28302-14_F43.EPS

Figure 43 Glass block bar.

out sacrificing noise control. Whether opaque or transparent, glass block allows a feeling of openness that is not possible with conventional partitions or walls (*Figure 44*).

The three categories of glass block are:

- Functional
- General purpose
- Decorative

The functional type of glass block controls both the distribution and diffusion or scattering of light. It is often used in schools, hospitals, and basement walls. The general-purpose type is installed where direct light transmission is required. Windows in bathrooms, entrance side panels, and interior partitions would require this type of glass block. When architectural design is most important, decorative block is used. Examples would include the front entrance of an office building, decorative screens, and corridor partitions or stairwell walls.

6.2.0 Understanding Variations of Glass Block

Glass block units are made in two sections that are later pressed together and sealed. The edges of the block are wrapped with a gritty, textured material that makes mortar bind to them. Glass block is available in a variety of sizes and thicknesses. The standard nominal sizes are 6-, 8-, or 12-inches square × 3⅞-inches thick. They weigh about 4 pounds, 6 pounds, and 16 pounds each, respectively. The actual face dimensions of the block are

5¾-inches square, 7¾-inches square, or 11¾-inches square. A ¼-inch mortar joint is standard with glass block. This results in less joint exposure, thereby increasing the block's weathering capabilities and allowing for modular dimensioning.

Within these standard shapes are various innovative differences. The basic unit is smooth on both sides, with no pattern. This block provides maximum light transmission and maximum visibility. Design variations are easily achieved by using patterned glass. If the outer faces are smooth, but a pattern is pressed on the inner faces, light is transmitted but images are distorted. If the pattern is pressed into the outer surfaces, images are distorted and glare is reduced. With patterns, the architect is able to control the direction of light and image distortion.

Several special types of glass block are also available. Three of the more common types include the following:

- The 3⅛-inch-thick units offer the light transmission, image distortion, and insulation of the standard units, but weigh about 20 percent less. These thin-line glass block units are widely used in interior applications. Rectangular glass block is also available.
- Solid glass block, sometimes called glass brick, is virtually indestructible. Bullets fired from a high-powered rifle at a distance of 25 feet are unable to penetrate this block. Because of this ruggedness and reduced sound transmission, solid glass units are often used in exterior structures. Solid glass block is 3-inches thick and available in a 7⅝-inch square.
- Solar-reflective glass block has a thermally bonded oxide surface coating that reduces both solar heat gain and transmitted light. Compared to conventional glass block, solar-reflective units reflect about twice as much light and reduce the interior heat gain by about 70 percent. The combination of reduced radiant heat gain and insulating value can mean a substantial reduction in air-conditioning costs.

28302-14_F44.EPS

Figure 44 Glass block partition.

No Assembly Required

Fully assembled glass block windows are available. They are faster to install than building them block by block, but are generally more expensive. Regardless of how you choose to install them, glass block is four times heavier than other windows of equal size. Check the instructions that come with the window kit for the weight specifications. Extra support for the window may be required.

6.3.0 Understanding Detail Procedures of Glass Block

Glass block are not designed as loadbearing units. When used in masonry walls, glass block panels are supported by steel lintels and recesses in the supporting structure. Glass block units may be set as panels in masonry, concrete, and steel- or wood-framed walls, both interior and exterior. Glass block panels cannot support structural loads above the panel, however. Unlike many other types of masonry units, glass block cannot be trimmed or cut.

There are two methods for installing glass block in a masonry wall: the chase method and the panel-anchor method. The chase method is normally preferred because it ensures a watertight and airtight joint. The maximum single-panel area using chase construction is 144 square feet, with a maximum height of 20 feet and a maximum width of 25 feet. The maximum single-panel area using panel-anchor construction is 100 square feet, with a maximum height of 10 feet and maximum width of 10 feet. Multipanel and curtain-wall sections may be erected up to 250 square feet if properly braced to limit movement and settlement.

Construction using the chase method requires placing the glass block in a chased or recessed opening to ensure weather tightness and lateral support, while allowing for movement in the wall or panel. This technique is shown in *Figure 45*.

Where the chase method is not possible, panel-anchor construction may be employed. This is used when the glass block must be placed against a flush wall. Anchoring strips 24 inches long and 1¾ inches wide of 20-gauge perforated and galvanized steel are used to provide lateral support for the glass block, yet allow slight differential movement. *Figure 46* shows the typical arrangement for a panel anchor between a wall and the glass block course.

When using panel anchors, they must be securely attached to the side jambs before starting the laying procedure. The method of construction used depends in part on the size of the area to receive the glass block. If the area is no more than 25 square feet, with a maximum width of 5 feet and a maximum height of 7 feet, a simple laying procedure can be used without expansion strips or anchors at the jambs. The side jambs of these small panels are simply mortared-in solid. A small space must be left at the head to allow for expansion and lintel deflection.

Glass block panels up to 144 square feet in area can be placed as a simple panel with no interior support points or joints. Care should be taken to

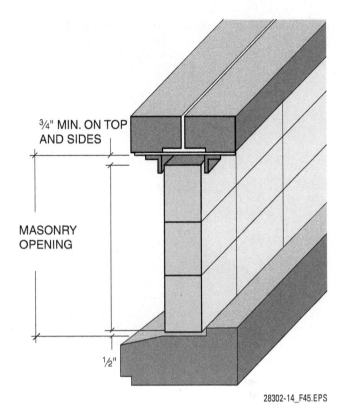

Figure 45 Chase method for glass block.

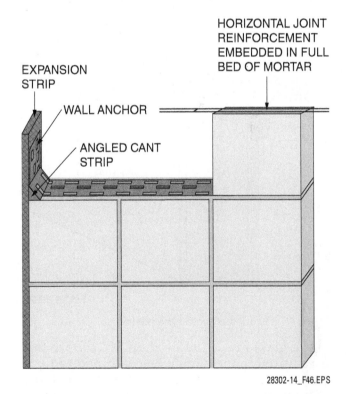

Figure 46 Panel-anchor construction.

provide adequate support and expansion space at the sill, jamb, and head. Typical installation details for these areas are shown in *Figures 47, 48,* and *49*, respectively.

 28302-14 *Specialized Materials and Techniques*   Module Two 41

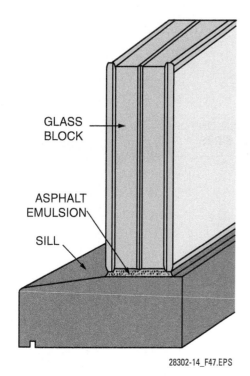

Figure 47 Elevation of typical sill detail.

GLASS BLOCK

ASPHALT EMULSION

SILL

28302-14_F47.EPS

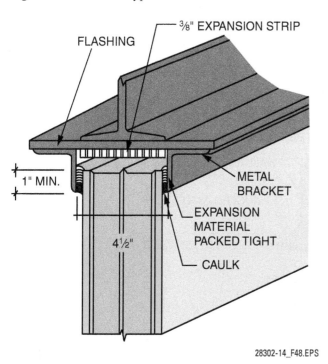

Figure 48 Elevation of typical head detail.

3⁄8" EXPANSION STRIP

FLASHING

1" MIN.

4½"

METAL BRACKET

EXPANSION MATERIAL PACKED TIGHT

CAULK

28302-14_F48.EPS

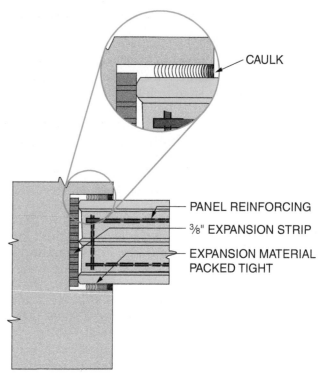

CHASE CONSTRUCTION

CAULK

PANEL REINFORCING

3⁄8" EXPANSION STRIP

EXPANSION MATERIAL PACKED TIGHT

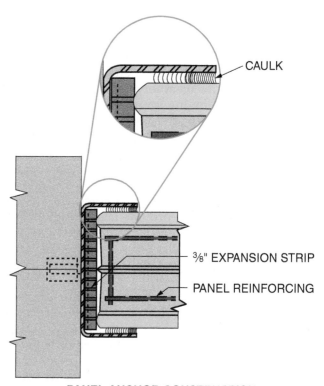

PANEL-ANCHOR CONSTRUCTION

CAULK

3⁄8" EXPANSION STRIP

PANEL REINFORCING

28302-14_F49.EPS

Figure 49 Plan view of typical jamb detail.

If a glass area of more than 144 square feet is required, panels must be combined to form the larger area. The panels must meet the same dimensional requirements as simple panels, and must be supported at all panel joints with an appropriate stiffener or mullion. *Figure 50* provides an example detail for installation of a stiffener for a large continuous panel. *Figure 51* gives a detail for the installation of a mullion.

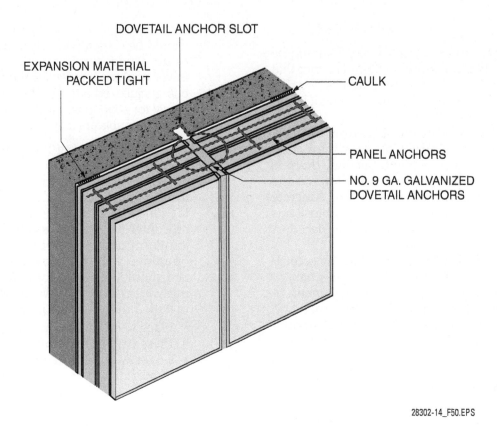

DOVETAIL ANCHOR SLOT

EXPANSION MATERIAL
PACKED TIGHT

CAULK

PANEL ANCHORS

NO. 9 GA. GALVANIZED
DOVETAIL ANCHORS

28302-14_F50.EPS

Figure 50 Plan view of typical stiffener detail.

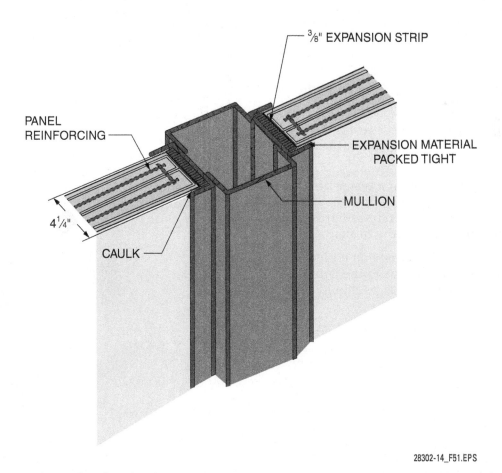

$\frac{3}{8}$" EXPANSION STRIP

PANEL
REINFORCING

EXPANSION MATERIAL
PACKED TIGHT

MULLION

$4\frac{1}{4}$"

CAULK

28302-14_F51.EPS

Figure 51 Plan view of typical mullion detail.

6.3.1 Mortar

The mortar used in laying glass block should be as dry as possible, yet still workable. The fact that glass block does not absorb water creates two distinct differences between laying glass block and laying regular concrete masonry units. First, the block does not absorb any moisture from the mortar, so no excess moisture is necessary for complete hydration of the cement. Second, there is no strength of bond between the mortar and the glass block as there is with a standard concrete masonry unit. Because of this, the glass block has a tendency to slide on the mortar when placed. A dry mortar decreases this tendency.

Mortar specifications may vary slightly according to the product and manufacturer, but generally should be mixed using 1-part portland cement, ¼- to 1¼-parts lime, a waterproofing admixture, and sand equal to between 2¼ and 3 times the amount of the cement plus lime. If a waterproof portland cement is used, the waterproofer can be omitted. For interior panels, the waterproofer is typically omitted. Admixtures such as accelerators and antifreeze compounds should not be used.

A white mortar formulated for use with glass block is commercially available, either with or without an epoxy binder. This mortar can be used for both interior and exterior walls. It has good waterproofing qualities, as well as a high refractory index.

6.3.2 Installation Procedure for a Straight Wall

The basic installation procedure for laying a straight glass-block panel is given in the following steps:

Step 1 Apply a heavy coat of a water-based asphalt emulsion to the sill area. Asphalt emulsion can be purchased from any building material supply dealer.

Step 2 Use the water-based asphalt emulsion to attach expansion strips to the jambs and head. The expansion strips should extend to the sill.

Step 3 Allow the emulsion to dry 12 to 24 hours.

Step 4 Place a full mortar bed on the sill.

Step 5 Set the lower course of glass block. Block should be placed about ½ inch away from the shoulder of the sill. Each block should be tapped down with a rubber-tipped mallet or handle so that it is firmly bedded in the mortar. The joint at the front of the sill should be left open about ½ inch to allow

for caulking. All vertical joints are full and should be ¼ inch wide. It may be preferred to use wedges in the mortar joints of the lower courses to prevent the mortar from being squeezed out. Another option, particularly for solid glass block, is to use universal mortar spacers (*Figure 52*) in the joints.

Step 6 The wall reinforcing for glass block consists of parallel 9-gauge wires 2 inches on center with 14-gauge cross wires. The reinforcing is placed in the horizontal mortar joints 24 inches on center, 16 inches on center for solid glass block, and in all joints immediately above and below all openings within a panel. A full mortar bed should be placed for joints not requiring wall reinforcing. Wall reinforcing is installed in horizontal joints as follows:

- Place lower half of mortar in bed joint.
- Press wall reinforcing in place.
- Cover wall reinforcing with upper half of mortar bed and trowel smooth.

> **NOTE**
>
> Wall reinforcing must run from one end of the panel to the other. If the reinforcing length requires the use of two or more sections, the ends should be overlapped at least 6 inches. Wall reinforcing should not bridge expansion joints.

Step 7 Repeat the procedures in Steps 5 and 6 in setting succeeding courses.

Step 8 Tool the joints smooth using a clean non-metallic jointer (see *Figure 53*) while the mortar is still plastic. At this time, rake out all joints requiring caulking to a depth equal to the thickness of the joint. Remove the surplus mortar from the

28302-14_F52.EPS

Figure 52 Universal mortar spacers for use with glass block.

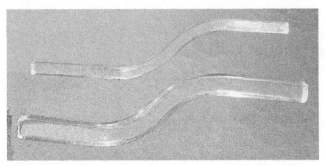

Figure 53 Nonmetallic jointers for tooling joints in glass block.

28302-14_F53.EPS

faces of the block and wipe dry. Before the mortar takes final set, tool the joints smooth and concave. Remove wedges, if used, from the lower courses, and point the voids with mortar.

Step 9 After final mortar set, place any gasket material tightly between the glass block panel and the jamb and head. Leave space for caulking.

Step 10 Caulk the interior and exterior perimeter of the panel with caulking compound.

Step 11 Remove surplus mortar and wipe the block faces dry at the time the joints are tooled. Clean the block by using a scrub brush with stiff bristles or soft rags. Final cleaning is done after the mortar has attained final set, but before becoming dry on the block surfaces. Abrasive cleaners such as steel wool, wire brushes, or acid should not be used at this time to remove mortar or dirt from the faces of the glass block.

Cleaning glass block during installation and regularly thereafter is important. During installation, a film of cement, lime, and water is certain to smear the surface of the block. This film is easily removed if allowed to dry; it is then wiped off with a dry cloth or scrubbed with a scrub brush.

The final cleaning after the mortar has set consists of a wash with a proprietary cleaner followed immediately by a thorough rinse with potable water. The usual deposits of airborne dust and dirt are removed by wiping with a damp cloth.

Where panels have stood for a long time without being cleaned, they should be washed with a mild detergent solution, then rinsed thoroughly with potable water.

6.3.3 Curved-Panel Construction

Curved panels of glass block can be used to produce a beautiful and dramatic wall. Curved panels are constructed by using a procedure similar to that for flat panels. The curvature of the wall is produced by using a different mortar-joint thickness on the outer surface of the curve than on the inner surface. The outside joint thickness is typically about ⅝ inches and the inside joint about ⅛ inches, but this will vary with the radius requirement. The inside joint thickness will become larger as the radius increases.

The first step in laying out a curved panel is to locate and mark the center of curvature. Measure from this point to the outer edges of the curved panel and clearly mark this arc. Dry-bond the first course of glass block and mark the position of each unit. The glass block for this course and for all other courses are laid using the procedure previously outlined. Normally, a vertical support and expansion joint will be located at each end of the curved panel.

Curved panels can be laid using any radius that is large enough to permit an inside mortar-joint thickness of at least ⅛ inch, and one that permits the use of whole block units. The minimum radius requirements for the most popular block widths are listed in *Table 1*. Note that 13 blocks are required to construct a 90 degree arc (quarter circle), regardless of the block width.

Table 1 Minimum Radius for Curved Glass-Block Panels

Block Width (Inches)	Outside Radius of Panel (Inches)	Joint Thickness (Inches)	
		Inside	Outside
4	34½	⅛	⅝
6	52½	⅛	⅝
8	69	⅛	⅝
12	102½	⅛	⅝

28302-14_T01.EPS

Additional Resources

Building with Glass Blocks. 1987. Bob Pennycock. Toronto, ON: Doubleday Canada.

Glass Block Handbook. 1995. Xavier G. Zeitoun. Canoga Park, CA: Builder's Book, Inc.

6.0.0 Section Review

1. Each of the following is a category of glass block *except* _____.

   a. general purpose
   b. decorative
   c. structural
   d. functional

2. To reduce joint exposure when laying glass block, the width of the mortar joint should be _____.

   a. ⅟₃₂ inch
   b. ⅟₁₆ inch
   c. ⅛ inch
   d. ¼ inch

3. To decrease the radius of a curved glass-block panel, _____.

   a. used specially made curved-glass block
   b. construct a curved centering frame
   c. decrease the mortar joint thickness on the inside joint
   d. increase the mortar joint thickness of the inside joint

---

SUMMARY

Masonry units are used to build sound-barrier walls along highways and other corridors where noise is a problem. Construction techniques for these structures conform to conventional masonry construction for long walls.

Arches are designed to be both decorative and functional. The structural efficiency of the arch is due to the fact that most of the arch will always be in compression under normal conditions. Arches other than jack arches, which are flat, must be built by using a form to support the units until the mortar has cured.

Acid brick is capable of resisting attack by harmful chemicals. Acid brick construction is done to close tolerances, and the workmanship must be very precise. Because the brick itself is not liquid-tight, it is usually used in conjunction with a membrane coating that is placed behind the brick surface.

Refractory brick is used in linings of blast furnaces, crucibles for melting metals, and other places where resistance to temperature and corrosion are required. Refractory brick is installed using special mortars that can also withstand high temperatures.

Structural glazed tile is made of burned clay or shale; glazed block is made of cementitious material. They are used in sewage- and water-treatment plants, bakeries, hospitals, food establishments, or wherever a hard, durable, stain-resistant surface is required. Tile and block are categorized by grade and type according to ASTM standards.

Glass block is a hollow, partially evacuated unit of clear, pressed glass. Glass block is available in a variety of sizes and thicknesses, with the standard units nominally being 6, 8, or 12 inches square × 3⅞ inches thick. Glass block is designed for use in nonbearing structures. Larger panel areas can be constructed, but only with additional bracing to limit movement and settlement.

1. In a pier-and-panel sound-barrier wall, the panel thickness is usually _____.
   a. 4 inches
   b. 6 inches
   c. 8 inches
   d. 12 inches

2. Panels in sound-barrier walls are supported by piles or _____.
   a. steel channels
   b. 1-inch-diameter rebar pins
   c. clip angles
   d. prestressed concrete beams

3. A major arch has a rise-to-span ratio that is greater than 0.15 and can support loads in excess of _____.
   a. 500 pounds per foot
   b. 1,000 pounds per foot
   c. 2,500 pounds per foot
   d. 5,000 pounds per foot

4. An arch that has a relatively high rise, and whose sides consist of arcs or circles, is the _____.
   a. semicircular arch
   b. multicentered arch
   c. jack arch
   d. Gothic arch

5. Buttresses are structures designed to resist an arch's _____.
   a. compressive force
   b. tensile stress
   c. thrust
   d. tangential expansion

6. In a major parabolic arch, the spring line is the intersection of the arch axis with the _____.
   a. skewback
   b. intrados
   c. camber
   d. soffit

7. The surface of the arch template that will support the arch until the masonry has cured is made from _____.
   a. ¾-inch plywood
   b. ³⁄₃₂-inch steel sheet
   c. ¼-inch plywood
   d. ⅛-inch tempered masonite

8. When building an arch, the height of each brick course is established by using a _____.
   a. mason's line
   b. protractor
   c. builder's level
   d. radius stick

9. A typical width of a small mortar joint used to lay arch brick is _____.
   a. less than ¼ inch
   b. ¼ inch to ⅜ inch
   c. ½ inch to ⅝ inch
   d. ¾ inch

28302-14_RQ01.EPS

Figure 1

10. The specialized mortar joint tool shown in Review Question *Figure 1* is a _____.
    a. plugging chisel
    b. joint cutter
    c. raker
    d. centering chisel

11. Steel angle iron will typically be used as a support for a jack arch with a span greater than _____.
    a. 18 inches
    b. 24 inches
    c. 36 inches
    d. 48 inches

12. A line stretched across the center of a jack arch to ensure it does not bulge out of position is called a _____.

    a. face line
    b. center line
    c. range line
    d. midpoint line

13. A concrete or steel substrate is often needed to protect loadbearing acid brick structures from _____.

    a. tensile forces
    b. skewing
    c. compressive loads
    d. shear stresses

14. Acid brick is exceptionally hard and may have a water absorption rate as low as _____.

    a. 9 percent
    b. 6 percent
    c. 3 percent
    d. 1 percent

15. Because it is very brittle, a type of brick that should not be used where thermal shock is likely is _____.

    a. fireclay brick
    b. common brick
    c. red-shale brick
    d. carbon brick

16. The best bonding for chemical-resistant brick structures is provided by _____.

    a. phenol-formaldehyde resin mortar
    b. epoxy mortar
    c. furan mortar
    d. polyester resin mortar

17. Because it is less effective at supporting structural material for acid-proof masonry, a material that should not be used as a substrate is _____.

    a. wood
    b. concrete
    c. reinforced fiberglass
    d. steel

18. For proper drainage, the concrete substrate for an acid brick floor should be sloped _____.

    a. 1 inch per foot
    b. ¾ inch per foot
    c. ½ inch per foot
    d. ¼ inch per foot

19. In acid brick floor construction, the maximum distance allowed between expansion joints is _____.

    a. 5 feet
    b. 10 feet
    c. 15 feet
    d. 20 feet

20. The numbering system for pyrometric ceramic cones used to measure heat resistance begins with 22 and ends with _____.

    a. 29
    b. 36
    c. 42
    d. 48

21. Low- and medium-duty refractory brick are used for _____.

    a. ceramic kilns
    b. fireplaces
    c. blast furnaces
    d. launchpads

22. The dimensions of the 9-inch straight or square type of refractory brick is _____.

    a. 9 × 4 × 2 inches
    b. 9 × 2½ × 4½ inches
    c. 9 × 6½ × 4½ inches
    d. 9 × 4½ × 2 ½ inches

23. Ready-to-use refractory mortar is supplied in containers ranging in capacity from 15 pounds to _____.

    a. 200 pounds
    b. 100 pounds
    c. 50 pounds
    d. 30 pounds

24. The two basic methods used to lay rotary-kiln linings are the kiln jack method and the _____.

    a. circular-bond method
    b. incremental method
    c. arch-form method
    d. dry-fit method

25. Ground-edge ceramic tile, which has very consistent face dimensions, is designated as Grade _____.

    a. GS
    b. S
    c. XS
    d. SS

26. In the standard coding system used to indicate variations in glazed block, the term *partially unglazed face* is represented by the letter(s) _____.

    a. FF
    b. X
    c. GG
    d. U

27. Special structural glazed-tile shapes with rounded corners are designated as _____.

    a. bull nose
    b. convex
    c. radiused
    d. snubnose

28. A 6-inch-square glass block weighs approximately _____.

    a. 16 pounds
    b. 6 pounds
    c. 4 pounds
    d. 2 pounds

29. The mortar used to lay glass block should have as little excess moisture as possible because _____.

    a. excess moisture would corrode joint reinforcement
    b. it will reduce setting time
    c. it will be less likely to squeeze out of joints
    d. glass block does not absorb water

30. Curved glass-block panels must allow an inside mortar thickness of no less than _____.

    a. ¹⁄₁₆ inch
    b. ⅛ inch
    c. ³⁄₁₆ inch
    d. ¼ inch

Trade Terms Quiz

Fill in the blank with the correct term that you learned from your study of this module.

1. Burned, pulverized refractory material such as broken pottery or firebrick, utilized in the preparation of refractory bodies, is called _____.

2. A(n) _____ is a specialized masonry unit that can withstand high temperatures; used to form an insulating layer where extreme heat would damage other components of the structure.

3. Glazed masonry units made from concrete are called _____.

4. A(n) _____ is a masonry unit that is manufactured with special properties to withstand harsh chemical environments without disintegrating.

5. The form of construction in which a number of units span an opening by transferring vertical loads laterally to adjacent units and thus to the supports is called a(n) _____.

6. _____ is a complex property of a material involving a combination of qualities of mobility and magnitude of yield value.

7. Glazed masonry units made from burned clay or shale are called _____.

8. A(n) _____ is a thin, vertical bar that divides lights in a window or panels in a door.

Trade Terms

Acid brick
Arch
Glazed block
Grog

Mullion
Plasticity
Refractory
Structural glazed tile

Trade Terms Introduced in This Module

Acid brick: Masonry units that are manufactured with special properties to withstand harsh chemical environments without disintegrating.

Arch: A form of construction in which a number of units span an opening by transferring vertical loads laterally to adjacent units and thus to the supports.

Glazed block: Glazed masonry units made from concrete.

Grog: Burned, pulverized refractory material such as broken pottery or firebrick, utilized in the preparation of refractory bodies.

Mullion: A thin, vertical bar that divides lights in a window or panels in a door.

Plasticity: A complex property of a material involving a combination of qualities of mobility and magnitude of yield value; a material's ability to be easily molded into various shapes.

Refractory: A specialized masonry unit that can withstand high temperatures; used to form an insulating layer where extreme heat would damage other components of the structure. Refractories require special mortars that can also withstand high temperatures.

Structural glazed tile: Glazed masonry units made from burned clay or shale.

Additional Resources

This module presents thorough resources for task training. The following resource material is suggested for further study.

ASTM C90, *Standard Specification for Loadbearing Concrete Masonry Units*, Latest Edition. West Conshohocken, PA: ASTM International.

ASTM C126, *Standard Specification for Ceramic Glazed Structural Clay Facing Tile, Facing Brick, and Solid Masonry Units*, Latest Edition. West Conshohocken, PA: ASTM International.

ASTM C652, *Standard Specification for Hollow Brick (Hollow Masonry Units Made from Clay or Shale)*, Latest Edition. West Conshohocken, PA: ASTM International.

ASTM C744, *Standard Specification for Prefaced Concrete and Calcium Silicate Masonry Units*, Latest Edition. West Conshohocken, PA: ASTM International.

Bricklaying: Brick and Block Masonry. 1988. Brick Industry Association. Orlando, FL: Harcourt Brace & Company.

Building with Glass Blocks. 1987. Bob Pennycock. Toronto, ON: Doubleday Canada.

The Complete Guide to Masonry & Stonework. 2010. Minneapolis, MN: Creative Publishing International (video).

Concrete Masonry Handbook: for Architects, Engineers, Builders, Fifth Edition. 1991. W. C. Panerese, S. K. Kosmatka, and F. A. Randall, Jr. Skokie, IL: Portland Cement Association.

Corrosion and Chemical Resistant Masonry Materials Handbook. 1987. Walter Lee Sheppard. Norwich, NY: William Andrew.

Dekker Mechanical Engineering Series, Book 95, *Refractory Linings: Thermomechanical Design and Applications*. 1995. Boca Raton, FL: CRC Press.

Glass Block Handbook. 1995. Xavier G. Zeitoun. Canoga Park, CA: Builder's Book, Inc.

Kiln Construction: A Brick by Brick Approach. 2006. Joe Finch. Philadelphia: University of Pennsylvania Press.

Technical Note TN45, *Brick Masonry Noise Barrier Walls—Introduction*. 2001. Reston, VA: The Brick Industry Association. **www.gobrick.com**

Technical Note TN45A, *Brick Masonry Noise Barrier Walls—Structural Design*. 1992. Reston, VA: The Brick Industry Association. **www.gobrick.com**

Figure Credits

Section Review Answers

Answer	Section Reference	Objective
Section One		
1. d	1.1.0	1a
2. c	1.2.0	1b
Section Two		
1. a	2.1.0	2a
2. b	2.2.1	2b
3. a	2.3.3	2c
4. c	2.4.3	2d
Section Three		
1. d	3.1.0	3a
2. a	3.2.2	3b
3. c	3.3.5	3c
Section Four		
1. c	4.1.0	4a
2. a	4.2.2	4b
3. b	4.3.0	4c
Section Five		
1. b	5.1.3	5a
2. d	5.2.1	5b
Section Six		
1. c	6.1.0	6a
2. d	6.2.0	6b
3. c	6.3.3	6c

NCCER CURRICULA — USER UPDATE

NCCER makes every effort to keep its textbooks up-to-date and free of technical errors. We appreciate your help in this process. If you find an error, a typographical mistake, or an inaccuracy in NCCER's curricula, please fill out this form (or a photocopy), or complete the online form at **www.nccer.org/olf**. Be sure to include the exact module ID number, page number, a detailed description, and your recommended correction. Your input will be brought to the attention of the Authoring Team. Thank you for your assistance.

Instructors – If you have an idea for improving this textbook, or have found that additional materials were necessary to teach this module effectively, please let us know so that we may present your suggestions to the Authoring Team.

NCCER Product Development and Revision

13614 Progress Blvd., Alachua, FL 32615

Email: curriculum@nccer.org
Online: www.nccer.org/olf

❏ Trainee Guide ❏ Lesson Plans ❏ Exam ❏ PowerPoints Other _____

Craft / Level: _____ Copyright Date: _____

Module ID Number / Title: _____

Section Number(s): _____

Description: _____

_____

_____

_____

_____

Recommended Correction: _____

_____

_____

_____

_____

Your Name: _____

Address: _____

_____

Email: _____ Phone: _____

Repair and Restoration

This module serves as a general introduction to the techniques and materials used to repair and restore masonry structures. Restoration of masonry structures typically involves cutting out and repairing mortar joints, removing efflorescence, repairing cracks, replacing brick, removing paint and stains, and repairing or rebuilding fireplaces and chimneys. Masons must be familiar with the equipment and materials used to perform these tasks, and also be able to recognize the causes of the deterioration.

Module Three

Trainees with successful module completions may be eligible for credentialing through the NCCER Registry. To learn more, go to **www.nccer.org** or contact us at **1.888.622.3720**. Our website has information on the latest product releases and training, as well as online versions of our *Cornerstone* magazine and Pearson's product catalog.

Your feedback is welcome. You may email your comments to **curriculum@nccer.org**, send general comments and inquiries to **info@nccer.org**, or fill in the User Update form at the back of this module.

This information is general in nature and intended for training purposes only. Actual performance of activities described in this manual requires compliance with all applicable operating, service, maintenance, and safety procedures under the direction of qualified personnel. References in this manual to patented or proprietary devices do not constitute a recommendation of their use.

REPAIR AND RESTORATION

Objectives

When you have completed this module, you will be able to do the following:

1. Explain how to inspect and identify common masonry problems.
   a. Identify common types of masonry deterioration and their causes.
   b. Explain how to inspect existing masonry structures.
2. Describe common masonry repair techniques.
   a. Explain how to tuck-point a masonry structure.
   b. Describe how to remove efflorescence.
   c. Describe how to repair masonry cracks.
3. Describe how to restore brick walls.
   a. Identify staining problems.
   b. Explain how to remove old paint.
   c. Describe how to clean brick.
   d. Explain how to replace brick and mortar joints.
4. Describe how to repair a foundation wall.
   a. Explain how to repair water intrusion.
   b. Explain how to repair cracks and localized problems.
5. Describe how to repair and rebuild chimneys and fireplaces.
   a. Explain how to repair chimneys.
   b. Explain how to repair fireplaces.

Performance Tasks

Under the supervision of your instructor, you should be able to do the following:

1. Repair mortar joints by tuckpointing.
2. Clean a masonry wall with a bucket and brush.
3. Replace a damaged masonry unit in a wall.

Trade Terms

Efflorescence	Portland cement paint
Parapet	Spall
Plugging chisel	Tuckpointing

Industry-Recognized Credentials

If you're training through an NCCER-accredited sponsor, you may be eligible for credentials from NCCER's Registry. The ID number for this module is 28303-14. Note that this module may have been used in other NCCER curricula and may apply to other level completions. Contact NCCER's Registry at 888.622.3720 or go to **www.nccer.org** for more information.

Code Note

Codes vary among jurisdictions. Because of the variations in code, consult the applicable code whenever regulations are in question. Referring to an incorrect set of codes can cause as much trouble as failing to reference codes altogether. Obtain, review, and familiarize yourself with your local adopted code.

Contents

Topics to be presented in this module include:

Figures and Tables

1.0.0 INSPECTION AND PROBLEM SOLVING

Objective

Explain how to inspect and identify common masonry problems.

    a. Identify common types of masonry deterioration and their causes.

    b. Explain how to inspect existing masonry structures.

Trade Terms

Efflorescence: A deposit or crust of white powder on the surface of brickwork, resulting when soluble salts in the mortar or brick are drawn to the surface by moisture.

Parapet: A low wall or railing.

Spall: A chip, fragment, or flake broken off from the edge or face of a stone masonry unit and having at least one thin edge.

Early deterioration of a masonry structure may be a result of improper design or poor workmanship. However, the principal causes of long-term deterioration are dirt, moisture, and temperature changes. These agents eventually erode mortar joints and crack other materials. While wind, rain, snow, and temperature changes are unavoidable, the damage they cause can often be prevented by regular inspections and repairs. Regular inspections can identify deterioration before it becomes a major problem in a structure.

1.1.0 Identifying Types of Deterioration and Their Causes

Most deterioration in masonry structures is caused by movement of the structure or by water. Proper design, construction methods, and materials will minimize most types of deterioration for the life of the structure. It is important to recognize the major types of deterioration and their causes as described in the following sections.

1.1.1 Weathering Damage

Weathering and frost damage is characterized by spalls, cracks, and surface erosion. Parapets, screen walls, fences, and fully exposed columns are especially vulnerable to this type of damage. A constant drip or flow of water can slowly erode brick, masonry units, or mortar.

Inspect masonry beneath recesses or in locations prone to collect water more frequently, especially if there is no flashing to deflect the water. Frost damage occurs when water that has accumulated in a masonry unit freezes. When water freezes, it expands and exerts tremendous pressure that can crack or severely damage the surrounding material.

Inspect mortar joints for weathering and frost damage. Deteriorated mortar often holds water, which can freeze and expand causing misalignment of brick or masonry units.

1.1.2 Cracking and Spalling

A building is made of materials and components that are constantly moving. The movement is caused by changes in temperature and moisture, loads, creep, and other factors. Restraint of these movements may cause stresses within the building elements that result in cracking or spalling.

Cracking is the distress that occurs most often in masonry walls. Cracks result from many different sources, but there are typical shapes and patterns of cracks. Often, the type and magnitude of cracking will indicate the cause. As a learning tool, it can be more beneficial to study what might happen if movement is not considered in a design than to study a properly designed and detailed project. The following locations, if poorly designed, are where cracking typically occurs:

- *Long walls* – Long walls, or walls with large distances between expansion joints, may create unrelieved distress within the wall. Expansion of the brickwork may force sealant material out of the expansion joint, or may crack the brickwork between expansion joints. Diagonal cracks often occur in piers between window or door openings. Cracks like the one shown in *Figure 1* usually extend from the head or sill at the jamb of the opening, depending on the direction of movement and the path of least resistance.
- *Corners* – An insufficient number of expansion joints or the improper location of these joints in a wall can lead to cracking at the corners. Perpendicular walls will expand in the direction of the corner, causing rotation, and will crack near that corner. This typically occurs at the first head joint from either side of the corner, as shown in *Figure 2*.
- *Offsets and setbacks* – Vertical cracks are quite common at wall setbacks and offsets if movement is not accommodated. *Figure 3* shows

Figure 1 Expansion cracking of a long wall.

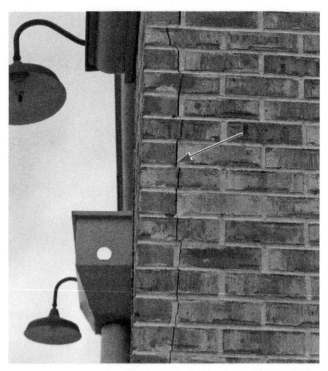

28303-14_F02.EPS

Figure 2 Crack at corner.

28303-14_F03.EPS

Figure 3 Crack at offset.

this effect at an offset. Typically, the header is twisted by forces within the offset. Often, the rotation of individual units will create more than one opening.

- *Shortening of structural frames* – In framed structures (predominantly concrete-frame buildings) vertical shortening due to creep or shrinkage of the structural elements may impose high stresses on the masonry. These stresses typically develop at window heads, shelf angles, and other points where stresses are concentrated. *Figure 4* shows brick veneer supported by a steel shelf angle on a concrete frame. Over time, the concrete frame has shrunk and caused the steel shelf angle to bear on the masonry below, resulting in spalling. Because a horizontal expansion joint was not provided, stresses became concentrated on the mortar joint directly below the angle, crushing the masonry below. This phenomenon can also cause bowing of the brickwork between floors if the veneer wall is not adequately attached to the backing, or if the backing is not sufficiently rigid.

- *Foundations* – Masonry walls built above grade on concrete foundations will expand, while the concrete foundation will shrink. This differential movement will cause shearing at the foundation interface if that area is tightly bonded together. Movement of the brick away from the corner or cracking of the mortar often results in cracking at the foundation corner. If the cracked corner of the foundation falls away, there will not be any structural support for the masonry resting on top.

- *Deflection and settlement* – Deflection and settlement cracks are identified by a tapered opening. *Figure 5* shows a deflection crack caused

28303-14_F04.EPS

Figure 4 Spalling due to shortening of the structural frame.

28303-14_F05.EPS

Figure 5 Crack due to deflection.

by insufficient support of the brickwork on a lintel. The crack shown in *Figure 6* is due to differential settlement of the foundation. If all settlement is equal, then little harm is done. Cracking occurs when one portion of a structure settles more than an adjacent part.

- *Encased columns* – When structural elements, such as columns, are rigidly encased in masonry, any movement of the element is transferred to the masonry, causing cracks and popouts (*Figure 7*). These movements may be due to drift of the building frame or lateral expansion resulting from creep. These cracks occur on the exterior as well as the interior of the building.

28303-14_F06.EPS

Figure 6 Crack due to differential settlement.

28303-14_F07.EPS

Figure 7 Encased column.

- *Parapet walls* – Parapet walls cause particular problems because there are two exposed surfaces subject to moisture and temperature extremes. Through-wall flashing is often necessary, but this creates a horizontal plane of weakness at the base of the parapet. In addition, the roof system may expand, making the situation worse. Under such conditions, the parapet may bow and crack horizontally at the roofline. Expansion joints spaced throughout the wall allow for expansion of the parapet wall.
- *Embedded items* – Items embedded in or attached to masonry walls may cause spalling or cracking when they move or expand. Reinforcement that is continuous across an expansion joint may buckle, pushing out adjacent mortar, which is shown in *Figure 8*. Corrosion of metal elements within masonry can cause volume increases of such a magnitude as to crack or spall the masonry.

28303-14_F08.EPS

Figure 8 Spalling in the bottom of an arch due to corrosion of joint reinforcement.

1.1.3 Water Penetration

Masonry structures are often subject to water penetration from groundwater, rainwater, and water vapor. Masonry foundations must be protected from groundwater penetration. Rainwater can penetrate exposed masonry surfaces where water can pool. The movement of water and air into and within masonry structures leads to other problems such as efflorescence, frost action, cracking, and some forms of staining.

Moisture may also accumulate within the wall as a result of condensation of water vapor. Frequently, efflorescence that appears on masonry walls protected from rain is caused by the accumulation of condensed moisture. This phenomenon is often called new-building bloom.

Condensation is usually due to moisture originating inside buildings. As it enters a heated building, cold outside air is typically low in moisture content. Moisture released from cooking, bathing, washing, and similar activities humidifies the air, as well as the breathing and perspiration of the occupants.

Higher moisture content increases the barometric pressure of the inside air to a level greater than that of the outside air. This increased pressure tends to drive the water vapor outward through any vapor-porous enclosing surfaces. When water vapor passes through porous materials, it may pass through without condensing into water. But if the flow of water vapor is impeded by a vapor-resistant surface at a lower temperature, the vapor will condense on these cold surfaces.

Another source of moisture that may contribute to future efflorescence is the water that enters the assembly during construction. Improper protection of a building during construction can cause efflorescence, as well as other problems. New construction is highly vulnerable to the entry of a considerable amount of moisture when interior assemblies are exposed and joints are open.

Areas that are consistently damp are also prone to the growth of mold. Water infiltration, excessive humidity, and condensation are key factors in the development of mold. There are thousands of different types of molds, and several are present in most buildings to varying degrees.

Molds thrive in an environment with moisture, a food source, oxygen, and moderate temperatures. The key to preventing mold is controlling moisture. All buildings have some organic-based components that can foster mold growth. The recent trends toward making buildings tight and energy efficient often limit the building's ability to breathe. This often traps moisture in the building, supporting the growth of mold.

Molds can cause an area to become slippery and unsightly. Mold allergies have been linked to a wide variety of flu-like symptoms, such as fever, cough, and runny nose. Some people have more severe reactions to specific kinds of molds and the toxins they produce.

Unlike other organic materials such as paper and wood, masonry does not provide a food source for molds. Masonry is not damaged by mold, and mold can be cleaned off the face of masonry.

1.1.4 Stains

Efflorescence is a white or gray crystalline deposit of water-soluble salts on the surface of brick masonry. The principal objection to efflorescence is its unsightly appearance. Although efflorescence is unattractive and a nuisance to remove, it is usually not harmful to the masonry.

Efflorescence occurs when water-soluble salts in solution are brought to the surface of the masonry and deposited there by evaporation. The salts may be present in the facing units, backup, mortar ingredients, or other building elements in contact with the masonry. A source of water must be in contact with the salts long enough for them to dissolve. The salt solution must then migrate to the surface. The water then evaporates, leaving the salt residue on the surface. *Figure 9* shows severe efflorescence on a brick sidewalk. The pattern indicates that the moisture is coming through at the joints.

28303-14_F09.EPS

Figure 9 Severe efflorescence on a brick sidewalk.

28303-14_F10.EPS

Figure 10 Stains on brick wall.

Efflorescence could be prevented if masonry could be constructed without water-soluble salts, or if no water were permitted to penetrate the masonry. However, in conventional masonry exposed to the weather, neither of these conditions are possible. The practical approach is to minimize the contributing factors.

Efflorescence will often disappear with normal weathering if the source of moisture is located and stopped. Efflorescence can also be dry-brushed, washed away by a thorough flushing with potable water, or scrubbed away with a brush.

Stains other than efflorescence will occasionally occur on the surfaces of masonry structures (see *Figure 10*). These are carbonate or silicate deposits. They appear as white scum, green or yellow stains, and brown stains.

Carbonate deposits usually appear as a gray-white crusty spot in the form of a vertical run-down shape on the face of the wall. These deposits are sometimes referred to as lime run. Carbonate deposits usually occur at a small hole or opening in the face of the masonry. This type of stain results from a great deal of water traveling the same path over an extended period of time. The process here is similar to the formation of stalactites in limestone caves. The water forms any of several calcium compounds into a solution and brings them to the surface of the masonry through the hole. The source of the calcium compounds may be trim, mortar, or backup material.

Carbonate stains can be removed using a weak solution of proprietary cleaner applied directly to the deposit, then scrubbing with a stiff fiber-bristle brush. Care must be taken to properly wet the wall area first and to rinse it thoroughly after cleaning. This is especially true when removing carbonate deposits from light-colored brick. The deposit is likely to reappear unless the water source is stopped.

> **WARNING!**
>
> Acids can cause chemical burns. Always wear personal protective equipment when using acids. Scrubbing with an acid solution causes a splash hazard. Protect yourself and others from chemical burns.

Silicate deposits, called scumming, sometimes occur as a general white or gray discoloration on the face of brick masonry. This discoloration may occur throughout the face of the masonry or as an irregular shape in specific locations of 100 to 200 square feet. Silicate stains may also occur adjacent to trim elements, precast concrete, and occasionally large expanses of glass. Silicate deposits on brick masonry should not be confused with the scumming that occasionally occurs on brick in the manufacturing process; that variety will be evident on the brick units in storage before they are placed in the structure.

There are several mechanisms that produce silicate deposits on brickwork. Many of these stains are related to the cleaning of brick masonry with hydrochloric acid solutions. Silicate deposits are very difficult to remove from brick masonry, as they are insoluble by most acid-based proprietary cleaners. Often, the only practical method of dealing with a silicate deposit is to conceal the stain and permit it to weather away over time.

Some structural clay products develop yellow or green efflorescent salts when they come in contact with water. These salts may be found on red, buff, or white clay products, but they are most apparent on lighter-colored units. The vanadium salts responsible for these stains come from the raw materials used in the manufacture of the clay products.

As moisture travels through the brick, it dissolves both the vanadium oxide and the sulfates. In this process, the solution may become quite acidic. As the solution evaporates from the surface of the product, the salts are deposited.

The chloride salts of vanadium require highly acidic leaching solutions, and are usually the result of washing brickwork with harsh cleaning compounds. Vanadyl chloride, one of the most prominent staining compounds, forms almost exclusively as a result of washing with hydrochloric acid. Preventing green stains caused by vanadium is important, as subsequent efforts at cleaning may turn it into a brown insoluble deposit that is difficult to remove.

The following guidelines are recommended to minimize the occurrence of green stain:

- Store brick off the ground and under protective cover.
- Never use or permit the use of acid solutions to clean brick.
- Seek and follow the recommendations of the brick manufacturer for cleaning procedures for all types and colors of brick.

Under certain conditions, tan, brown, and sometimes gray staining may occur on the mortar joints of brickwork. Occasionally, the brown stain which results from the use of manganese dioxide as a coloring agent in the units, will streak down onto the face of the brick. This type of stain is the result of the use of manganese dioxide as a coloring agent in the units. This staining problem is closely related to general efflorescence, because it is the sulfate and chloride salts of manganese that travel to the surface of the brick and are deposited on the mortar joints.

During the brick firing process, the manganese coloring agents undergo several chemical changes, resulting in manganese compounds that are insoluble in water and have varying degrees of solubility in weak acids. It is also possible that rainwater in some areas may be acidic. Always request and follow the advice of the brick manufacturer for cleaning a brown- or manganese-stained brick.

1.2.0 Inspecting Existing Masonry Structures

Inspections of masonry structures should be conducted at regular intervals, such as once a year, and in an established manner. A good procedure is to begin with the basement and work up to the walls, chimneys, and other parts of the structure. The inspector should pay particular attention to the masonry at floor lines and around all openings for signs of distress.

1.2.1 Basements

Foundation walls should be examined for cracks and for loose mortar in joints. Inspect floors for cracks or disintegration, and look for signs of leakage through walls or floors and between wall and floor intersections. If moisture is evident in the basement, the source of the moisture must be determined.

Checking for Condensation

A simple test can be made as a starting point. Place a thin sheet of bright tin, about 6 inches square, flat against the damp wall. The tin may be secured in place with adhesive at the corners. If moisture collects on the visible surface in one hour, condensation is the cause. If the metal remains dry, the dampness is the result of water penetration through the wall.

1.2.2 Exterior Walls

In masonry walls, look for cracks or broken masonry units, especially above door and window openings. Observe if joints need repointing. If a thin knife blade will pass between the mortar and masonry units, repairs are needed. Take note if there is efflorescence on the face of walls, particularly below windowsills and near downspouts.

1.2.3 Stucco

Evidence of stucco failure usually appears in the form of cracks or chipping, which appears most often over doors and windows and near the ground. These defects are most readily apparent after a rain, when moisture penetration becomes visible. Cracks may also be caused by uneven settling of the building or by temperature differences around chimneys.

1.2.4 Chimneys

Examine chimneys for defective joints and bonding. Crumbling mortar may leave openings into the flue, presenting a serious fire hazard. If joint failures are evident throughout, the chimney should be dismantled and rebuilt.

1.2.5 Wall Openings

Check the caulking around all doors and windows to determine if it satisfactorily seals all spaces. Inspect the condition of all flashing and determine if it is functioning properly.

1.2.6 Inspection Checklist

A prescribed checklist should be used to ensure a thorough inspection. The following items can be made into a checklist and modified for local conditions. As part of your inspection, check the following:

- Basement walls for cracks or faulty mortar joints
- Basement walls and floors for leaks that may require waterproofing or drainage
- Basement floors for cracks or settlement
- Grading around foundations for proper drainage
- Masonry walls for cracks or broken areas
- Mortar joints to see if pointing is needed
- Walls for leakage that may require damp-proofing
- Masonry walls for efflorescence, scum, or stains
- Painted surfaces to see if blistering, cracking, or peeling has occurred and if repainting is necessary
- Window caps to see if new flashing or joint repair is needed
- Eaves or tops of walls for leakage to see if repairs or copings are needed
- Porch walls and floors for defects
- Chimneys for defects and the need for pointing or replacement
- Need for chimney caps or pots
- Cracks between chimneys and adjoining walls
- Mortar joints and masonry units in and around fireplaces
- Stucco walls for cracks, discoloration, or damaged portions

A more formalized checklist is shown in *Figure 11*. Note that it is broken down into locations, such as North and Below Grade. While not all parts of the checklist may apply to every structure, it is easier to use one general reference than to try to think of everything that applies each time you need to do an inspection.

Following this or any other checklist provides a systematic approach so that nothing will be overlooked. A visual inspection is generally sufficient. An in-depth inspection can be made using instruments such as a fiber optic camera scope, shown in *Figure 12*.

MASONRY WALLS	NORTH	SOUTH	EAST	WEST
ABOVE GRADE				
Masonry				
Cracked Units				
Efflorescence				
Loose Units				
Missing/Clogged Weepholes				
Deteriorated Mortar Joints				
Plant Growth				
Deteriorated/Torn Sealants				
Out-of-Plumb				
Spalled Units				
Stains				
Water Penetration				
Flashing/Counterflashing				
Bent				
Missing				
Open Lap Joints				
Stains				
Caps/Copings				
Cracked Units				
Drips Needed				
Loose Joints				
Open Joints				
Out-of-Plumb				
BELOW GRADE				
Foundation Walls				
Cracks				
Deteriorated Mortar Joints				
Inadequate Drainage				
Differential Movement				
Water Penetration				
Retaining Walls				
Cracked Units				
Decayed Mortar Joints				
Damp				
Inadequate Drainage				
Out-of-Plumb				
Spalled Units				

28303-14_F11.EPS

Figure 11 Brick inspection checklist.

28303-14_F12.EPS

Figure 12 A fiber optic camera scope.

Additional Resources

Building Block Walls: A Basic Guide, Latest Edition. Herndon, VA: National Concrete Masonry Association.

Masonry Design and Detailing for Architects, Engineers and Contractors, Sixth Edition. 2012. Christine Beall. New York: McGraw-Hill.

1.0.0 Section Review

1. Condensation is usually caused by moisture that originates _____.

    a. in cracks and plugged weepholes
    b. from rain and humidity in the outside air
    c. inside a building
    d. between the wythes of a cavity wall

2. Cracks and chips in stucco do *not* typically appear _____.

    a. at wall intersections
    b. near the ground
    c. above doors
    d. above windows

2.0.0 REPAIR TECHNIQUES

Objective

Describe common masonry repair techniques.
   a. Explain how to tuck-point a masonry structure.
   b. Describe how to remove efflorescence.
   c. Describe how to repair masonry cracks.

Performance Task

Repair mortar joints by tuckpointing.

Trade Terms

Plugging chisel: A chisel with a tapered blade used for removing mortar from joints.

Tuckpointing: The process of cutting away defective mortar and refilling the joints with fresh mortar.

Masonry materials are some of the most durable of all construction materials. As such, masonry structures can be expected to have many years of serviceable life. However, even the most durable materials require repair or restoration work after a period of time. Most repair work on masonry structures involves surface blemishes, minor cracks, or moisture control. If these problems are corrected when first discovered, the serviceable life of the structure should not be affected. If neglected, the damage may escalate and seriously damage the structure.

When the damage becomes so severe that many of the masonry units must be cut out and replaced, it may be better to tear down the structure to a solid base and rebuild. However, repair may be possible, depending on the type and extent of the damage. Several types of repairs described below are common to many different types of structures and problems. Repair techniques are described in this section in general terms; specific applications are presented later in this module.

The first priority when making repairs should be identifying and treating the cause of deterioration and damage, rather than the effects. Before surfaces cracked by settlement are repaired or replaced, the foundation itself must be stabilized. Roof leaks must be stopped before repairing moisture-damaged walls.

2.1.0 Using Tuckpointing to Repair Masonry Structures

As mortar cures, some shrinkage occurs. Shrinkage forms small cracks in which water collects. If this water is exposed to freezing and thawing, the mortar joints will deteriorate over time. Tuckpointing is the process of chiseling or grinding out the old, deteriorated mortar to a uniform depth and refilling the joint with new mortar. Tuckpointing is a maintenance procedure that helps prevent water from re-entering the wall.

Prior to undertaking a tuckpointing project, you must decide whether or not to use power tools for cutting out old mortar. Tuckpointing should be done only by a qualified, properly trained craftsperson. Tuckpointing requires a great deal of practice with specific tools, so even an accomplished mason may not be skilled at tuckpointing. Improper use of tuckpointing tools can cause cracks in masonry structures.

Old mortar can be removed with a plugging chisel or a tuckpoint grinder. The plugging chisel, shown in *Figure 13*, is used for small areas. It has a cutting edge of at least ½ inch. The powered tuckpoint grinder, shown in *Figure 14*, is used when large areas are to be repointed.

> **WARNING!**
>
> Wear eye protection when using a plugging chisel. Wear full face protection when using a tuckpoint grinder.

Practice is necessary with either of these tools to avoid accidental damage to the surrounding masonry. How to operate a tuckpoint grinder was described in *Masonry Level One*. The general procedure for tuckpointing is as follows:

Step 1 Remove the old mortar, cutting out to a uniform depth of at least ¾ inch and until sound mortar is reached. Use a plugging chisel or a tuckpoint grinder, according

28303-14_F13.EPS

Figure 13 Plugging chisels.

28303-14_F14.EPS

Figure 14 Tuckpoint grinder.

PROPERLY PREPARED
JOINT

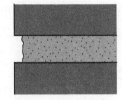

IMPROPERLY PREPARED
JOINT—TOO SHALLOW

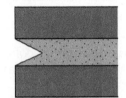

IMPROPERLY PREPARED
JOINT—FURROW SHOULD
BE ELIMINATED

28303-14_F15.EPS

Figure 15 Preparation of joints for tuckpointing.

to the requirements of the job. The joint should have a square bottom after the deteriorated mortar is removed. See *Figure 15* for proper and improper joint preparation.

Step 2 Once the deteriorated mortar is removed, the joints should be cleaned with a wire brush or pressurized air. All dust and debris must be removed from the joint by brushing, blowing with air, or rinsing with water. Any masonry units that are broken or badly spalled must be removed and replaced. Be careful not to damage the brick edges.

Step 3 Fill the joint as shown in *Figure 16*. The joints to be tuck-pointed should be dampened to ensure a good bond, but be sure that the brickwork absorbs all surface water. Water should be added to the prehydrated mortar to bring it to a workable consistency, which will be somewhat drier than conventional mortar. Use a grout bag or mortar pump to precisely place the mortar in thin layers. Tightly pack the mortar into the joints in thin layers (¼ inch maximum). Each layer should become thumbprint hard before applying the next layer.

Step 4 Tool the joint. The joints should be tooled in the same way as the original job, so they will match the original profile, after the last layer of mortar is thumbprint hard. This procedure brings the pointing mortar to the same condition as the old material that has already gone through the hardening and shrinking stages.

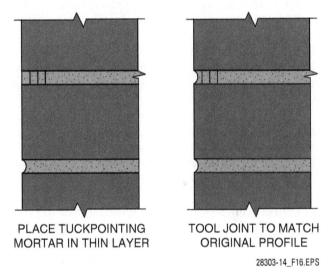

PLACE TUCKPOINTING
MORTAR IN THIN LAYER

TOOL JOINT TO MATCH
ORIGINAL PROFILE

28303-14_F16.EPS

Figure 16 Filling a joint.

Recommendations for repointing mortar vary greatly among preservationists. In any case, the mortar should be carefully selected and properly proportioned. For best results, the original mortar proportions should be duplicated. If this is not possible, Type N or O mortar should be used, as mortars with higher cement contents may not properly bond to the original mortar. *Table 1* lists the proper proportions for Types N and O mortars.

Non-preblended mortar should be prehydrated to reduce excessive shrinkage. To properly prehydrate mortar, mix all dry ingredients thoroughly, then add only enough potable water to produce a damp, workable consistency that will retain its shape when formed into a ball. The mortar should

Table 1 Mixing Proportions for Type N and Type O Mortar

Mortar Type	Parts by Volume		
	Portland Cement	Type S Hydrated Lime	Aggregate, Measured in a Damp, Loose Condition
N	1	1	4½ to 6
O	1	2	6¾ to 9

28303-14_T01.EPS

then stand in this dampened condition for 1 to 1½ hours. After this time, water can be added to bring it to a workable consistency.

2.2.0 Removing Efflorescence from Masonry Structures

Efflorescence occurs when soluble salts in the masonry units or mortar are taken into solution by water entering through joint separation cracks, faulty copings, leaking window flashing, or other construction defects. As the wall begins to dry, the salt solution migrates towards the surface through capillary pores. When the water evaporates, the salts are deposited on the face of the wall.

Efflorescence is more likely to appear in late fall, winter, and early spring, when evaporation is slower and temperatures cooler. This is particularly true for locations that experience long rainy or damp periods when temperatures are above freezing. Hot summer months are not favorable for efflorescence because the wetting and drying of the wall is generally quite rapid.

The best way to remedy efflorescence is to take measures to prevent it. Keeping materials dry before laying and during construction goes a long way towards preventing efflorescence, and good workmanship helps considerably. Masonry construction should be cleaned only after the masonry has had sufficient time to dry. If pressurized cleaning is used, be sure to repoint as necessary. Gutters and downspouts should be checked and any leaks repaired. Windowsills, copings, and overhangs should have drip grooves to keep water away from the wall.

Prevention of efflorescence is often impossible in areas of excessive rainfall. If efflorescence cannot be prevented, it can sometimes be controlled by brushing with a stiff fiber or wire brush. For best results, brush when the masonry units are dry. If the problem is severe, washing with a chemical solution may be the only alternative.

WARNING!

Always read the safety data sheet (SDS) and the instructions when using proprietary cleaners. Many chemical cleaners contain acids that can cause chemical burns. Wear personal protective equipment as directed on the SDS. Scrubbing with an acid solution causes a splash hazard. Protect yourself and others from chemical burns.

Proprietary cleaners are available for both block and brick. A typical solution consists of 1 part proprietary cleaner and 10 parts water. Some cleaning solutions are acid based; they should be used with extreme care. Acid-based solutions should never be used on concrete masonry units. They can be used on brickwork only after the wall has been presoaked with potable water. Presoaking fills the pores of the brick and mortar, preventing the solution from entering and discoloring the material.

WARNING!

Never pour a noncaustic solution, such as water, into a caustic solution, such as acid. The water could cause the acid to splash back into your face or onto your body. Always pour caustic solutions gently into noncaustic solutions so splash-back is minimized. Always wear appropriate personal protective equipment when working with acid-based solutions.

Efflorescence spots are scrubbed using the cleaner solution and a fiber brush (*Figure 17*). The process is repeated as often as necessary. After scrubbing, the wall is thoroughly rinsed with potable water. A pint of ammonia can be added for every 2 gallons of rinse water to help remove all traces of cleaner.

2.3.0 Repairing Cracks in Masonry Structures

Cracking is the most common problem in masonry structures. It has several causes; some are avoidable, and some are not. Moisture is responsible for many problems, causing cracking if it is allowed to penetrate the structure and freeze. Water can also pass through the wall between the units and the mortar if the mortar bond is inadequate. The original bond may be unsatisfactory, or a worker may disturb the position of a unit after it is laid, breaking the bond. This invites large-scale moisture penetration and subsequent cracking.

Figure 17 Fiber brushes.

Other causes of cracking are not so easily prevented. Poor soil conditions or inadequate foundation designs are often responsible for extensive cracking in some structures (*Figure 18*). Ultimately, a certain amount of cracking is to be expected in any masonry structure, but accepted and effective methods of repair can be used to remedy the problem.

2.3.1 Exterior Concrete Masonry Walls

Cracks should not be repaired as soon as they appear. For instance, if a crack occurs because of a foundation settlement, then immediate repair would be wasted since the stresses causing the problem are still present. Generally, cracks in a new wall should not be filled until a year after completion of the work.

Cracks should be repaired with material that matches the original construction. Plastic compounds are sometimes used, but are not recommended because they create an undesired appearance. The first step in crack repair is to clean the crack area and moisten the surrounding masonry. Use mortar that matches the original as closely as possible. When done, clean the wall using the techniques that were presented in *Masonry Level One*. Check the masonry unit manufacturer's SDSs, for recommended chemical cleaners and cleaning procedures. If not used properly, proprietary cleaners can be very destructive and must be used with protective countermeasures.

Follow the manufacturers' directions for mixing, using, and storing any chemical solution. Any cleaner used should first be tested on an inconspicuous 4-foot × 5-foot section of wall. Repairs on painted walls should be allowed to thoroughly cure before the wall is repainted.

Figure 18 Extensive cracking in masonry foundation wall.

If no cleaning procedures are given, use the bucket-and-brush method, described later in this module.

Cracks over ⅛ inch should be raked out to provide a firm key for the new mortar. Form a square notch at least ½-inch deep. Use 1 part cement and 2 parts sand to prepare a prehydrated mortar with the consistency of damp earth. Allow the mortar to stand for about an hour and temper it with water immediately before using. Tamp the mixture solidly into the crack to ensure a good bond. After the mortar has had time to set up, tool any repaired joints to match the existing surface.

2.3.2 Stucco

Stucco is typically laid over an insulating layer, especially in cold climates. Today, some type of proprietary exterior insulation foam system (EIFS) is most commonly used. When repairing cracks in stucco, it is important to avoid damaging any undamaged insulation beneath it.

The first step to repairing stucco is to chip out any cracks in a square notch shape so that the new material will remain firmly in place. Be careful not to chip out the insulation. If the insulation is damaged, refer to the manufacturer's instructions for proper repair procedures. Make sure the stucco surface is clean, without particles or dust.

Dampen the stucco before applying the new material.

A mixture identical to the original specifications should be used for all repair work. If these specifications are not known, a mixture of 1 part cement, 3 parts sand, and $\frac{1}{10}$ part hydrated lime is satisfactory. Any mineral pigments needed for coloring should also be added. Add enough water to form a mixture with the consistency of putty. The material is tamped firmly into the crack and kept damp for several days. If cracks appear in the new work, the entire surface may have to be covered with cement paint to adequately repair the defect.

2.3.3 Foundation Walls

Another common type of cracking occurs in foundation walls in a horizontal plane. This is usually caused by backfill being pushed against the foundation before it is secured with other parts of the structure. The wall normally cracks across the bed joint where the pressure is applied. This can be prevented by either bracing the foundation wall before backfilling or waiting until the structure is built on the foundation so the weight of the structure can hold the walls in place during the backfilling process.

The only correct way to repair this type of crack is to dig out the earth from around the foundation wall near the crack and re-lay the concrete block in fresh mortar. This is an expensive, time-consuming procedure, but if a crack of this type is visible, it is because the wall has been pushed out of alignment. Merely pointing mortar in the cracked joints does not solve the problem.

Additional Resources

Bricklaying: Brick and Block Masonry. 1988. Brick Industry Association. Orlando, FL: Harcourt Brace & Company.

Pocket Guide to Brick Construction. 1990. Reston, VA: Brick Industry Association.

Principles of Brick Masonry. Reston, VA: Brick Industry Association.

2.0.0 Section Review

1. The width of the cutting edge of a plugging chisel is at least _____.

   a. 1¼ inches
   b. 1 inch
   c. ¾ inch
   d. ½ inch

2. When cleaning efflorescence from a masonry surface, a typical cleaning solution consists of 1 part proprietary cleaner and _____.

   a. 20 parts water
   b. 15 parts water
   c. 10 parts water
   d. 5 parts water

3. EIFS stands for _____.

   a. exterior insulation foam system
   b. exterior inspection form sheet
   c. external iron foundation support
   d. efflorescence inhibitor fluid solution

3.0.0 RESTORATION OF BRICK WALLS

Objective

Describe how to restore brick walls.
  a. Identify staining problems.
  b. Explain how to remove old paint.
  c. Describe how to clean brick.
  d. Explain how to replace brick and mortar joints.

Performance Tasks

Clean a masonry wall with a bucket and brush.
Replace a damaged masonry unit in a wall.

Restoration of a brick wall may involve one or more repair techniques. Thoroughly examine the wall to determine what actually caused the problem. In some cases, the mortar has simply worn out; old mortar often contains no portland cement, just lime and sand with animal hair as the binder. Over time, this type of mortar will soften and break down, losing its ability to seal joints or adhere to the brick.

In other cases, improper flashing or structural damage to the roof has allowed water penetration behind the brick. This water can be absorbed by the brick and mortar. As the wall dries out, the moisture escaping can cause the brick to chip and mortar to pop out. Sometimes, the ground beneath the wall has settled, which can cause cracks in the wall that might eventually require rebuilding. Before any actual repointing is done, find out why the cracks developed and correct these conditions first.

3.1.0 Removing Stains

Stains may be caused by materials in masonry units or mortar, or from outside contaminants such as cleaners. Because each type of stain has

Restoration Rules

According to the Masonry Contractors Association of America, recent statistics show that over half of all new construction contracts are for restoration or adaptive re-use. This is roughly 17 percent of the masonry contracting dollars.

its own chemical composition, the means of removing it will vary. This section provides a broad overview of common types of stains in brick masonry structures and how to remove them. These include lime run, white scum, green or yellow stains, brown stains, and stains from external sources. Always refer to the manufacturer's instructions before cleaning block or brick. Using the wrong stain-removal techniques could cause more stains or even damage the masonry unit and mortar. *Table 2* provides additional information on cleaning specific types of stains.

3.1.1 Lime Run

Calcium is a common component in mortar, cement, and other construction materials. When a hairline crack or small hole opens up in a masonry surface, water flowing along the crack or through the hole for an extended period of time can cause the calcium to leach out of the masonry unit or mortar and react with carbon dioxide in the air to become calcium carbonate. The calcium carbonate then leaves streaking stains on the masonry surface as the running water carries it away (*Figure 19*). The white streaks are called lime run.

Before the masonry surface can be cleaned of lime-run stains, the source of the water flow will have to be identified and either stopped or redirected away from the masonry surface. Repair the crack or hole using the appropriate repair techniques for the materials. Then use a heavy-duty alkaline-based proprietary cleaner designed for use with the masonry materials. Depending on the extent of the staining, repeated applications of the cleaner and scrubbing may be required. Be sure to protect wood, metal, and painted surfaces from exposure to the proprietary cleaner, as they may be stained or damaged by coming into contact with the cleaner.

> **CAUTION**
>
> Disposal of excess chemicals is regulated by federal and state laws. Check the SDS or your safety coordinator for proper disposal. Only mix the amount of chemicals that you need for a project.

3.1.2 White Scum

If a masonry surface has been cleaned with a cleaning solution other than a proprietary cleaner designed for use with that type of surface, or if the surface was insufficiently prewetted prior to cleaning or rinsed after cleaning, whitish-gray silicate deposits may form on the surface of the masonry units (*Figure 20*). These deposits are

Table 2 Cleaning Guide for Masonry

Brick Category	Cleaning Method	Remarks
Red and Red Flashed	Bucket-and-Brush Hand Cleaning Pressurized Water Abrasive Blasting	Water, detergents, emulsifying agents, or suitable proprietary compounds may be used.
White, Tan, Buff, Gray, Pink, Brown, Black Specks and Spots	Bucket-and-Brush Hand Cleaning Pressurized Water Abrasive Blasting	Clean with water, detergents, emulsifying agents, or suitable proprietary compounds. Unbuffered muriatic acid solutions tend to cause stains in brick containing manganese and vanadium. Light-colored brick are more susceptible to "acid burn" and stains, compared to darker units.
Sand Finish or Surface Coating	Bucket-and-Brush Hand Cleaning	Clean with water and scrub brush using light pressure. Stubborn mortar stains may require use of cleaning solutions. Abrasive blasting is not recommended. Cleaning may affect appearance. See Brick Category for additional remarks based on brick color.
Glazed Brick	Bucket-and-Brush Hand Cleaning Pressurized Water	Wipe glazed surface with soft cloth within a few minutes of laying units. Use a soft sponge or brush plus ample water supply for final washing. Use detergents where necessary and proprietary cleaners only for very difficult mortar stain. Consult brick and cleaner manufacturer before use of proprietary cleaners on salt-glazed or metallic-glazed brick. Do not use abrasive powders. Do not use metal cleaning tools or brushes.
Colored mortars	Method is generally controlled by Brick Category	Many manufacturers of colored mortars do not recommend chemical cleaning solutions. Unbuffered acids and some proprietary cleaners tend to bleach colored mortars. Mild detergent solutions are generally recommended.

28303-14_T02.EPS

28303-14_F19.EPS

Figure 19 Lime run.

commonly called white scum. White scum may coat the face of masonry units or appear as small spots. It can also appear on trim and glass.

White scum can be treated by using proprietary cleaners designed to remove it from the specific type of masonry surface where it appears. White-scum remover consists of a blend of inorganic and mild organic acids plus special wetting agents and inhibitors. Following the application of the cleaning agent, wash the brick according to the manufacturer's instructions.

> **NOTE**
> White scum resembles the scumming that can appear on brick during the manufacturing process. This can be distinguished from white scum because it will be present on the surface of masonry units before they are laid; white scum appears after laying.

NCCER – *Masonry Level Three* 28303-14

Figure 20 White scum.

WARNING!

Sodium hydroxide, a chemical commonly used in many proprietary cleaners, is a corrosive material. Wear appropriate personal protective equipment when using sodium hydroxide.

3.1.3 Green or Yellow Stains

If unbuffered muriatic acid is used to clean brick, the acid can cause green or yellow stains on brick (*Figure 21*). These stains are caused by sulfate and chloride salts that have been leached out of the brick by exposure to the acid. Green and yellow stains can also be caused by prolonged exposure to excessive moisture. These stains are difficult to clean, and if cleaned improperly may cause even more staining.

To remove green or yellow stains from masonry surfaces, begin by allowing the masonry to dry thoroughly. Patch any cracks or holes that may allow water to seep or collect on the masonry surface. Because green and yellow stains tend to occur on lighter-color brick or mortar, use a proprietary cleaner designed for use on such brick. After application of the cleaning solution, rinse the masonry surface with water according to the manufacturer's instructions. Protect polished surfaces and adjacent nonmasonry materials from exposure to the proprietary cleaner. Fumes from the cleaner may be harmful; observe all recommended safety precautions.

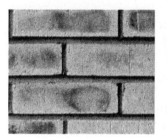

GREEN STAINS   YELLOW STAINS

Figure 21 Green and yellow stains.

WARNING!

Breathing small amounts of trichloroethylene, a chemical used in some proprietary cleaners, may cause headaches, lung irritation, dizziness, poor coordination, and difficulty concentrating. Ventilate the area well and/or wear a respirator. Read and follow safety precautions on the product label.

3.1.4 Brown Stains

Brick and mortar that has been exposed to unbuffered cleaning agents such as muriatic acid, or occasionally to rain with a high acid content, may develop tan or brown discolorations in the mortar joints or on the surface of the brick (*Figure 22*). These stains, called brown stains, are caused by manganese coloring agents in the brick react-

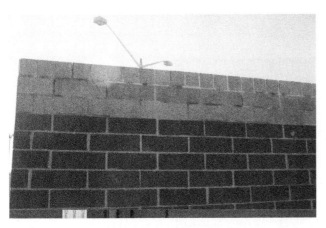

BROWN STAINS
ON MORTAR JOINTS

BROWN STAINS
ON BRICK SURFACE

28303-14_F22.EPS

Figure 22 Brown stains.

ing to the acid. The chemical processes that cause brown stains are similar to those that cause efflorescence.

To treat brown stains on brick and mortar, allow the masonry to dry thoroughly. Apply a proprietary cleaning solution that has been designed to remove brown stains from the specific type of masonry surface. The cleaner may be the same type as that used to clean green and yellow stains. After applying the cleaning solution for the required amount of time, rinse the masonry surface with water according to the manufacturer's instructions. Protect polished surfaces and adjacent nonmasonry materials from exposure to the proprietary cleaner. Fumes from the cleaner may be harmful; observe all recommended safety precautions.

> **WARNING!**
>
> Hydrofluoric acid is a hazardous, corrosive acid. Observe proper safety precautions when using and handling it. The liquid and vapor forms of the acid can cause severe burns. Specialized medical treatment is required to treat any exposure to hydrofluoric acid.

3.1.5 Stains from External Sources

Masonry surfaces and mortar joints can also be stained by exposure to pollution; organic growth such as mold; hard water; runoff of stormwater and paint. Chemical reactions with metals that are adjacent to or in contact with the masonry structure can also cause staining. If the cause of the staining is not immediately apparent from a visual inspection, tests will have to be conducted to determine the composition of the stain and identify the proper method of treatment.

Stains caused by acids can be removed using proprietary cleaners designed for use on green, yellow, and brown stains. Surprisingly, mud is the most difficult stain to remove from a masonry surface. Special restoration cleaners are available for mud, but they may require longer exposure on the surface than other types of cleaners. The Brick Industry Association (BIA) recommends the use of scouring powder and a stiff bristle brush for brick with smooth texture. Use pressurized-water cleaning on brick with rougher textures.

Stains from smoke and other atmospheric pollutants can be washed off using a bleach scouring powder or proprietary smoke-removal products followed by an application of a proprietary cleaner and rinsing.

Specially formulated cleaners are available to remove oil and tar from masonry surfaces. The BIA recommends the use of household bleach, ammonium sulfate, or weed killer to remove mold stains.

3.2.0 Removing Old Paint

Old paint must be removed before brickwork can be repaired. Chemical compounds are usually needed to remove paint, shellac, or wax coatings. Alkaline cleaners, or strippers, are very effective in removing multilayered applications of paint. Depending on the type of cleaner used to remove

Environmental Laws

It is illegal in many places in the United States to intentionally pour or spill many types of chemicals onto the ground. In most instances, the runoff from chemical solutions must be controlled and not allowed to enter storm drains or sewers. Check with the local authorities before using chemicals and allowing the rinse water to seep into the ground. Check for a nontoxic alternative to better ensure greater worker and environmental safety.

paint, the cleaning runoff may need to be captured and disposed of according to procedures specified in the local applicable code.

3.3.0 Using Appropriate Cleaning Techniques

There are more than just cosmetic reasons to clean and maintain older masonry structures. In fact, cosmetic reasons alone may not be a good reason for a full-scale cleaning program. The weathered patina of masonry often becomes a part of a building's character. The unnecessary cleaning of undamaged or lightly soiled walls may do more harm than good. The harsh chemicals or abrasive action can remove the protective crust that has formed on the masonry surface. It should not be disturbed, as long as it does not contribute to or conceal deterioration. The body of the brick or stone underneath may be too soft to withstand the attack of modern urban pollutants.

On the other hand, excessive soiling can disguise or even contribute to physical damage of the masonry. A heavy dirt buildup may conceal cracks and other signs of deterioration that warrant investigation and repair. Sometimes the surface must be cleaned before a thorough investigation can be performed.

Dirt may also cause or aggravate deterioration of the masonry, because it significantly increases the amount of moisture that is attracted to and held on a wall surface. Dirt impedes natural drying after a rain.

Prolonged dampness increases the chemical reactivity of the masonry with common air pollutants. It also increases the risk of freeze-thaw damage in the winter and the growth of micro-vegetation in the summer. Water that gets into the wall from other sources is trapped because it cannot evaporate at the surface. This moisture contributes to the accelerated corrosion and failure of concealed metal components and structural supports. Moisture damage can go beyond the masonry wall itself to interior finishes and other adjacent elements.

Start a cleaning program with carefully planned, on-site testing of specific materials and cleaning methods. Begin testing well in advance of necessary work. Choose test patches that represent the different types of substrates involved and the substances to be removed.

Remove as much of the dirt or stain as possible before test cleaning. Scrape the area by hand with a brush with nonmetallic bristles, ensuring the most accurate test results. Follow the same procedure when full-scale cleaning begins.

> **CAUTION**
>
> Do not use brushes with ferrous iron or nonstainless steel bristles to clean masonry surfaces. The iron in the bristles can transfer chemically to the mortar and cause stains.

Test patches serve as a standard to assess full-scale cleaning. Do not evaluate the test areas until they are dry and have weathered as much as possible. Ideally, exposure to a complete one-year weathering cycle will give the most accurate and reliable information. When this is not feasible, allow a minimum of one month. During this time there should be several wetting cycles and many temperature variations.

All Gloves Are Not Created Equal

When using acids, oxidizers, or caustic chemicals, the gloves you use are important to your safety. Always check that the gloves are designed to protect against the chemical you are using. Some gloves, like latex, cotton, or leather, will be destroyed by certain solutions. Refer to the manufacturer's specifications and to the SDSs for the chemicals being used. Wear gloves over your clothes, and use tape or rubber bands to seal them against your clothing.

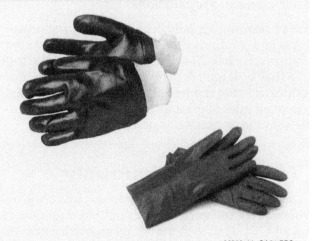

28303-14_SA01.EPS

3.3.1 Common Cleaning Failures

Because of the different approaches to cleaning masonry structures, there are many misconceptions and misunderstandings about cleaning requirements. This often creates ineffective results. Cleaning failures generally fall into one of the following three categories:

- *Failure to saturate the masonry surface thoroughly with water before and after application of chemical or detergent cleaning solutions* – Dry masonry will absorb the cleaning solution and may result in mortar smear, white scum, or the development of efflorescence or green stain. Saturation of the surface prior to cleaning reduces the absorption rate. The cleaning solution will stay on the surface of the masonry rather than being absorbed.
- *Failure to use chemical cleaning solutions properly* – Improperly mixed or overly concentrated chemical solutions can etch or wash out cementitious materials from the mortar joints. They have a tendency to discolor masonry units, particularly the lighter shades. They can produce discoloration and promote the development of green and brown stains.

> **CAUTION**
>
> Do not overmix cleaning chemicals in an attempt to be more efficient. Overmixed chemicals can cause damage to masonry surfaces.

- *Failure to protect windows, doors, and trim* – Many cleaning agents will corrode metal. Some cleaning solutions may cause pitting if they contact metal frames. They can stain the masonry surface as well as trim materials such as limestone and cast stone.

3.3.2 General Preparations

Before cleaning actually begins, all cleaning procedures and solutions should be applied to a test area of approximately 20 square feet. The size of the test area may be larger, depending on the cleaning procedure. Inspect the test area after waiting at least one week after application, then modify the cleaning solution or process as necessary.

The indiscriminate use of proprietary cleaners or the wrong cleaning compound can cause difficult-to-remove stains. Reactions of brick to cleaning solutions are not always predictable. It is safer to use a trial-and-error method on a small test area before committing the entire project to a set procedure. Minute quantities of certain minerals found in some fired-clay masonry units may react with some solutions and cause staining.

> **WARNING!**
>
> Acids and other cleaning solutions can cause chemical burns. Protect yourself and others. Wear personal protective equipment. Use barriers to keep people away from your work area.

Sample testing should be performed under temperature and humidity conditions that closely approximate the conditions under which the brickwork will be cleaned. Chemical cleaning solutions are generally more effective when the outdoor temperature is 50°F or above. Keeping brickwork free of mortar smears is advisable, although in modern construction where speed is important, this may be difficult.

Good practice dictates that the cleaner you maintain your work, the less cleanup you have to perform when the job is completed. Follow these precautions to minimize cleanup when laying brick:

- Protect the base of the wall from rain-splashed mud and mortar splatter. Use straw, sand, sawdust, or plastic sheeting spread out on the ground, extending 3 to 4 feet from the wall surface and 2 to 3 feet up the wall.
- Turn scaffold boards near the wall on edge at the end of the day to prevent possible rainfall from splashing mortar and dirt directly onto the completed masonry.
- Cover walls with a waterproof membrane at the end of the workday to prevent mortar-joint washout and entry of water into the completed masonry.
- Protect site-stored brick from mud. Store brick off the ground and under protective covering.
- Practice careful craftwork to prevent excessive mortar droppings. Excess mortar should be cut

Test Inconspicuously

Locate test patches in an inconspicuous area of the building. Select areas that are not easily seen, for example a back wall or corner. There are many unforeseeable factors and the results are not always predictable. You do not want a failed test to be the first thing people notice. Paint removal testing, however, should be done near the front entrance to the building where the most layers of paint are likely to be found.

Figure 23 Trim excess mortar.

Figure 24 Brush off excess mortar after tooling.

off with the trowel as the brick is laid (*Figure 23*). Joints should be tooled when thumbprint hard. After tooling, excess mortar and dust should be brushed from the surface (*Figure 24*). Avoid any motion that will result in rubbing or pressing mortar particles into the brick faces. A medium-soft bristle brush is preferable.

• Retool joints as needed.

> **WARNING!**
>
> The solutions and fumes of some chemicals or chemical compounds used to clean brickwork may be harmful to your health. Use protective clothing and accessories, proper ventilation, and safe handling procedures. The use and disposal of some chemicals or chemical compounds is regulated by federal, state, or local laws and should be researched before use. The manufacturer's material and handling requirements must be strictly observed when using any proprietary compound.

> **NOTE**
>
> Always read the SDS and the instructions when using cleaning chemicals. Personal protection, first-aid, cleanup, and disposal requirements are listed on every SDS.

3.3.3 Bucket-and-Brush Hand Cleaning

Simple bucket-and-brush cleaning is probably the most popular, but most misunderstood of all the methods used for cleaning brick masonry. There are many different proprietary cleaning compounds and brushes available for this method of cleaning. Select a brush designed to be used with the proprietary cleaner that you are using, particularly if the cleaning compound contains acid. Acid can damage or destroy some types of brushes. Different styles of cleaning brushes are shown in *Figure 25*.

Work on a small area at a time. The size of the wash-down area should be determined after a trial run. The following is a recommended general procedure for using proprietary cleaners:

Step 1 Wear appropriate personal protective equipment.

Step 2 Select the proper solution.

• For proprietary compounds, check that the product is suitable for brick. Follow the manufacturer's dilution instructions. Many cleaning solutions work well for their intended cleaning jobs, but the formula may change. Therefore, test each product on a panel or inconspicuous wall area before use.

• Use TSP (trisodium phosphate) laundry detergent or soap solutions to remove mud, dirt, and soil accumulation. A suggested solution is one-half cup dry measure of dishwashing-machine detergent and one-half cup dry measure of laundry detergent dissolved in 1 gallon of potable water.

- For solutions, mix a 10 percent solution of proprietary cleaner in a nonmetallic container. This is 9 parts potable water to 1 part cleaner. Pour the cleaner into the water. Do not permit metal tools to contact the cleaner solution.

> **NOTE**
> Every brick bundle has a tag that specifies the proprietary cleaners that are safe for use with that brick.

Step 3   Remove larger dirt particles by hand with nonmetallic scraper hoes or chisels.

Step 4   Protect metal, glass, wood, limestone, and cast-stone surfaces. Mask or otherwise protect windows, doors, and ornamental trim from the cleaning solution.

Step 5   Pre-soak or saturate the area to be cleaned. Rinse with water from the top down. Saturated brick masonry will not absorb the cleaning solution or dissolved mortar particles. Saturate areas below the area to be cleaned to prevent absorption of the runoff.

Figure 25   Different styles of cleaning brushes.

Use the Correct Mixture

It is tempting to mix a cleaning solution stronger than recommended in order to clean stubborn stains. Overuse of a cleaning solution can cause further stains, especially on light or colored brick. Read and follow the manufacturer's instructions.

Step 6   Apply the cleaning solution starting at the top. Use a long-handled, stiff fiber or bristle brush as recommended by the cleaning-solution manufacturer. Allow the solution to remain on the brickwork for 5 to 10 minutes, and then begin washing from the bottom up. This allows for the gradual construction of scaffold when cleaning tall surfaces. For proprietary compounds, follow the manufacturer's instructions for application and scrubbing. Nonmetallic tools may be used to remove stubborn particles. Do not use metal scrapers or chisels, because metal marks will oxidize and cause staining.

Step 7   Heat, direct sunlight, warm masonry, and winds will affect the drying time and reaction rate of cleaning solutions. Ideally, work in shaded areas to avoid rapid evaporation that causes streaking.

Step 8   Rinse all areas thoroughly. Flush walls with large amounts of potable water from top to bottom before they can dry. Failure to completely rinse the wall of cleaning solution and dissolved matter from top to bottom may result in the formation of white scum.

> **WARNING!**
> Always pour proprietary cleaner into a container of water; never pour water into a container of proprietary cleaner. Pour carefully to minimize the splashing of the cleaner out of the container. Always use the proper personal protective equipment when handling proprietary cleaners.

> **NOTE**
> Cleaning solutions and acids can damage grass and other plants.

28303-14_F25.EPS

28303-14_F26.EPS

Figure 26 Typical pressure-washing operation.

3.3.4 Pressurized-Water Cleaning

Pressurized-water washing of masonry structures (*Figure 26*) involves the use of water and a detergent to clean the surface by causing dirt to absorb water and swell, thereby loosening it so that it can be washed away. To prevent efflorescence, plan to use the least amount of water and detergent that is required to clean the wall completely.

Begin by removing clay and dirt with a dry fiber or nylon brush. Do not use steel brushes, as they can leave behind metal particles that rust and discolor the surface. Use heated water on greasy surfaces or when cleaning during cold weather.

The National Concrete Masonry Association (NCMA) recommends that masons follow these guidelines when using pressure-washing equipment:

- The water pressure should be limited to 400 to 600 pounds per square inch (psi)
- Use a wide-flange tip only, not a pointed tip
- Keep the tip at least 12 inches away from the masonry while cleaning
- Direct the spray at a 45-degree angle to the wall, never perpendicular to the wall

Muriatic Acid

In the past, masons used muriatic acid to clean masonry surfaces. However, muriatic acid is no longer used because it is highly caustic and hard to neutralize. Also, the appropriate concentration to use in cleaning is not regulated. Today, masons use proprietary cleaners that are more controllable and cost effective.

Easy Does It

Do not use high pressure to apply cleaning solutions. If high pressure is used, the cleaning solution can be driven into the masonry. That can become the source of future staining. Reduce the risk of penetration by saturating the walls sufficiently with water before application.

Pressure washing has the best results when the operator uses a fan-type tip, dispersing the water through 25 to 50 degrees of arc. The amount or volume of water has more effect than the amount of pressure. The minimum flow should be 4 to 6 gallons per minute (gpm). Usually, the compressor should develop from 400 to 600 psi water pressure for the most effective washing. It is important to keep the water stream moving to avoid damaging the wall.

Pressure washing can be used in combination with various cleaning compounds. Training and practice are necessary to properly control the mix of proprietary cleaners, pressure, and spray pattern. *Figure 27* shows what can happen when pressure washing is done incorrectly. Improper

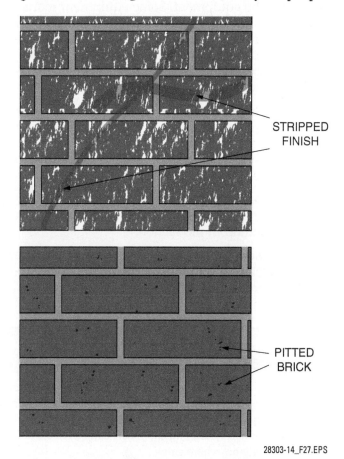

STRIPPED FINISH

PITTED BRICK

28303-14_F27.EPS

Figure 27 Results of improper pressure washing.

28303-14_F28.EPS

Figure 28 Portable pressure washer.

pressure-washing technique can remove finish and even score brick, resulting in costly repairs.

For ease of use, equipment should be as portable as possible. Units may be on wheels (*Figure 28*), skids, trailers, or pickup-truck beds. More elaborate systems include pumps, engines, cleaner containers, and water storage tanks fixed on truck beds.

Check that the cleaning compound is compatible with the equipment. Some equipment manufacturers allow only specific cleaning compounds to be used with their equipment. Some pumps will resist different types of solutions.

The following procedure is suggested when using a pressure system:

Step 1 Select and test the cleaning solution on a sample area. Check that the equipment is compatible with the cleaning solution. Mix proprietary compounds according to the manufacturer's instructions.

Step 2 Prepare the wall for cleaning. Follow Steps 3 through 5 of the bucket-and-brush method in the section.

Step 3 Use the pressure system to apply the cleaning solution, following the manufacturer's instructions.

Step 4 Permit the cleaning solution to remain on the wall for approximately five minutes.

Step 5 Starting at the top, rinse the wall with the pressure nozzle using water only. Use minimal pressure for the rinse cycle.

This cleaning method is likely to change the appearance of certain types of masonry. Pressure cleaning can damage sand-molded brick, sand-faced extruded brick, and brick with glazed coatings or slurries applied to the finished faces. Check with the brick manufacturer or an expert in brick-structure preservation before using this technique on these types of masonry.

> **WARNING!**
>
> Pressurized cleaning equipment creates dangerous conditions. Read the manufacturer's operating manual before using. Make sure that you know how to use all pressure-release valves and safety switches. Wear safety glasses and other appropriate personal protective equipment.

3.4.0 Replacing Brick and Mortar Joints

Moisture may penetrate brick units that are broken or heavily spalled. When this occurs, replacement of the affected units may be necessary. The following procedure is suggested for the removal and replacement of masonry units:

Step 1 Remove the damaged brick. Drill several holes in the middle of the masonry unit to weaken the unit. Break the unit by chiseling between the holes for easier removal. Use a toothing chisel (*Figure 29*) to cut out the mortar that surrounds the affected unit.

Step 2 Clean out the area. Carefully chisel out all of the old mortar. Sweep out all dust and debris with a brush. If the units are located in the exterior wythe of a cavity wall, do not allow debris to fall into the cavity.

Protect Yourself

It is nearly impossible not to splash chemicals when scrubbing with chemical cleaning compounds. Read the instructions and wear proper personal protective equipment. This can include goggles, a face shield, chemical-resistant gloves, and a chemical-resistant apron or overalls. Splashing certain acids or chemicals in your eyes can cause blindness. Cleaning chemicals can cause severe skin burns. Review first-aid and other safety precautions before using any chemicals.

Step 3 Dampen the brick surfaces in the wall and the new replacement brick before new units are placed. Allow the masonry to absorb all surface moisture to ensure a good bond.

Step 4 Butter the appropriate surfaces of the surrounding brickwork and the replacement brick with mortar. Prehydrated mortar may be necessary to match the existing joints.

Step 5 Center the replacement brick in the opening and press it into position. Remove the excess mortar with a trowel. Point around the replacement brick to ensure full head and bed joints.

Step 6 Tool the joints to match the original profile when the mortar becomes thumbprint hard.

Step 7 Remove mortar crumbs with a stiff bristle brush after mortar has set up.

Deteriorated mortar joints in existing masonry can be cleaned out and repointed with fresh mortar. If the joints are bad, but the masonry units are in good shape, there is no need to remove that whole portion of the wall; only the damaged joints should be repaired. The procedure is the same as that for normal joint tuckpointing.

The mortar used for repointing should match the existing material as closely as possible in color and texture, and should match or exceed the strength and hardness of the existing mortar. Modern mortars containing portland cement

are much harder than older mixtures. In some cases, they are harder than the brick or stone being used. Ideally, the new mortar should have the same density and absorbency as the stone or brick in the wall.

A hard mortar used with soft brick or stone can cause deterioration of the masonry. The two components will not respond to temperature and moisture changes at the same rate or to the same degree. The softer material will absorb more movement stress and more moisture. The hard mortar can act as a wedge, breaking the edges off the units. Many buildings have been irreparably damaged in this manner. Strong portland cement mortars may also shrink, leaving minute cracks at the mortar-to-unit interface.

To ensure a good bond with brick and stone, dampen the cleaned joints with water just before beginning work. Mortar is placed by using a grout bag or a tuckpointer's trowel. Fill the joint with mortar applied in thin, 1/4-inch layers. These must become thumbprint hard before the next layer is placed. Fill the vertical joints before the horizontal joints.

Joints should not be overfilled to the point where mortar hides the edges of the units. This makes the joint appear too wide, and the edges break off too easily, leaving voids through which moisture can penetrate. For brick or stone that has weathered to a rounded profile, the new mortar joint should be slightly recessed from the unit surface and tooled concave to avoid feathered edges. Stippling joints with the bristles of a stiff, nonmetallic brush while the mortar is still soft will give it a worn appearance. Moist curing may be necessary in hot, dry weather to ensure proper hydration and good mortar bond to the units.

28303-14_F29.EPS

Figure 29 Toothing chisel.

Clean and Green

Harsh chemicals are not always required to remove tough stains. There are many nontoxic products designed to remove paint and other masonry stains. These products are effective stain removers, but do not have the problems associated with harsh chemicals. They do not require special protective equipment or expensive disposal. Many contracts require the use of nontoxic cleaners.

Additional Resources

Good Practice for Cleaning New Brick Work. 2009. Charlotte, NC: Brick Industry Association Southeast Region. **http://www.gobricksoutheast.com**

Sure Klean Guide to Common Brick Staining in New Construction. 2013. Lawrence, KS: PROSOCO, Inc. **http://www.prosoco.com**

Technical Note TN20, *Cleaning Brickwork*. 2006. Reston, VA: The Brick Industry Association. **www.gobrick.com**

Technical Note TN23, *Stains—Identification and Prevention*. 2006. Reston, VA: The Brick Industry Association. **www.gobrick.com**

3.0.0 Section Review

1. When calcium leaches out of masonry unit or mortar through a hairline crack or small hole and reacts with carbon dioxide in the air to become calcium carbonate, the resulting stain is called _____.

   a. green or yellow stain
   b. brown stain
   c. white scum
   d. lime run

2. Multilayered applications of paint can be effectively removed using _____.

   a. tuckpointing
   b. pressure washing
   c. stripper
   d. paint thinner

3. Probably the most popular method for cleaning brick masonry is _____.

   a. bucket-and-brush cleaning
   b. pressure washing
   c. sandblasting
   d. repainting

4. When replacing brick and mortar joints, before placing new units the brick surfaces in the wall and the new replacement brick should be _____.

   a. dried
   b. dampened
   c. roughed
   d. smoothed

4.0.0 REPAIR OF FOUNDATION WALLS

Objective

Describe how to repair a foundation wall.

a. Explain how to repair water intrusion.
b. Explain how to repair cracks and localized problems.

Trade Term

Portland cement paint: Cement-based paint. Type I, containing 65 percent portland cement by weight, is for general use; Type II, containing 80 percent portland cement by weight, is used where maximum durability is needed. Within each type there are two classes: Class A contains no aggregate filler and is for general use; Class B contains 20 to 40 percent sand filler and is used on open-textured surfaces.

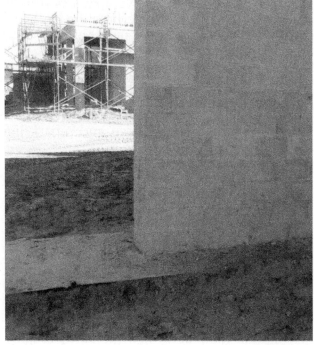

28303-14_F30.EPS

Figure 30 Concrete block foundation on concrete footer.

Foundation and basement walls are usually built of 8- or 12-inch-wide hollow concrete block, as shown in *Figure 30*. They are built on either masonry or concrete footers. Some building codes require that a solid concrete block cap be used on the top of the wall instead of placing mortar in all the cells of the top course. Any part of the foundation or basement wall that is underground must be parged from the top of the footer to the ground level to prevent water from getting into the wall. Usually, an asphalt mixture is brushed, sprayed, or troweled on top of the parge coat.

Foundation walls usually require repair for cracking and water intrusion. Cracking can result from many different causes, ranging from uneven settlement of the footings to hydrostatic pressure on the wall exterior caused by higher-than-expected groundwater. Inadequate sealing and caulking during initial construction may allow water intrusion through the joints or through the block itself.

Foundation walls are normally backfilled on the outside, so accessibility may be a problem. Making repairs from the inside should only be done to correct problems with the inside surface. Any structural repairs or heavy waterproofing should be done on both sides of the wall. This means the backfill around the outside must be excavated to allow access to the wall.

4.1.0 Repairing Water Intrusion

The cause of a leaky basement is either water buildup against the exterior wall or groundwater from under and around the footings and wall edges. The best guard against water penetration is good workmanship. Slope the grade away from the wall to keep water away from walls. Reduce surface water by providing gutters, downspouts, or drain tile around the building.

Even with sound workmanship, water buildup sometimes occurs. To remedy the problem of wet basement walls, the area around the wall should be excavated and the wall waterproofed. When

---

Old Mortar

Historic mortars were generally soft, and may have been mixed from clay, gypsum, lime, and natural cement. Some earlier mortars may contain lumps of partially burned lime or dirty lime, shell, natural cements, clay, lampblack, or even animal hair. Some later mortars were mixed with portland cement. Use laboratory analysis to determine the exact ingredients and proportions of old mortar.

---

the problem is not extensive, use less drastic methods. These include waterproofing the joints or applying portland cement paints or other coatings to the interior wall.

Small openings and cracks in the mortar joints can be sealed with a grout coating, which may reduce water penetration. A typical grout coating consists of ¾-part portland cement, 1-part sand that passes a No. 30 sieve, and ¼-part fine hydrated lime.

Mix all the ingredients with water shortly before use to obtain a fluid consistency. Wet the joints thoroughly, and allow the masonry to absorb surface water before the grout is applied. Then, brush the grout into the joints with a fiber brush. Normally, two coats of grout are necessary for adequate protection.

Chip out large cracks to form a square notch with rough edges. Remove all loose material from the defective joint. Brush out and dampen the joints to prevent moisture absorption from the new mortar. Tamp in the new mortar thoroughly to form a complete bond with all surfaces. After the material has set, it should be kept moist for several days to assist curing.

4.1.1 Portland Cement Paint

Portland cement-based masonry paints can also reduce water penetration. The paints are widely available in standard and heavy-duty types. The standard type, called Type I, has a minimum of 65 percent portland cement by weight. It is suitable for general use. The heavy-duty type, called Type II, contains 80 percent portland cement. Each type is available with a siliceous sand additive for use as a filler on a porous surface.

Mix the paints with water just before use, as they set by hydration of the cement. Apply them with a brush to a moist surface. Keep the area damp using a fine spray of water for 48 to 72 hours until the paint cures. These paints are very successful when mixed properly and applied following the manufacturer's instructions.

The newest generation of portland cement paints includes many proprietary products that mix chemical modifiers and plastics with the cement and sand. These newer products are used for applications such as reservoirs, tunnels, swimming pools, fountains, and fish tanks. They provide a tanking skin that waterproofs concrete and masonry. These products are also used as a waterproof coating in showers, bathrooms, and basements. *Figure 31* shows a dock being cleaned and sealed with a penetrating water-dispersed thermoplastic polyester polymer.

28303-14_F31.EPS

Figure 31 Cleaning and sealing a dock.

Application instructions vary with the manufacturer. Application may require several coats, with curing intervals. Be sure to read and follow manufacturers' instructions for these new products.

4.1.2 Exterior Repairs

Water intrusion through building exteriors can create serious problems. Moisture trapped inside a structure can cause general deterioration of materials. It can also cause structural problems created by the freeze-thaw cycle. The primary defense against this problem is to keep the moisture from entering in the first place. Where this is not possible, use other measures such as weepholes, sumps, and air vents to allow drainage and drying.

One source of moisture penetration is missing or deteriorated sealant in contact areas between brickwork and other materials (*Figure 32*). Inspect the sealant joints in window and door frames, expansion joints, and sill plates. If the sealant is missing, apply a full bead of high-quality, permanently elastic sealant compound in the open joints.

If a sealant material was installed but has torn, deteriorated, or lost elasticity, carefully cut it out and replace it. Clean the opening of all old sealant material. Place a new sealant in the cleaned joint if required by the manufacturer's instructions.

Surface grouting is an effective way to seal small hairline cracks in mortar joints. One recommended grout mixture is 1-part portland cement, ⅓-part hydrated lime and 1⅓-parts fine sand passing a No. 30 sieve. Dampen the joints to be grouted. Allow the brickwork to absorb all surface water to ensure a good bond. Add potable water to the dry ingredients to obtain a fluid consistency. Apply the grout mixture to the joints with a stiff fiber brush to force

DETERIORATING SEALANT

Figure 32 Deteriorating sealant between masonry and door frame.

28303-14_F32.EPS

the grout into the cracks. Use two coats to effectively reduce moisture penetration. Tool the joints after the grout application to force the grout into the cracks. Use a template or masking tape to help keep the brick faces clean.

In some instances, moisture may penetrate walls but be unable to escape to the exterior. Instead, it finds its way to the interior of the wall. If this is the case, inspect the weepholes to determine if they are clogged or improperly spaced. If the weepholes are clogged, clean them out by probing them with a thin wooden dowel or stiff wire. If the weepholes were not properly spaced, drilling new weepholes at closer intervals may be necessary.

4.2.0 Repairing Cracks and Localized Problems

Cracks, spalls, wall deflections, and bulging are all signs of uncontrolled or unaccommodated movement. Complex settlement and lateral pressure problems cause cracking in foundations. This can require extensive structural analysis to arrive at an engineered solution. In these cases, a structural engineer evaluates the problem and recommends a solution.

Common causes for differential settlement include poor soil compaction under the footings, improper footing size, hydrostatic pressure, and excessive loading from above. Repairs include mud jacking, additional bracing, or a complete rebuild of part or all of the foundation.

Cracks can develop in foundation walls for several reasons other than settlement or lateral pressure. In these cases, some localized repairs are possible and desirable. Cracks and spalls often accompany deflection or bulging due to expansion of the masonry units. Proper spacing and construction of control joints will accommodate movement resulting from moisture and temperature changes. Correct a lack of proper jointing by cutting new joints into the wall.

Sometimes, an existing joint does not perform properly because it is inadvertently bridged with mortar. This condition is easily corrected: clean the hardened mortar from the joint. Correct any other conditions that restrain joint movement.

Cracks can develop where a change in material occurs. Vertical cracks tend to form in the center of walls that are overloaded. Cracks also occur around improperly designed openings in masonry and near corners that are twisting due to structural restraint. Install bond-breaking flashing between the masonry and the foundation to prevent spalls and cracks at foundation corners.

Cracks will form due to extensive rusting of lintels and shelf angles. In repairing shelf angles, relieve stress from the top down to avoid damage to headers or transverse bond within mortar joints. Relieve stress by cutting new movement joints or relieving the restraint at existing movement joints.

Wall deflection or bulging due to a lack of ties can be corrected by installing special retrofit repair anchors or ties. Space these properly throughout the wall. Straighten and re-anchor bulged masonry before making other repairs. Replace spalled or cracked masonry units. Tuckpoint damaged mortar joints. The degree of repair depends on the exposure condition and the amount of damage incurred. Structural problems that may be causing the cracking can be difficult to identify and are not covered in this module.

Additional Resources

Building Block Walls: A Basic Guide, Latest Edition. Herndon, VA: National Concrete Masonry Association.

Pocket Guide to Brick Construction. 1990. Reston, VA: Brick Industry Association.

4.0.0 Section Review

1. The minimum amount of portland cement by weight in Type I portland cement paint is _____.

   a. 65 percent
   b. 50 percent
   c. 35 percent
   d. 20 percent

2. Poor soil compaction under footings is a common cause of _____.

   a. spalling
   b. shortening of the structural frame
   c. differential settlement
   d. bulging

5.0.0 REPAIRING AND REBUILDING CHIMNEYS AND FIREPLACES

Objective

Describe how to repair and rebuild chimneys and fireplaces.

  a. Explain how to repair chimneys.
  b. Explain how to repair fireplaces.

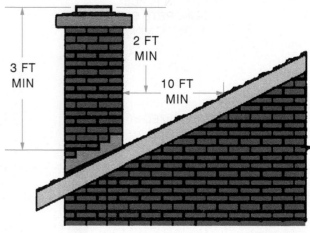

28303-14_F33.EPS

Figure 33 Minimum chimney clearance.

Repairing and rebuilding old fireplaces and chimneys is generally more difficult than the original construction. Fireplaces and chimneys must be properly maintained, as they are particularly susceptible to weather damage and can deteriorate rapidly. If they deteriorate even slightly, they may become hazardous to operate.

5.1.0 Repairing Chimneys

A fireplace chimney serves a dual purpose: to create a draft and to dispose of the products of combustion. If there is damage to a chimney that creates a problem with either of these functions, the chimney must be repaired or replaced. For example, if the chimney does not draw well, increasing its height may improve the draft.

A general guideline for chimney height states that the chimney opening should be at least 3 feet above a flat roof, 2 feet above the ridge of a pitched roof, or 2 feet above any part of the roof within a 10-foot radius of the chimney, as shown in *Figure 33*.

Repairs for chimneys fall into several categories. If the problem involves deteriorated mortar joints or spalling of the brick face, the repairs can be made by tuckpointing the joints and chipping out and replacing individual brick. Other types of repairs include rebuilding the chimney top, replacing racked-face surfaces, and resealing openings in appliance chimneys.

5.1.1 Chimney Tops

The top of the chimney is subject to moisture, gases, and extreme fluctuations in temperature. As the chimney ages, it may deteriorate. Perform a complete inspection of the chimney before any repair or rebuilding work is started. *Figure 34* shows

the design details for the top of a chimney. If the only problem is deteriorated mortar joints, they can be cut out and repaired. However, if the chimney is badly damaged, from the roof to the top, or if the brick is burned out, the best procedure is to tear down the brickwork to the roofline and rebuild.

The first steps in the repair process are to build a scaffold and to cover the roof. These steps often involve more work than the actual chimney repair. Erect a strong scaffold fastened to the building. Set the legs on a strong, flat base. Tie it to the structure with wire attached to the scaffold frame.

If the scaffold must be built on the roof, use scaffold frames that fit the slope of the roof. The top of the working platform must be level. Tie the scaffold to the base of the chimney so it cannot slip. Lay an old rug, a canvas tarpaulin, or other nonslip coating over the roof in the work area. This will protect the roof against mortar stains or other damage to the shingles. Remove this covering after the job is complete.

Before tearing down any old brickwork, make provisions to collect any debris that may fall into the chimney. Fill a burlap bag with straw and attach it to the end of a rope. Push the bag snugly down inside the chimney, just past the point where it will be torn down for repairs. The bag will catch any pieces of brick or mortar that may drop down the flue. Falling pieces of brick or mortar can damage the flue or damper. They can also spread soot into the room below. After the work is complete, pull the bag up and remove the debris.

As the chimney is torn down, load the old brick and mortar into buckets. Lower them to the ground with a rope and pulley. Build the new chimney the required height to meet the top of the flue opening. Cap off the chimney and seal off around the flue and brick, as shown in *Figure 35*.

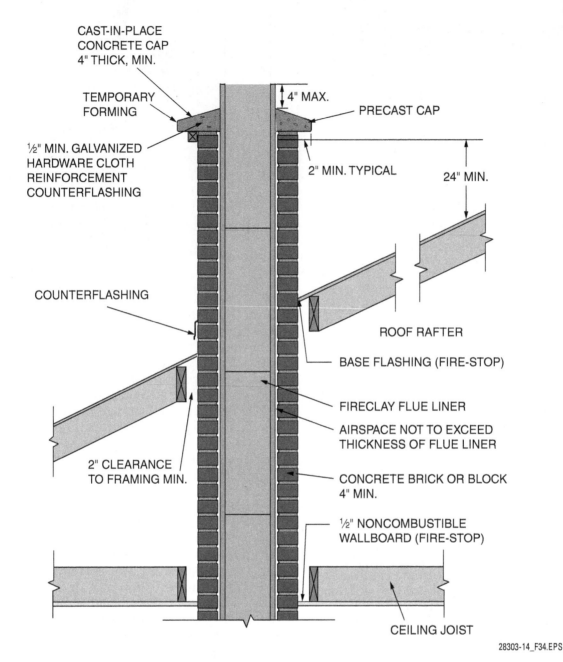

CAST-IN-PLACE
CONCRETE CAP
4" THICK, MIN.

TEMPORARY
FORMING

½" MIN. GALVANIZED
HARDWARE CLOTH
REINFORCEMENT
COUNTERFLASHING

COUNTERFLASHING

2" CLEARANCE
TO FRAMING MIN.

4" MAX.

PRECAST CAP

2" MIN. TYPICAL

24" MIN.

ROOF RAFTER

BASE FLASHING (FIRE-STOP)

FIRECLAY FLUE LINER

AIRSPACE NOT TO EXCEED
THICKNESS OF FLUE LINER

CONCRETE BRICK OR BLOCK
4" MIN.

½" NONCOMBUSTIBLE
WALLBOARD (FIRE-STOP)

CEILING JOIST

28303-14_F34.EPS

Figure 34 Design details for a chimney top.

Apply a portland cement-lime mortar wash coat. The wash, sloped upwards from the cap toward the top of the flue, deflects air currents and drains water from the cap. Type M or S mortar is normally recommended for the wash. The flue linings should extend at least 4 inches above the top of the finished cove.

Minor repairs are usually limited to tuckpointing the brick and resealing between the flue liner and the chimney cap. When the job is complete, carefully remove the scaffold so that the roof is not damaged.

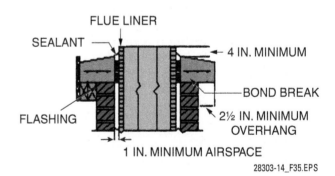

FLUE LINER

SEALANT

FLASHING

4 IN. MINIMUM

BOND BREAK

2½ IN. MINIMUM
OVERHANG

1 IN. MINIMUM AIRSPACE

28303-14_F35.EPS

Figure 35 Chimney cap details.

5.1.2 Chimney Racked Faces

Chimneys are generally not as wide as the body of the fireplace below. The transition between the wide fireplace and the narrower chimney with its exposed brick is called the racked face. The preferred method of construction consists of setting a bed of uncored or paving brick over the racked face to provide a weather-resistant surface. Mortar washes may also be used for the surface; however, they may not be as durable as the brick.

Inspect these areas carefully and repair them as necessary. When repairing a mortar wash, take care not to bridge over the racked face. Make sure to fill each step individually. Both methods of repairing racked faces are shown in *Figure 36*.

Uneven Settling

Soils must be compacted before pouring foundations, slabs, or floors. Improperly prepared soils can settle, causing foundations or slabs to crack. However, some structures are built so solidly that they lean instead of cracking. The most famous example of settling is the Leaning Tower of Pisa. It was built plumb, but now leans at an angle of 12 degrees.

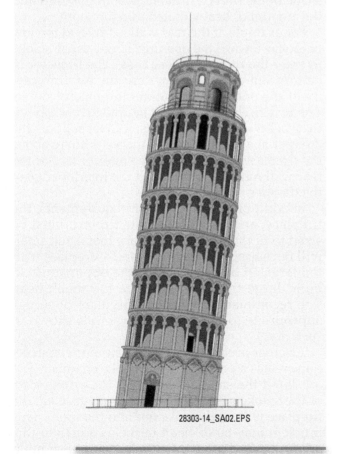

28303-14_SA02.EPS

5.1.3 Appliance Chimneys

Fireplace and appliance chimneys are very similar. Most repairs to an appliance chimney are made in the same way as repairs to a fireplace chimney. However, appliance chimneys have two specific items that should be maintained—the cleanout door and the thimble.

- *Cleanout door* – Newer chimneys may not have a cleanout door. They are not needed where venting appliances use clean-burning fuels, such as natural gas. Other fuels may produce combustion by-products that accumulate at the bottom of the chimney. These by-products require periodic removal. Check the cleanout door for tightness and corrosion. If the door must be replaced, the replacement door should be made of a ferrous metal. Make sure that it forms an airtight seal at the base of the chimney.
- *Thimble* – The thimble is the lined opening through the chimney wall that receives the smoke pipe connector. It should be built integrally with the chimney and made as airtight as possible using either boiler putty or high-temperature grout to install it. Set the thimble flush with the interior face of the flue liners. Repairs are normally limited to inspecting and repairing any loose material around the thimble.

5.2.0 Repairing Fireplaces

The more common repairs required for fireplaces include tuckpointing the front facing, replacing the firebrick in the firebox, and removing smoke stains from the front facing and mantle.

Tuckpointing the mortar joints uses the same techniques described earlier. Because the area being repaired is easily seen and exposed to close examination, make a special effort to do the best

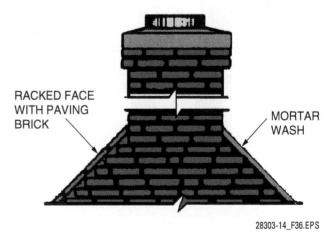

RACKED FACE WITH PAVING BRICK

MORTAR WASH

28303-14_F36.EPS

Figure 36 Repairing racked faces.

job possible. The new joints should match the old in both color and shape. This may take some practice on other surfaces before the final work is done on the fireplace itself.

Removal of smoke stains was also addressed previously. Again, take special care because of the location of the repair, as some cleaning techniques can create a mess. Make sure to put down drop cloths and cover the floor area where the work will be performed.

In addition to these common problems, a fireplace may have more serious defects. *Figure 37* shows a cross section of a typical residential fireplace and chimney. Areas where problems could occur include the smoke chamber, damper, throat, and firebox.

A fireplace is subjected to several forces that tend to speed up the deterioration process. Temperature changes that occur during use of the fireplace tend to loosen mortar and brick within the structure. Buildup of soot, ash, and creosote in the chimney can create poor ventilation. This may cause chimney fires that damage both the

fireplace and chimney. Other problems may be caused by poor design or craftsmanship during the initial construction.

Continuous heating and cooling may cause the firebrick to deteriorate. Replace the firebrick if the damage is extensive. Replacing the firebrick on the floor of the firebox can be easy or difficult, depending on how the original firebrick was laid. If the joints between the brick are mortared, removal is more difficult. However, if the units were laid very tightly and the joints not mortared, the only bond that needs to be broken is between the back of the unit and the supporting surface.

Broken and cracked firebrick on the walls can be drilled and chipped out in the same manner as replacing regular brick. When replacing individual units or repointing, the original width of the joints must be maintained for uniformity of appearance. However, when replacing the complete firebox, it is important to lay the head and bed joints as tightly as possible because the fire will burn out large joints over time.

The proper functioning of a fireplace depends on the shape and relative dimensions of the firebox, the proper location of the fireplace throat in relation to the smoke shelf, and the ratio of the flue area to the fireplace opening. The size and shape of the firebox influences both the draft and the amount of heat radiated into the room.

For example, if the rear wall is finished too low or too far toward the opening, it may cause smoke to enter the room. In these cases, the firebox can be torn out and rebuilt within the existing cavity of the outer walls without too much trouble. The rear firebox wall must be finished at a point that meets two requirements: it must support the damper at a height that is 6 inches or more above the fireplace opening; and the interior face of the rear wall must be in line with the interior edge of the damper.

In addition to these design requirements, the fireplace, and particularly the firebox, must be sized to fit the room. A fireplace that is too small will not supply enough heat, and a fireplace with too large an opening allows the opportunity to build large fires that generate too much heat. The recommended width of fireplace openings appropriate to the size of the room is shown in *Table 3*.

Another area where poor design or construction creates a problem is the smoke chamber (*Figure 38*). If the smoke chamber is the wrong size, or the smoke shelf is improperly constructed, the fireplace will send smoke out into the room. This problem may be difficult to fix, as access to this area is limited after the fireplace has been built.

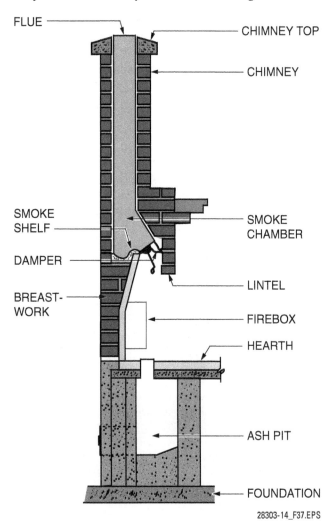

FLUE
CHIMNEY TOP
CHIMNEY
SMOKE SHELF
SMOKE CHAMBER
DAMPER
LINTEL
BREAST-WORK
FIREBOX
HEARTH
ASH PIT
FOUNDATION

28303-14_F37.EPS

Figure 37 Cross section of fireplace and chimney.

Table 3 Fireplace Widths for Different Room Sizes

Suggested Width of Fireplace Openings Appropriate to Size of a Room		
Size of Room in Feet	Width of Fireplace Opening in Inches	
	in Short Wall	in Long Wall
10 × 14	24	24 to 32
12 × 16	26 to 36	32 to 36
12 × 20	32 to 36	36 to 40
12 × 24	32 to 36	36 to 48
14 × 28	32 to 40	40 to 48
16 × 30	36 to 40	48 to 60
20 × 36	40 to 48	48 to 72

28303-14_T03.EPS

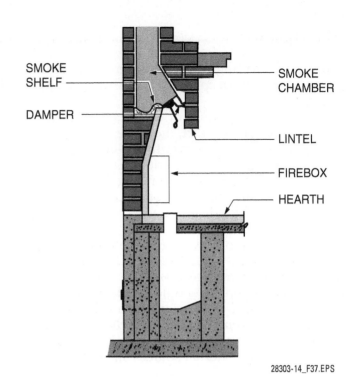

Figure 38 Fireplace firebox and smoke chamber.

28303-14_F37.EPS

If the throat is large enough, it may be possible to remove the damper assembly and gain access for filling any voids or holes on top of the smoke shelf or remolding its shape.

Metal components that fit in openings of the fireplace, such as the damper, may require periodic maintenance or repair. They are usually set into the brick and fastened with anchors and grout. Over time, the grout can weaken and crumble and the anchors become loose.

To repair these problems, remove the component from the opening for examination and refitting. Clean the area around the opening of any loose material and repoint it if necessary. Create new insets for the anchors or bolts by filling in the old holes with mortar and drilling new holes that will tightly hold the inset. Replace any seals around the edge of the opening before reinstalling the component.

Additional Resources

Technical Note TN19, *Residential Fireplace Design*. 1993. Reston, VA: The Brick Industry Association. **www.gobrick.com**

Technical Note TN19A, *Residential Fireplaces, Details and Construction*. 2000. Reston, VA: The Brick Industry Association. **www.gobrick.com**

Technical Note TN19C, *Contemporary Brick Masonry Fireplaces*. 2001. Reston, VA: The Brick Industry Association. **www.gobrick.com**

5.0.0 Section Review

1. The transition between the fireplace and the chimney is called a _____.

   a. flue
   b. smoke shelf
   c. firebox
   d. racked face

2. Each of the following elements of a fireplace is located above the firebox *except* for the _____.

   a. smoke chamber
   b. lintel
   c. hearth
   d. damper

Summary

The art of repairing and restoring masonry structures has become a specialty craft within the masonry trade.

The principal causes of deterioration in masonry structures are dirt, moisture, and temperature changes, which eventually erode mortar joints and crack other materials. Cracking is the distress that occurs most often in masonry walls. Efflorescence and stains are other common problems that can degrade the appearance of masonry structures.

Most of the repair work required on masonry structures involves surface blemishes, minor cracks, or moisture control. The first priority when making repairs should be identifying and treating the cause of deterioration and damage.

Stains in masonry structures include lime run, white scum, green or yellow stains, brown stains, and stains from external sources such as mud or acid rain. These can be treated with proprietary cleaners formulated for the particular type of stain.

Foundation walls usually require repair for cracking and water intrusion. Replace torn, deteriorated, or inelastic sealant material in exterior walls, and perform surface grouting to seal small hairline cracks in mortar joints.

Fireplace chimneys must be repaired or replaced if they are damaged in such a way as to prevent them from forming a draft or disposing of the products of combustion. The more common repairs required for fireplaces include tuckpointing the front facing, replacing the firebrick in the firebox, and removing smoke stains from the front facing and mantle.

Take the time to learn these specialized techniques for repair and restoration of block and brick masonry. You may decide to specialize in repair and restoration during your masonry career; if so, these skills will serve as a solid foundation.

1. Improper location of expansion joints can cause cracking _____.

   a. above window openings
   b. at building corners
   c. at the foundation interface
   d. at offsets

2. Cracks exhibiting a tapered opening are typically caused by _____.

   a. freezing
   b. improper reinforcement
   c. poor workmanship
   d. settlement or deflection

3. Masonry walls may exhibit new-building bloom as a result of _____.

   a. rain penetration
   b. groundwater exposure
   c. accumulation of condensed moisture
   d. plumbing leaks

4. Efflorescence appears on brick masonry surfaces in the form of _____.

   a. tiny black specks
   b. white or gray crystalline deposits
   c. a yellow film
   d. a brownish stain

5. Green stains on light-colored brick are typically caused by the _____.

   a. raw materials in the brick
   b. residue from pressure washers
   b. alkaline solutions used in proprietary cleaners
   d. iron content in water

6. Inspection of masonry structures should be conducted _____.

   a. beginning at the north side
   b. from the basement upward
   c. from the chimney cap downward
   d. outward from the main entrance

7. If a thin knife blade can be slipped between mortar and a masonry unit, the affected wall section should be _____.

   a. torn out and rebuilt
   b. caulked
   c. repointed
   d. sealed

8. When preparing a joint for tuckpointing, remove old mortar to a depth of at least _____.

   a. 1 inch
   b. ¾ inch
   c. ½ inch
   d. ¼ inch

9. Plastic compounds should not be used to repair cracks in exterior masonry walls because the results are _____.

   a. unsightly in appearance
   b. less durable than mortar
   c. more costly than mortar
   d. difficult to paint

10. For repair work on stucco, use a mixture of 1-part cement, 3-parts sand, and _____.

    a. ½-part hydrated lime
    b. ¼-part hydrated lime
    c. ⅛-part hydrated lime
    d. ⅒-part hydrated lime

11. White streaks on masonry called lime run are caused by _____.

    a. sodium hydroxide
    b. carbon compounds
    c. calcium carbonate
    d. muriatic acid

12. Exposure of brick and mortar to rainwater with a high acid content can result in _____.

    a. efflorescence
    b. brown stains
    c. green or yellow stains
    d. white scum

13. Besides being unsightly, dirt on masonry surface can aggravate deterioration by _____.

    a. clogging drainage channels
    b. forming acidic compounds with moisture
    c. promoting surface erosion
    d. impeding natural drying after a rain

14. Before a cleaning material or procedure is used, try it on a sample test area of approximately _____.

 a. 5 square feet
 b. 10 square feet
 c. 20 square feet
 d. 50 square feet

15. When using the bucket-and-brush cleaning method, begin scrubbing after the solution has been on the brickwork for _____.

 a. at least five minutes
 b. 5 to 10 minutes
 c. 12 to 15 minutes
 d. no more than 15 minutes

16. For most effective cleaning with the pressure-washing method, choose equipment that will provide a water flow at a minimum rate of between _____.

 a. 4-6 gpm
 b. 6-8 gpm
 c. 8-10 gpm
 d. 10-12 gpm

17. Modern portland cement mortars should not be used with older soft brick because _____.

 a. they are not chemically compatible
 b. different expansion rates can cause brick damage
 c. the mortars are waterproof
 d. they will stain the brick

18. To better match the appearance of older mortar and brick, replaced mortar can be _____.

 a. brushed with a staining material
 b. wire-brushed
 c. washed with hydrochloric acid
 d. stippled with a stiff-bristle nonmetallic brush

19. The best guard against basement water penetration is the use of _____.

 a. gutters and downspouts
 b. waterproofing on foundation walls
 c. good workmanship
 d. sump pumps

20. When grouting basement walls to seal small cracks, grout should be applied in _____.

 a. a single coat
 b. two coats
 c. three coats
 d. four coats

21. To allow moisture to escape, clogged weep-holes should be cleared by using _____.

 a. stiff wire
 b. compressed air
 c. a hammer and chisel
 d. an electric drill

22. Before making other repairs, areas of bulged masonry should be _____.

 a. tuck-pointed
 b. torn out and replaced
 c. straightened and re-anchored
 d. reinforced

23. Bricks in a chimney that exhibit spalling should be _____.

 a. mortared over
 b. chipped out and replaced
 c. repaired with epoxy compound
 d. repointed

24. When erecting a scaffold on a roof for use in chimney repair, _____.

 a. nail it into the roof deck
 b. avoid placing it in contact with the chimney
 c. tie it to the base of the chimney
 d. fasten all scaffold joints with carriage bolts

25. Substances that can build up in a chimney, causing poor ventilation and a potential fire hazard, include soot, ash, and _____.

 a. pine tar
 b. carbohydrates
 c. creosote
 d. incomplete combustion residues

Trade Terms Quiz

Fill in the blank with the correct term that you learned from your study of this module.

1. A(n) _____ is a low wall or railing.

2. A deposit or crust of white powder on the surface of brickwork resulting when soluble salts in the mortar or brick are drawn to the surface by moisture is called _____.

3. A chip, fragment, or flake broken off from the edge or face of a stone masonry unit and having at least one thin edge is called _____.

4. _____ is cement-based paint containing either 65 percent or 80 percent portland cement by weight.

5. The process of cutting away defective mortar and refilling the joints with fresh mortar is called _____.

6. A(n) _____ is a chisel with a tapered blade used for removing mortar from joints.

Trade Terms

Efflorescence
Parapet

Plugging chisel
Portland cement paint

Spall
Tuckpointing

Trade Terms Introduced in This Module

Efflorescence: A deposit or crust of white powder on the surface of brickwork, resulting when soluble salts in the mortar or brick are drawn to the surface by moisture.

Parapet: A low wall or railing.

Plugging chisel: A chisel with a tapered blade used for removing mortar from joints.

Portland cement paint: Cement-based paint. Type I, containing 65 percent portland cement by weight, is for general use; Type II, containing 80 percent portland cement by weight, is used where maximum durability is needed. Within each type there are two classes: Class A contains no aggregate filler and is for general use; Class B contains 20 to 40 percent sand filler and is used on open-textured surfaces.

Spall: A chip, fragment, or flake broken off from the edge or face of a stone masonry unit and having at least one thin edge.

Tuckpointing: The process of cutting away defective mortar and refilling the joints with fresh mortar.

Additional Resources

This module presents thorough resources for task training. The following resource material is suggested for further study.

Bricklaying: Brick and Block Masonry. 1988. Brick Industry Association. Orlando, FL: Harcourt Brace & Company.

Building Block Walls: A Basic Guide, Latest Edition. Herndon, VA: National Concrete Masonry Association.

Good Practice for Cleaning New Brick Work. 2009. Charlotte, NC: Brick Industry Association Southeast Region. **www.gobricksoutheast.com**

Masonry Design and Detailing for Architects, Engineers and Contractors, Sixth Edition. 2012. Christine Beall. New York: McGraw-Hill.

Pocket Guide to Brick Construction. 1990. Reston, VA: Brick Industry Association.

Principles of Brick Masonry. Reston, VA: Brick Industry Association.

Sure Klean Guide to Common Brick Staining in New Construction. 2013. Lawrence, KS: PROSOCO, Inc. **www.prosoco.com**

Technical Note TN19, *Residential Fireplace Design.* 1993. Reston, VA: The Brick Industry Association. **www.gobrick.com**

Technical Note TN19A, *Residential Fireplaces, Details and Construction.* 2000. Reston, VA: The Brick Industry Association. **www.gobrick.com**

Technical Note TN19C, *Contemporary Brick Masonry Fireplaces.* 2001. Reston, VA: The Brick Industry Association. **www.gobrick.com**

Technical Note TN20, *Cleaning Brickwork.* 2006. Reston, VA: The Brick Industry Association. **www.gobrick.com**

Technical Note TN23, *Stains—Identification and Prevention.* 2006. Reston, VA: The Brick Industry Association. **www.gobrick.com**

Figure Credits

Section Review Answers

Answer	Section Reference	Objective
Section One		
1. c	1.1.3	1a
2. a	1.2.3	1b
Section Two		
1. d	2.1.0	2a
2. c	2.2.0	2b
3. a	2.3.2	2c
Section Three		
1. d	3.1.1	3a
2. c	3.2.0	3b
3. a	3.3.3	3c
4. b	3.4.0	3d
Section Four		
1. a	4.1.1	4a
2. c	4.2.0	4b
Section Five		
1. d	5.1.2	5a
2. c	5.2.0	5b

NCCER CURRICULA — USER UPDATE

NCCER makes every effort to keep its textbooks up-to-date and free of technical errors. We appreciate your help in this process. If you find an error, a typographical mistake, or an inaccuracy in NCCER's curricula, please fill out this form (or a photocopy), or complete the online form at **www.nccer.org/olf**. Be sure to include the exact module ID number, page number, a detailed description, and your recommended correction. Your input will be brought to the attention of the Authoring Team. Thank you for your assistance.

Instructors – If you have an idea for improving this textbook, or have found that additional materials were necessary to teach this module effectively, please let us know so that we may present your suggestions to the Authoring Team.

NCCER Product Development and Revision
13614 Progress Blvd., Alachua, FL 32615

Email: curriculum@nccer.org
Online: www.nccer.org/olf

❏ Trainee Guide   ❏ Lesson Plans   ❏ Exam   ❏ PowerPoints   Other _____

Craft / Level: _____ Copyright Date: _____

Module ID Number / Title: _____

Section Number(s): _____

Description: _____

_____

_____

_____

_____

Recommended Correction: _____

_____

_____

_____

_____

Your Name: _____

Address: _____

_____

Email: _____ Phone: _____

28304-14

Commercial Drawings

Commercial and industrial construction is a large part of the construction market in the United States. In order to build commercial structures, you must be able to understand and interpret the plans and drawings. The basic principles are the same as those that apply to residential drawings; however, commercial structures have more dimensions to interpret. Additional section and detail drawings mean there will be more changes in scale. This module covers the requirements for and the contents of commercial drawings, how to read and interpret them, and how to interpret specifications.

Module Four

Trainees with successful module completions may be eligible for credentialing through the NCCER Registry. To learn more, go to **www.nccer.org** or contact us at **1.888.622.3720**. Our website has information on the latest product releases and training, as well as online versions of our *Cornerstone* magazine and Pearson's product catalog.

Your feedback is welcome. You may email your comments to **curriculum@nccer.org**, send general comments and inquiries to **info@nccer.org**, or fill in the User Update form at the back of this module.

This information is general in nature and intended for training purposes only. Actual performance of activities described in this manual requires compliance with all applicable operating, service, maintenance, and safety procedures under the direction of qualified personnel. References in this manual to patented or proprietary devices do not constitute a recommendation of their use.

Objectives

When you have completed this module, you will be able to do the following:

1. Identify the requirements for and contents of commercial drawings.
   a. Explain the requirements for commercial drawings.
   b. List the contents of commercial plans and describe the purpose of each.
2. Read and interpret commercial drawings.
   a. Identify common views used in commercial drawings.
   b. Explain how to read and interpret architectural drawings.
   c. Explain how to read and interpret structural drawings.
   d. Explain how to read and interpret shop drawings.
   e. Define building information modeling and describe its applications.
3. Explain the purpose of written specifications.
   a. Describe how specifications are written.
   b. Explain the format of specifications.

Performance Tasks

Under the supervision of your instructor, you should be able to do the following:

1. Locate 10 items contained in a set of instructor-chosen commercial drawings, including all of the following:
   - Wall height from finished floor
   - Entire wall elevation length
   - Wall composition
   - Wall-reinforcement size and spacing
   - Section view
2. Locate the section of a set of specifications that shows the type of mortar to be used.

Trade Terms

As-built drawing	Easement	Isometric drawing	Plan view
Beam	Elevation view	Joist	Property line
Benchmark	Fabricator	Landscape drawing	Reflected ceiling plan
Callout	Front setback	MEP drawings	Riser diagram
Civil drawing	Girder	Monument	

Industry-Recognized Credentials

If you're training through an NCCER-accredited sponsor, you may be eligible for credentials from NCCER's Registry. The ID number for this module is 28304-14. Note that this module may have been used in other NCCER curricula and may apply to other level completions. Contact NCCER's Registry at 888.622.3720 or go to **www.nccer.org** for more information.

Code Note

Codes vary among jurisdictions. Because of the variations in code, consult the applicable code whenever regulations are in question. Referring to an incorrect set of codes can cause as much trouble as failing to reference codes altogether. Obtain, review, and familiarize yourself with your local adopted code.

Contents

Topics to be presented in this module include:

Figures

SECTION ONE

1.0.0 REQUIREMENTS AND CONTENTS OF COMMERCIAL DRAWINGS

Objective

Identify the requirements for and contents of commercial drawings.

   a. Explain the requirements for commercial drawings.
   b. List the contents of commercial plans and describe the purpose of each.

Trade Terms

Benchmark: A point established by the surveyor on or close to the building site and used as a reference for determining elevations during the construction of a building.

Civil drawing: A drawing that shows the overall shape of the building site. Also called a site plan.

Easement: A legal right-of-way provision on another person's property (for example, the right of a neighbor to build a road or a public utility to install water and gas lines on the property). A property owner cannot build on an area where an easement has been identified.

Fabricator: A person who provides detailed shop drawings for the fabrication of components and who fabricates them in a shop for later installation at the job site.

Front setback: The distance from the property line to the front of the building.

Isometric drawing: A three-dimensional drawing in which the object is tilted so that three faces are equally inclined to the picture plane.

Landscape drawing: A drawing that shows proposed plantings and other landscape features.

MEP drawings: The set of construction drawings that consists of mechanical, electrical, and plumbing drawings.

Monument: A physical structure that marks the location of a survey point.

Plan view: A drawing that represents a view looking down on an object.

Property line: The recorded legal boundary of a piece of property.

Riser diagram: A type of isometric drawing that depicts the layout, components, and connections of a piping system.

Commercial or industrial construction work is often called heavy construction, because heavy equipment and heavy materials are used for these jobs. Cranes, hoists, and graders are heavy equipment. Heavy materials include steel and concrete. Most commercial projects are larger and more complicated than residential projects. They require a greater variety of construction techniques, equipment, and materials.

Consequently, the plans and drawings for a commercial project are also more complicated than residential plans and drawings, though the basic principles, such as scaling and dimensioning, are the same. Commercial structures have many different uses. Safety and environmental requirements must also be considered. The complexity of the commercial project is reflected in the complexity of the plans.

1.1.0 Requirements for Commercial Plans

There are several reasons why commercial construction plans are more detailed than residential plans. First, the structures are usually larger and more expensive to build. Another major consideration is legal liability. The contractor's legal liability is far greater in commercial construction because there are more applicable codes, ordinances, and regulations. Other reasons that commercial plans are more complex include the following:

- The architectural plans, drawings, schedules, and specifications are legal documents. Many state, local, and federal agencies demand greater detail in construction drawings, to substantiate any legal disputes.
- Code restrictions and safety requirements for commercial and industrial buildings are far more complicated than for residential construction. More detailed drawings are used to make certain that all codes and local ordinances are met.
- The size of a commercial building requires a greater number of drawings, sections, details, and schedules, with more detail required to correlate the various parts of the structure.
- The materials used in commercial construction call for more detailed information on construction techniques, especially for structural steel.

A major commercial project may have hundreds of digital drawings in its plan set, plus associated schedules. Specialty subcontractors will generally get a partial set of the plans relating to their work. At least one complete set of plans will be kept in the general contractor's

---

field office for reference. The plans are sent to the contractor electronically, or posted online so that the contractor can download them. The contractor then prints out the plans at the contractor's expense.

1.2.0 Commercial Plan Contents

Construction drawings consist of several different kinds of drawings assembled into a set (*Figure 1*). Each type of drawing is assigned a letter. For each type, there may be several drawings, which are then numbered. For example, the first three electrical drawings would be numbered E1, E2, and E3. A complete set of commercial construction plans typically includes the following drawing types:

- Architectural – A
- Structural – S
- Mechanical – M
- Electrical – E
- Plumbing – P

The exact content of the plan set will vary, depending on the type and size of the job and local code requirements. For example, a drawing set for a commercial office building may include landscape drawings. These would be denoted with the letter *L*. In some commercial drawing sets, the site plans are called civil drawings. They are marked C1, C2, C3, and so forth. Specifications and schedules may be referenced or included with each type of drawing. These two components are similar in content, if not in quantity, to architectural drawings for residences.

Reading and interpreting residential drawings was covered in Masonry Level Two. The same line and symbol conventions are used in both residential and commercial plans. As a memory refresher, *Figure 2* shows common drawing lines and *Figure 3* shows common symbols for various materials.

Object lines show the main outline of the structure. Dimension lines, extension lines, and leader lines show the size of an object. Arrows indicate the extent of the measurement. The measurement is written above the dimension line.

Section lines, or cutting-plane lines, show the location of a section view of that particular part of the structure. Designers use various formats to indicate sections. Usually, they are in the form of a U with arrows at the tips. A letter identifies the section. The section drawing is marked with a corresponding letter. The section view is placed either on the same page as the main drawing or on a separate page for all sections.

Symbols provide another graphic indication for different types of materials. There are many commonly used symbols. However, there are no standardized symbols for specific materials. The best-known reference for standard symbols, lines, and abbreviations is *Architectural Graphic Standards* prepared by the American Institute of

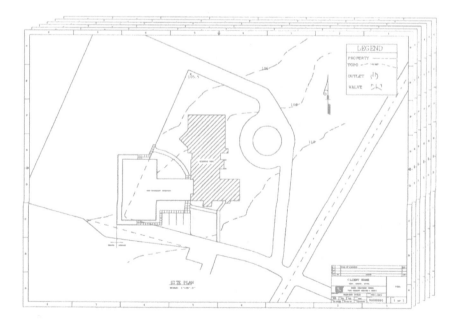

TITLE SHEET(S)
ARCHITECTURAL DRAWINGS
- SITE (PLOT) PLAN
- FOUNDATION PLAN
- FLOOR PLANS
- INTERIOR/EXTERIOR ELEVATIONS
- SECTIONS
- DETAILS
- SCHEDULES

STRUCTURAL DRAWINGS

MECHANICAL PLANS

ELECTRICAL PLANS

PLUMBING PLANS

28304-14_F01.EPS

Figure 1 A typical drawing set.

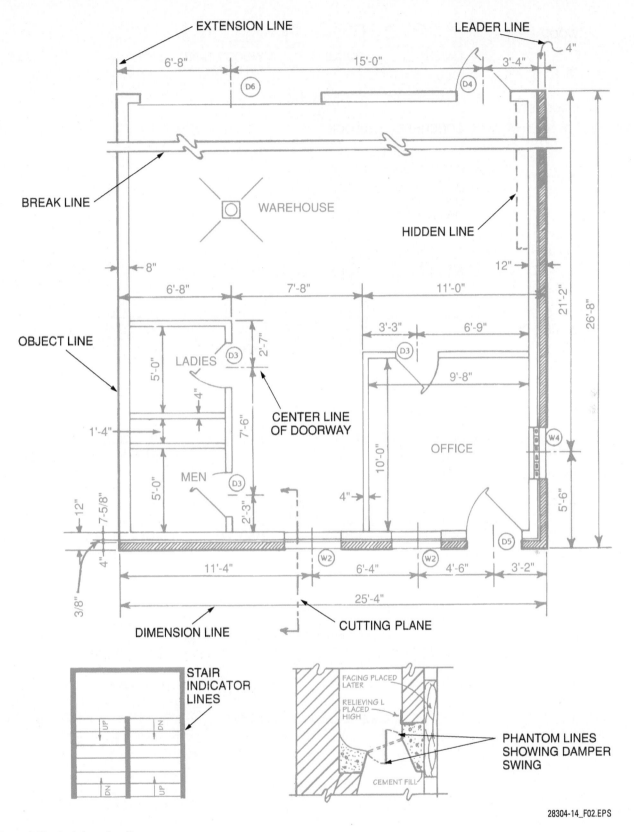

Figure 2 Typical drawing lines.

28304-14_F02.EPS

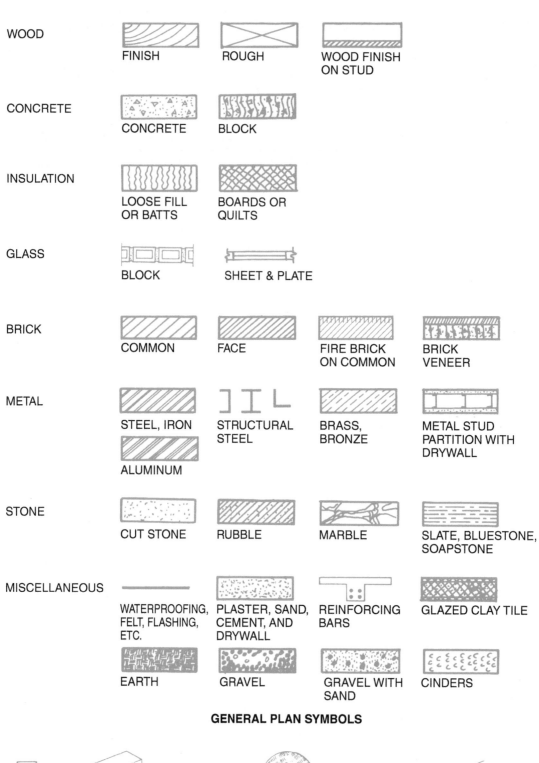

GENERAL PLAN SYMBOLS

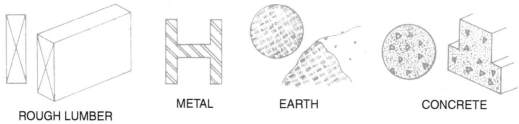

SECTION-VIEW SYMBOLS

28304-14_F03.EPS

Figure 3 Typical material symbols.

Brick upon Brick

For nearly 40 years, the Empire State Building was the tallest building in the world. Approximately 10 million brick masonry units were used in its construction. It was completed in 1931. Today it is likely that curtain walls containing brick facades would be used in place of individual brick masonry units.

Despite the enormity of the project, construction of the Empire State Building was completed in about 15 months. One of the methods used to speed up construction was to have trucks dump brick down a chute, instead of dumping them in the street. The chute led to a large hopper from which brick were then dumped into carts, and hoisted to the location where they were needed. The innovative technique eliminated the backbreaking work of moving brick from the pile to the bricklayer using a wheelbarrow.

Architects. Large commercial projects will have a key block or legend. It may be on the index page or a separate page that lists the symbols and abbreviations used throughout the plan set.

1.2.1 Site Plans

Man-made and topographic (natural) features and other relevant project information are shown on a site plan, including the information needed to properly locate the structure on the site. Man-made features include roads, sidewalks, utilities, and buildings. Topographic features include trees, streams, springs, and existing ground contours. Project information includes the building outline, general utility information, proposed sidewalks, parking areas, roads, landscape information, proposed contours, and any other information that will convey what is to be constructed or changed on the site. A prominently displayed North arrow is included for orientation purposes on site plans. Sometimes a site plan contains a large-scale map of the overall area that indicates where the project is located on the site. *Figures 4* and *5* show examples of basic site plans.

Typically, site plans include the following types of detailed information:

- Coordinates of control points or property corners
- Direction and length of property lines or control lines
- Description, or reference to a description, for all control and property monuments
- Location, dimensions, and elevation of the structure on site
- Finish and existing grade contours
- Location of utilities
- Location of existing elements such as trees and other structures
- Location and dimensions of roadways, driveways, and sidewalks
- Names of all roads shown on the plan

- Locations and dimensions of any easements

Like other drawings, site plans are drawn to scale. The scale used depends on the size of the project. A project covering a large area typically will have a small scale, such as 1" = 100', while a project on a small site might have a large scale, such as 1" = 10'.

The dimensions shown on site plans are typically expressed in feet and tenths of a foot (engineer's scale). However, some site plans state the dimensions in feet, inches, and fractions of an inch (architect's scale). Dimensions to the property lines are shown to establish code requirements. Frequently, building codes require that nothing be built on certain portions of the land. For example, a local building code may have a front setback requirement that dictates the minimum distance that must be maintained between the street and the front of a structure.

Typically, side yards have an easement to allow for access to rear yards and to reduce the possibility of fire spreading to adjacent buildings. A property owner cannot build on an area where an easement has been identified. Examples of typical easements are the right of a neighbor to build a road; a public utility to install water, gas, or elec-

Safeguarding the Drawing Set

Always treat a drawing set with care. It is best to keep two sets: one for the office, and one for field use. Never remove the bid set (the set of drawings that were submitted in response to the request for proposal) from the office. Removal of the bid set from the office may result in their loss or damage. It can also cause delays and lost time for others who need to use or reproduce the drawings.

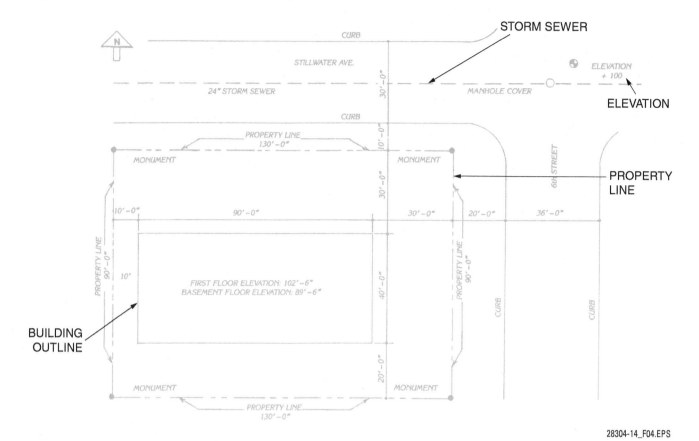

Figure 4 Typical site plan.

tric lines on the property; or an area set aside for groundwater drainage.

Site plans show finish grades (also called elevations) for the site, based on data provided by a surveyor or engineer. It is necessary to know these elevations for grading the lot and for construction of the structure. Finish grades are typically shown for all four corners of the lot as well as other points within the lot. Finish grades or elevations are also shown for the corners of the structure and relevant points within the building.

All the finish-grade references shown are keyed to a reference point, called a benchmark. A benchmark, or job datum, is a reference point established by the surveyor on or close to the property, usually at one corner of the lot. At the site, this point may be marked by a plugged pipe driven into the ground, a brass marker, or a wood stake. The location of the benchmark is shown on the site plan with a grade figure next to it. This grade figure may indicate the actual elevation relative to sea level, or it may be an arbitrary elevation of 100.00', 500.00', etc. All other grade points shown on the site plan will be relative to the benchmark.

A site plan usually shows the finish floor elevation of the building. This is the level of the first floor of the building relative to the job-site bench-

mark. For example, if the benchmark is labeled 100.00' and the finish floor elevation indicated on the plan is marked 105.00', the finish floor elevation is 5' above the benchmark. During construction, many important measurements are taken from the finish floor elevation point.

1.2.2 Architectural Drawings

The architectural drawings are usually labeled with page numbers beginning with the letter *A*. These contain general design features of the building, room layouts, construction details, and materials requirements. The architectural drawings include the following:

- A site plan and/or other location plans
- Floor plans
- Wall sections
- Door and window details and schedules
- Elevations
- Special application details including finish details and schedules

The site plan shows topographic features including trees, bodies of water, and ground cover. It will also show man-made features such as roads, railroad tracks, and utility lines. Typical topographic symbols are shown in *Figure 6*.

NCCER – *Masonry Level Three* 28304-14

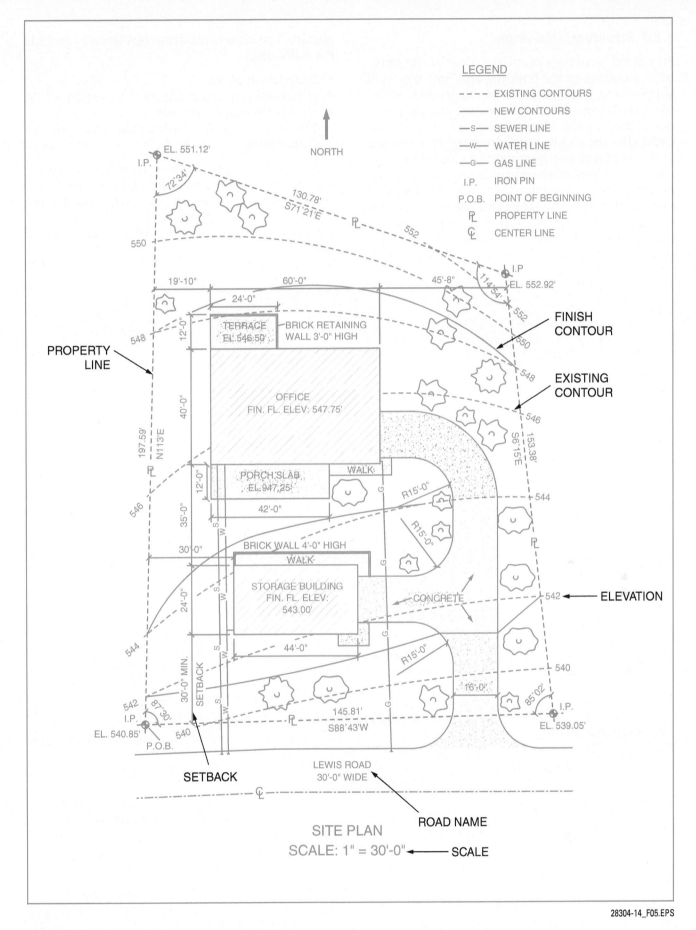

Figure 5 Typical site plan showing topographic features.

28304-14_F05.EPS

1.2.3 Structural Drawings

Structural drawings provide a view of the structural members of the building and how they will support and transmit loads to the ground. Structural drawings are numbered sequentially and designated by the letter *S*. They are normally located after the architectural drawings in a plan set.

A structural engineer prepares the structural drawings. They must calculate the forces on the building and the load that each structural member must withstand. The structural support information includes the foundation, size, and reinforcing requirements; the structural frame type and size of each member; and details on all connections re-

quired. The structural drawings usually include the following:

- Foundation plans
- Structural framing plans for floors and roofing
- Structural support details
- Notes to describe construction and code requirements

Structural support for a commercial building may be steel framing, precast concrete structural elements, or cast-in-place concrete. Unlike residences, modern commercial buildings do not have wooden frames, except in unusual circumstances.

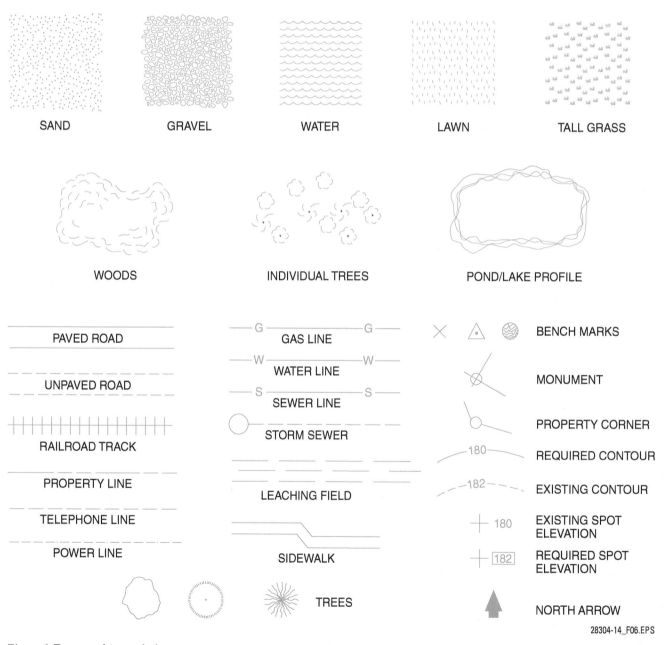

Figure 6 Topographic symbols.

28304-14_F06.EPS

Structural drawings provide useful information. They can stand alone for craftworkers such as framers and erectors. The structural drawings show the main building members. They also show how they relate to the interior and exterior finishes. They do not include information that is unnecessary at the structural stage of construction.

Structural drawings start with the foundation plans. Foundation plans are followed by ground-floor or first-floor plans, upper-floor plans, and the roof plan. Only information essential to the structural systems is shown. For example, a second-floor structural plan would show the steel or concrete framing and the configuration and spacing of the loadbearing members. The walls, ceilings, or floors would not be shown.

1.2.4 Shop Drawings

Shop drawings, also called shops, are specialized drawings that show how to fabricate and install components of a construction project. One type of shop drawing that may be created after an engineer designs the structure is a detail drawing showing the locations of all holes and openings and providing notes specifying how the component is to be made. Assembly instructions may be included. This type of drawing is used principally for structural steel members.

Another type of shop drawing, commonly referred to as a submittal drawing, pertains to special items or equipment purchased for installation in a building. Submittal drawings are usually prepared by equipment manufacturers. They show overall sizes, details of construction, methods of securing the equipment to the structure, and all pertinent data that the architect and contractor need to know for the final placement and installation of the equipment.

Shop drawings produced by a contractor or fabricator are usually submitted to the owner or architect for approval and revisions or corrections.

Location, Location, Location

Some site plans include a small map, called a locus, showing the general location of the property in respect to local highways, routes, and roads.

1.2.5 Mechanical, Electrical, and Plumbing Drawings

Mechanical, electrical, and plumbing drawings are often referred to collectively as MEP drawings. Masons refer to MEP drawings because they show how masonry walls interact with the MEP systems. Penetrations, chases, and embedded items necessary for these systems should be shown on MEP drawings.

Mechanical drawings show the different mechanical systems of the building. Specifically, they include the heating, ventilating, and air conditioning (HVAC) drawings, plumbing drawings, and fire-protection drawings. These drawings have a prefix of H, P, and FP respectively. The plan view format is commonly used on these drawings. This view offers the best illustration of the location and configuration of the work. The drawings serve as a diagram of the system layout.

Figure 7 provides an example of a typical mechanical drawing for an HVAC system. This involves laying out all the piping, controllers, and air handlers to scale on a basic floor plan. Most of the details that are normally on a floor plan have been removed so that the details of the HVAC system can be seen.

Information shown in *Figure 7* includes the following:

- Layout of all of the water supply and return and air-handling units
- Piping diagrams for the hot-water and chilled-water coils
- Room numbers

Mechanical plans typically contain schedules that identify the different types of HVAC equipment. As appropriate, the plans include a detailed view describing the installation of the HVAC equipment. Depending on the nature of the project, these views can include a refrigeration piping schematic, chilled-water coil and hot-water coil piping schematics, and piping runs for other HVAC equipment.

A large amount of information is required for mechanical work. There is limited space on the drawing to show the piping, valves, and connections, so special symbols are used for clarity. *Figure 8* contains some common HVAC symbols.

Detail drawings are sometimes used on mechanical drawings. Unlike the details shown on architectural drawings, these detail drawings are not

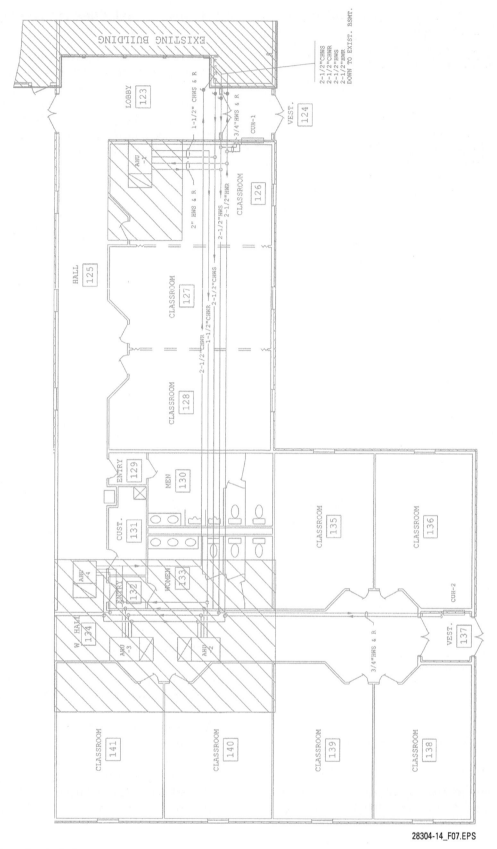

Figure 7 HVAC mechanical plan.

28304-14_F07.EPS

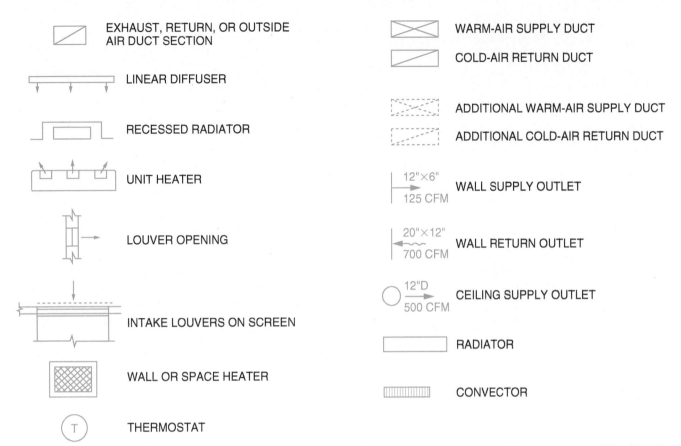

| | EXHAUST, RETURN, OR OUTSIDE AIR DUCT SECTION | | WARM-AIR SUPPLY DUCT |

LINEAR DIFFUSER

RECESSED RADIATOR

UNIT HEATER

LOUVER OPENING

INTAKE LOUVERS ON SCREEN

WALL OR SPACE HEATER

THERMOSTAT

WARM-AIR SUPPLY DUCT

COLD-AIR RETURN DUCT

ADDITIONAL WARM-AIR SUPPLY DUCT

ADDITIONAL COLD-AIR RETURN DUCT

12"×6"
125 CFM   WALL SUPPLY OUTLET

20"×12"
700 CFM   WALL RETURN OUTLET

12"D
500 CFM   CEILING SUPPLY OUTLET

RADIATOR

CONVECTOR

28304-14_F08.EPS

Figure 8 HVAC symbols.

normally drawn to scale. Usually, they are drawn as an elevation or perspective view. They show details about the configuration of the equipment.

The last part of the plan set usually contains the electrical drawings. They show the various electrical and communications systems of the building.

The electrical drawings are labeled with page numbers beginning with the letter *E*. They contain information on the electrical service requirements for the building. They also show the location of all outlets, switches, and fixtures. In addition to schematics of the branch circuits for the building, electrical drawings include the following:

- Site plan for electrical service requirements
- Floor plans for the outlet and switch locations and the branch-circuit requirements
- Lighting plans
- Emergency power and lighting systems
- Life safety systems
- Any backup power-generation facilities
- Notes and details to describe other parts of the electrical system

Figure 9 shows a wiring diagram for the lighting system on the main floor of a building. It mainly consists of overhead fluorescent lighting controlled by wall switches. Photocells control the individual exterior lights over the doorways.

Like mechanical drawings, electrical drawings use the plan view to show system layout. Details and schedules provide clarification. One drawing may include power, lighting, and telecommunications layouts, but in more complex structures the systems are shown separately. There are many different symbols for electrical connections and fixtures. Commonly used symbols are shown in *Figure 10*. Special symbols for components such as power supplies, security systems, and circuit boards are usually designated by the manufacturer.

Plumbing drawings are considered part of the mechanical plans. However, they are usually placed on their own set of drawings for clarity. Unless the building is very basic, placing them on the same sheets as the HVAC drawings would cause confusion.

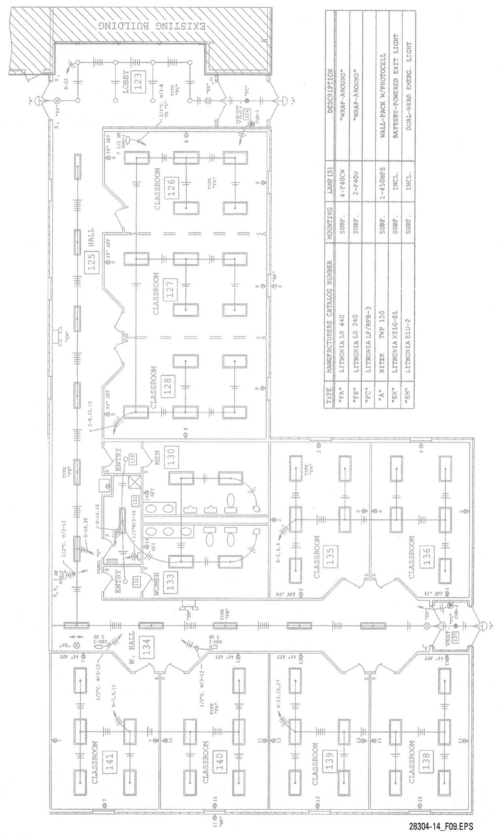

The table within the figure reads:

TYPE	MANUFACTURERS CATALOG NUMBER	MOUNTING	LAMP(S)	DESCRIPTION
"FA"	LITHONIA LB 440	SURF.	4-F40CW	"WRAP-AROUND"
"FB"	LITHONIA LB 240	SURF.	2-F40U	"WRAP-AROUND"
"FC"	LITHONIA LP/RPB-3	SURF.		
"A"	HITEK TWP 150	SURF.	1-450HPS	WALL-PACK W/PHOTOCELL
"EX"	LITHONIA XSIG-BL	SURF.	INCL.	BATTERY-POWERED EXIT LIGHT
"EM"	LITHONIA ELU-2	SURF.	INCL.	DUAL-HEAD EMERG. LIGHT

Figure 9 Electrical plan.

28304-14_F09.EPS

GENERAL OUTLETS

Junction Box, Ceiling

Fan, Ceiling

Recessed Incandescent, Wall

Surface Incandescent, Ceiling

Surface or Pendant Single
Fluorescent Fixture

SWITCH OUTLETS

Single-Pole Switch

Double-Pole Switch

Three-Way Switch

Four-Way Switch

Key-Operated Switch

Switch w/Pilot

Low-Voltage Switch

Door Switch

Momentary Contact Switch

Weatherproof Switch

Fused Switch

Circuit-Breaker Switch

RECEPTACLE OUTLETS

Single Receptacle

Duplex Receptacle

Triplex Receptacle

Split-Wired Duplex Recep.

Single Special Purpose Recep.

Duplex Special Purpose Recep.

Range Receptacle

Switch & Single Receptacle

Grounded Receptacle

Duplex Weatherproof Receptacle

AUXILIARY SYSTEMS

Telephone Jack

Meter

Vacuum Outlet

Electric Door Opener

Chime

Pushbutton (Doorbell)

Bell and Buzzer Combination

Kitchen Ventilating Fan

Lighting Panel

Power Panel

Television Outlet

28304-14_F10.EPS

Figure 10 Electrical symbols.

Lighting the New York Skyline

The first searchlight on top of the Empire State Building heralded the election of Franklin D. Roosevelt in 1932. A series of floodlights were installed in 1964 to illuminate the top 30 floors of the building. Today, the color of the lights is changed to mark various events; for example, yellow marks the US Open and red, white, and blue marks Independence Day.

The building is lit from the 72nd floor to the base of the TV antenna by 204 metal halide lamps and 310 fluorescent lamps. In 1984, a color-changing apparatus was added in the uppermost mooring mast. There are 880 vertical and 220 horizontal fluorescent lights. The colors can be changed with the flick of a switch.

Plumbing drawings usually include the following:

- Site plan for water-supply and sewage-disposal systems
- Floor plans for the fire system, including hydrant connections and sprinkler systems
- Floor plans for the water supply system and fixture locations
- Floor plans for the waste disposal system
- Riser diagrams to describe the vertical piping features
- Floor plans and riser diagrams for the gas lines

Plumbing drawings usually appear as plan-view drawings (*Figure 11*) and as isometric drawings called riser diagrams. The plan views show the horizontal distances or piping runs. The riser diagram shows vertical pipes in the walls. Plumbing drawings for water systems usually show two separate systems: a water source or distribution system, and a waste collection and disposal system.

The most common symbols used to designate the water and gas systems on plumbing drawings are shown in *Figure 12*. In addition to these symbols, there will be specific graphic symbols for the layout of such items as toilets, sinks, water heaters, and sump pumps. These will vary from structure to structure, depending on the specific design.

Isometric Drawings

Isometric means *equal measurement*. A designer uses the true dimension of an object to construct the drawing. An isometric drawing shows a three-dimensional view of where the pipes should be installed. The piping may be drawn to scale or to dimension, or both. In an isometric drawing, vertical pipes are drawn vertically on the sketch, and horizontal pipes are drawn at an angle to the vertical lines.

The Empire State Building

The Empire State Building has 102 stories. It has 70 miles of water pipe that provide water to tanks at various levels. The highest tank is on the 101st floor. There are two public restrooms on each floor and a number of private bathrooms. There are, however, no water fountains.

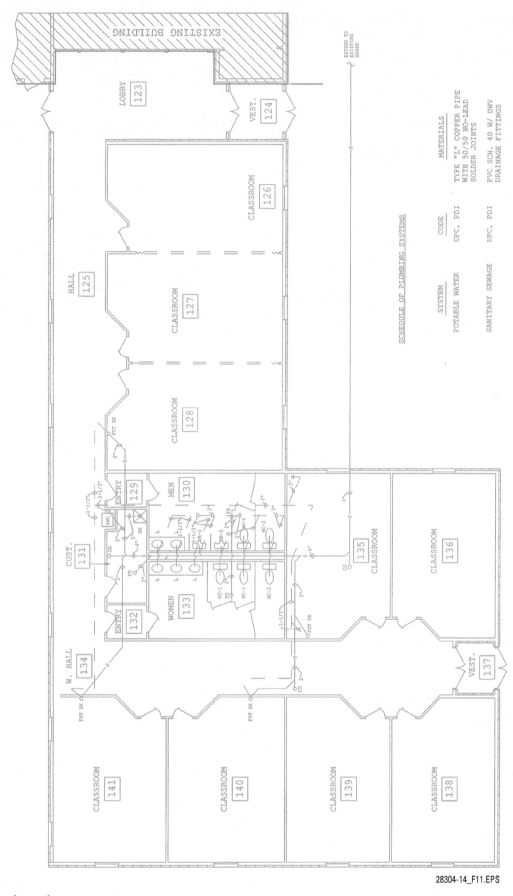

Figure 11 Plumbing plan.

28304-14_F11.EPS

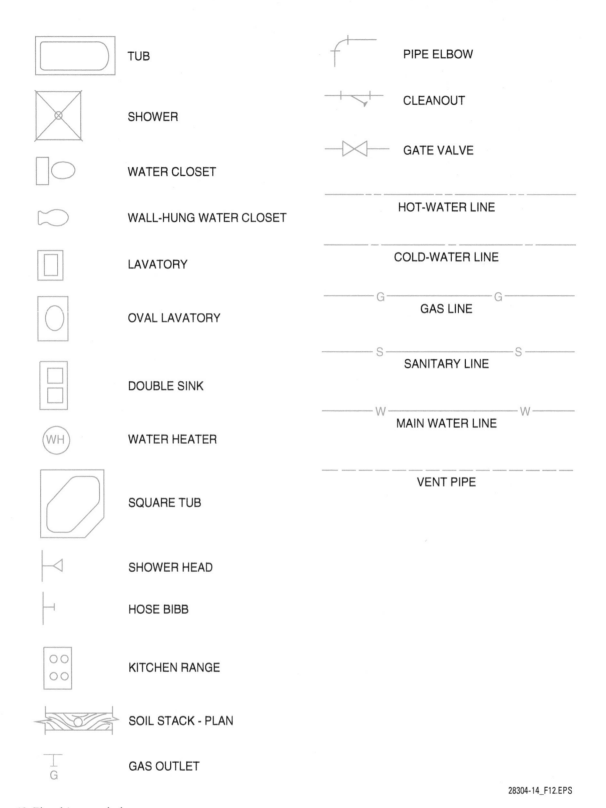

TUB

SHOWER

WATER CLOSET

WALL-HUNG WATER CLOSET

LAVATORY

OVAL LAVATORY

DOUBLE SINK

WATER HEATER

SQUARE TUB

SHOWER HEAD

HOSE BIBB

KITCHEN RANGE

SOIL STACK - PLAN

GAS OUTLET

PIPE ELBOW

CLEANOUT

GATE VALVE

HOT-WATER LINE

COLD-WATER LINE

GAS LINE

SANITARY LINE

MAIN WATER LINE

VENT PIPE

28304-14_F12.EPS

Figure 12 Plumbing symbols.

Additional Resources

Architectural Graphic Standards. 1998. The American Institute of Architects. New York: John Wiley & Sons, Inc.

A Manual of Construction Documentation: An Illustrated Guide to Preparing Construction Drawings. 1989. Glenn E. Wiggins. New York: Whitney Library of Design.

Masonry Design and Detailing for Architects, Engineers and Contractors, Sixth Edition. 2012. Christine Beall. New York: McGraw-Hill.

1.0.0 Section Review

1. For the reference use of specialty subcontractors, at least one complete set of plans should be kept _____.
   a. near the location of the work being done
   b. in the field office
   c. by the authority having jurisdiction
   d. at the local records office

2. A foundation plan is an example of a(n) _____.
   a. architectural drawing
   b. shop drawing
   c. MEP drawing
   d. structural drawing

SECTION TWO

2.0.0 READING AND INTERPRETING COMMERCIAL DRAWINGS

Objective

Read and interpret commercial drawings.
  a. Identify common views used in commercial drawings.
  b. Explain how to read and interpret architectural drawings.
  c. Explain how to read and interpret structural drawings.
  d. Explain how to read and interpret shop drawings.
  e. Define building information modeling and describe its applications.

Performance Task

Locate 10 items contained in a set of instructor-chosen commercial drawings, including all of the following:

- Wall height from finished floor
- Entire wall elevation length
- Wall composition
- Wall-reinforcement size and spacing
- Section view

Trade Terms

As-built drawing: A construction drawing that shows a project as it was completed, including all changes incorporated into the design during the construction process.

Beam: Loadbearing horizontal framing element supported by walls or columns and girders.

Callout: Marking or identifying tag describing parts of a drawing on detail drawings, schedules, or other drawings.

Elevation view: A drawing giving a view from the front or side of a structure.

Girder: Large steel or wooden beam supporting a building, usually around the perimeter.

Joist: Horizontal member of wood or steel supported by beams and holding up the planks of floors or the lathes of ceilings. Joists are laid edgewise to form the floor support.

Reflected ceiling plan: A drawing that shows the details of the ceiling as though the ceiling were reflected by a mirror on the floor.

When reading commercial plans, it is best to follow a step-by-step process to avoid confusion and to catch all the important details. Use the following list as a guide:

Step 1 Begin by reading the project specifications to pick up details not found on the drawings.

Step 2 Quickly review all drawings that give a general impression of the shape, size, and appearance of the structure, including the site plan, floor plans, and exterior elevation views.

Step 3 Begin correlating the floor plans with the exterior elevations, making certain that the parts appear to fit together logically.

Step 4 Next, look at the wall sections and determine the wall types (materials, loadbearing or nonbearing) and construction details or procedures.

Step 5 Review the structural plans. Determine the foundation requirements and the type of structural system. Turn back to the site plan, floor plans, wall sections, and elevations as often as needed to relate the structural and architectural drawings.

Step 6 Review all details on the architectural and structural drawings. Carefully consider all items that may require special construction procedures.

Step 7 Review all interior elevations and try to get a clear picture of what the interior of the building will look like.

Step 8 Review the finish schedule.

Step 9 Review the mechanical and electrical plans. In particular, look for work items that will be done by the masons or those that will affect the work of the masons.

If some part of a drawing is not clear, check with your supervisor. There may be a logical explanation for the item in question, or the plans may be incorrect. It is best to find an answer before construction begins. The architect should clarify and resolve all conflicts in writing so there will not be any confusion later in the project.

2.1.0 Reading and Interpreting Views

A set of construction drawings shows the planned structure from a variety of perspectives. These perspectives are called views. The drawings show elements of the structure viewed from above and looking down (plan views) and viewed from the side (elevation view). They also show the interior features of structural elements (section view) and enlargements of special features (detail views).

2.1.1 Plan Views

Plan views show the structure as if it were being viewed from above, looking straight down. Floor plans and roof plans show the structure in plan view.

The floor plan is the main drawing of the entire set. For a floor-plan view, an imaginary line is cut horizontally across the structure at varying heights so all the important features such as windows, doors, and plumbing fixtures can be shown (*Figure 13*). For multistory buildings, separate floor plans are typically provided for each floor. However, if several floors have the same layout (such as a hotel), one drawing may be used to show all the floors that are similar. *Figure 14* shows an example of a basic floor plan. The types of information commonly shown on floor plans include:

- Outside walls, including the location and dimensions of all exterior openings
- Types of construction materials
- Location of interior walls and partitions
- Location and swing of doors
- Stairways

- Location of windows
- Location of cabinets, electrical and mechanical equipment, and fixtures
- Location of cutting-plane lines

Each door or window shown on a floor plan for a commercial building is typically accompanied by a number, letter, or both. This number/letter is an identifier, which refers to a door or window schedule that describes the corresponding size,

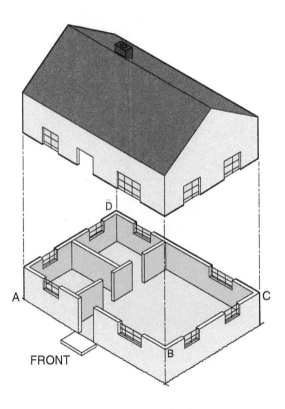

28304-14_F13.EPS

Figure 13 Visualizing a simple floor plan.

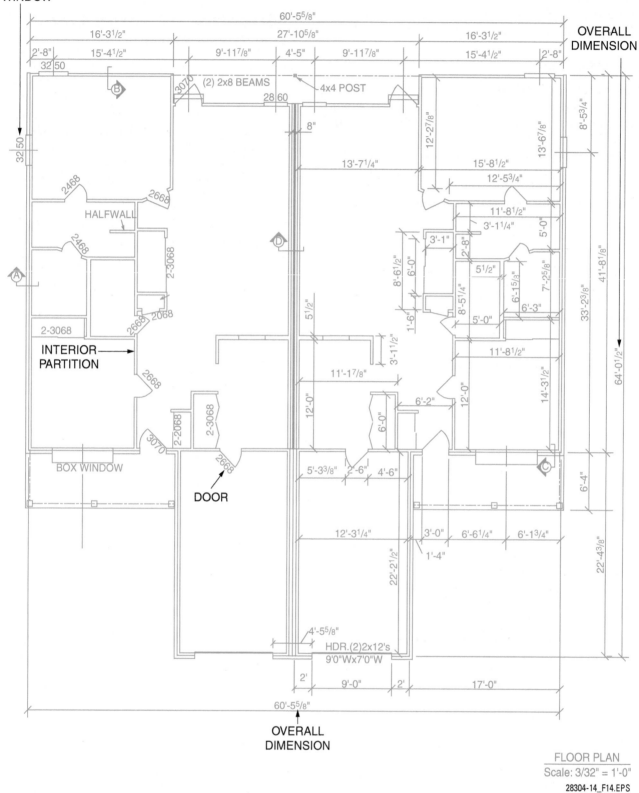

Figure 14 Basic floor plan (not to scale).

Drawing Revisions

When a set of drawings has been revised, always make certain that the most up-to-date set is used for all future work. Either destroy the old, obsolete drawing or else clearly mark on the affected sheets Obsolete Drawing – Do Not Use. A good practice is to remove the obsolete drawing from the set and file it as a historical copy for possible future reference.

Also, when working with a set of construction drawings and written specifications for the first time, thoroughly check each page to see if any revisions or modifications have been made to the original. Doing so can save time and expense for all concerned.

type of material, model number, etc., for the specific door or window. Residential floor plans often show door sizes directly on the floor plan.

Roof plans provide information about the roof slope, roof drain placement, and other pertinent information regarding ornamental sheet metalwork, gutters and downspouts, etc. Where applicable, the roof plan may also show information on the location of air conditioning units, exhaust fans, and other ventilation equipment. Not all sets of drawings include roof plans.

Some drawing sets have reflected ceiling plans, which show the details of the ceiling as though it were reflected by a mirror on the floor. Reflected ceiling plans show features of the ceiling while keeping those features in proper relation to the floor plan. For example, if a vertical pipe runs from floor to ceiling in a room and is drawn in the upper-left corner of the floor plan, it is also shown in the upper-left corner of the re-

flected ceiling plan of that same room. Reflected ceiling plans also show the height of the ceiling.

2.1.2 Elevation Views

Elevation views provide a view of the side of a structure. The structure's features are projected on a vertical plane. Typically, elevation views are used to show exterior features so the general size and shape of the structure can be determined. Elevation views clarify much of the information on the floor plan. For example, a floor plan shows where the doors and windows are located in the outside walls; an elevation view of the same wall shows actual representations of these doors and windows. *Figure 15* shows an example of a basic elevation drawing. The types of information normally shown on elevation drawings include:

- Floor height
- Window and door types
- Rooflines and slope, roofing material, vents, gravel stops, and projection of eaves
- Exterior finish materials and trim

Unless one or more views are identical, four exterior elevation drawings are generally used to show the exterior of a building. More than four elevation drawing views may be required for complex buildings. Commercial exterior elevation drawings are typically designated by compass direction. For example, if the front of the building faces north, then this becomes the north elevation. The other elevations are then labeled accordingly (east, south, and west).

Drawing sets may also include interior elevation drawings for the walls in each partitioned area, especially for walls that have special features such as fireplaces.

GOING GREEN

Insulated Concrete Masonry Units

Insulated concrete masonry units incorporate a polystyrene thermal energy barrier material that increases the thermal mass of the masonry structure. This allows them to absorb more heat and radiate it more slowly than other types of masonry units. By reducing air temperature fluctuations, the use of insulated concrete masonry units permits the installation of smaller HVAC units, helps to reduce energy consumption, and lowers the owner's operating costs.

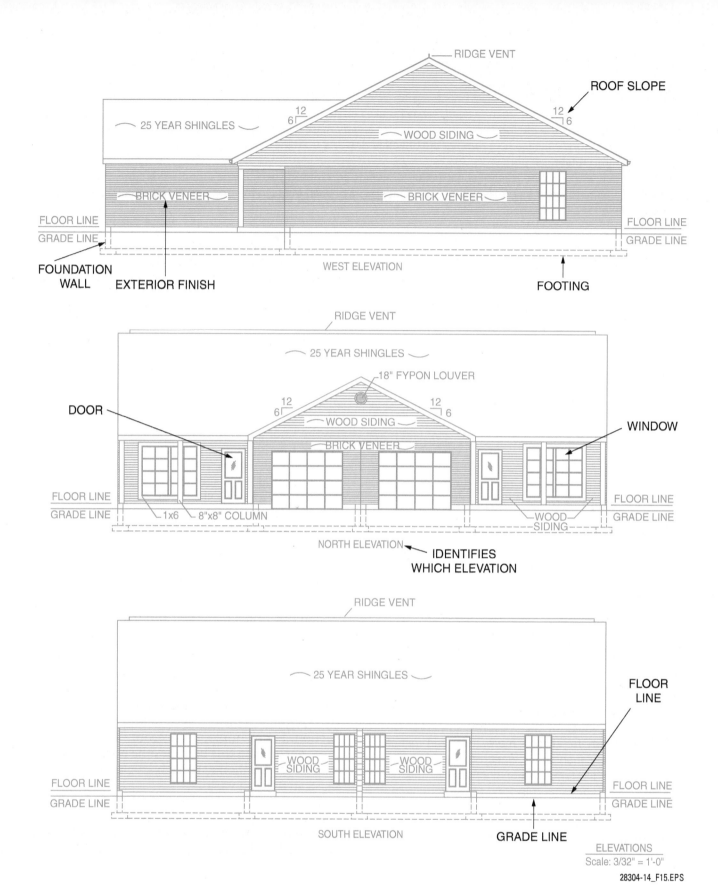

Figure 15 Elevation drawing (not to scale).

2.1.3 Section Views

A section drawing or section view (*Figure 16*) shows the interior features of a wall or other structural feature. Section drawings are drawn as if a cut has been made through a feature at a certain location. The location of the cut and the direction from which the section is to be viewed are shown on the related plan view.

Section drawings that show a view made by cutting through the length of a structure are referred to as building sections, while those showing the view of a cut through specific walls or partitions within the structure are referred to as wall sections.

To show greater detail, section drawings are normally drawn to a larger scale than plan views. The types of information commonly shown by a section drawing include:

- Details of construction and information about stairs, walls, chimneys, or other parts of construction that may not show clearly on a plan view
- Floor levels in relation to grade
- Wall thickness at various locations
- Anchors and reinforcing steel

2.1.4 Detail Views

Detail drawings are enlargements of special features of a building or of equipment installed in a building. They are drawn to a larger scale in order to make the details clearer. *Figure 17* shows a series of detail drawings.

Typically, detail drawings are used for the following objects or situations:

- Footings and foundations, including anchor bolts, reinforcing, and control joints
- Beams, floor joists, bridging, and other support members
- Sills, floor framing, exterior walls, and vapor barriers
- Floor heights, thickness, expansion, and reinforcing

- Masonry wall thickness, reinforcing, and intersections with other building elements
- Windows, exterior and interior doors, and door frames
- Roofs, cornices, soffits, and parapets
- Gravel stops, fascia, and flashing
- Fireplaces and chimneys
- Elevators and stair assemblies
- Shelf angles and lintel details
- Millwork, trim, ornamental iron, and specialty items

2.2.0 Reading and Interpreting Architectural Drawings

Architectural drawings are the core drawings of any plan set. They are sequentially numbered, usually starting with the site plan or the basement. In some cases, the exterior drawings, such as the site plan and landscaping plans, are numbered separately from the architectural drawings. If this is the case, the architectural drawings are numbered in order of basement or ground-floor plans; upper-level floor plans; exterior elevations; sections; interior elevations; details; and window, door, and room finish schedules.

2.2.1 Floor Plans

Floor plans of both commercial and residential structures show various floor levels as if a horizontal plane had been cut through the structure. This results in an overhead view of each floor.

The most noticeable difference between commercial and residential floor plans is the amount of detail. Commercial plans contain more details of room use, finishes, wall types, sound transmission, and fire retardation. In fact, most commercial floor plans incorporate a legend or chart to specify the various interior wall types shown on the plan. The drawing scale is typically ⅛ inch per foot for commercial drawings. The detailing instructions usually include the following types of information:

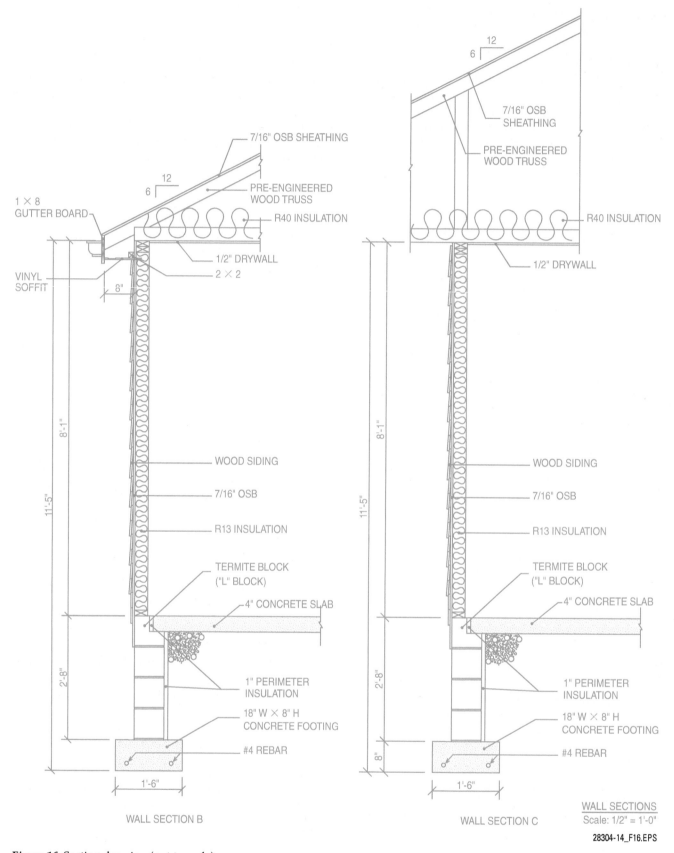

Figure 16 Section drawing (not to scale).

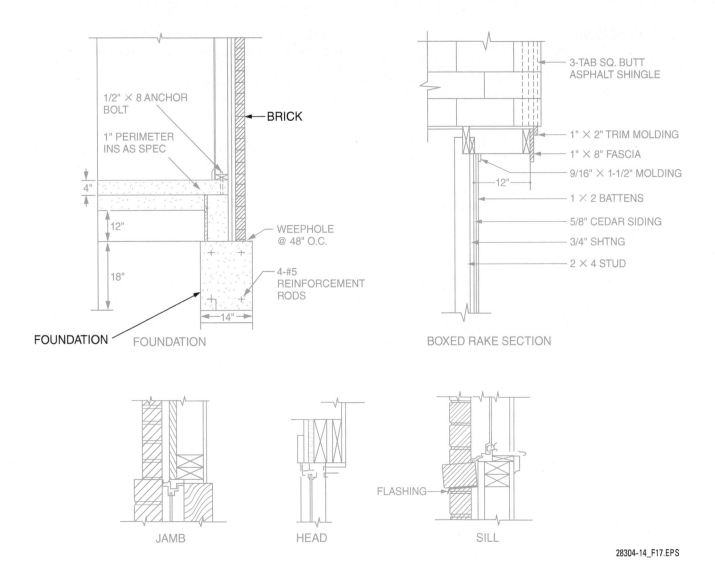

Figure 17 Detail drawings.

28304-14_F17.EPS

- Numerous callouts specifying section views and details
- Room assignment designations by function or number
- Detailed dimensioning of all visible parts of the structure
- Finish designations referenced to schedules

On commercial plans, little is left to chance for several reasons. First, the construction itself is varied and complex. Second, construction must meet all code specifications. Finally, different contractors will be working on the same job. Since there are many ways to accomplish the same task, the requirements are specified in detail to achieve consistency throughout the structure.

Each door or window on a floor plan for a commercial building is typically accompanied by a number, letter, or both. This number/letter is an identifier that refers to a door or window schedule that describes the corresponding door by size, type of materials, or model number for the specific door.

To help clarify ambiguous parts of the drawing, notes may be written on commercial floor plans. This is particularly true when any feature differs from one area to another. In most instances, these notes are important not only because they show variations, but also because they detail the responsibilities of those involved in the construction. A plan note may read, "Furnished by owner," "Refer to structural drawings," or "Not in contract."

For large buildings, the architect often will divide the floor plan into sections by grid lines. The grid is the same for all floors of the building. Using the grid allows the architect and engineer to locate or place features very specifically anywhere in the building. The grid lines are useful for locating features that are repeated on one floor or from one floor to another. The grid markings on the floor plan also reappear on the structural drawings where they locate structural elements such as footings and columns.

2.2.2 Roof Plans

When supplied, roof plans (*Figure 18*) provide information about the roof slope, roof drain placement, and other pertinent information. Where applicable, the roof plan may also show information on the location of air conditioning units, exhaust fans, and other ventilation equipment. The top or bottom elevation of the roof deck is usually given in the roof plans.

2.2.3 Exterior and Interior Elevations

Elevation views on commercial construction drawings are similar to residential elevations, but provide more information. Exterior elevations (refer to *Figure 15*) provide views of the building from each major orientation, as well as references for section views and other exterior elements such as backfilled retaining walls. Elevations are normally drawn to the same scale as the floor plans. Masons refer to them for such items as finishes, patterns, and bands, and also for related items such as canopies.

Some interior elevations provide vertical dimensioning for interior work, materials lists, and construction details such as landings and stairways. They are important for built-in cabinets, shelving, finish carpentry, millwork items, and walls that are hidden from other views. The scale of the drawing depends on the detail required. This may be as small as ¼ inch to 1 foot, or as large as ¾ inch to 1 foot.

2.2.4 Building and Wall Sections

There are many wall sections to show the different types of exterior and interior walls. The interior sections will detail the construction of each wall type, such as curtain walls, partitions, loadbearing walls, fire-resistant walls, and noise-reduction walls. The floor plans incorporate a legend that specifies the interior wall type. The wall section drawing provides the necessary detail. Wall sections usually provide the following details:

- Construction techniques and materials types
- Stud types and placement

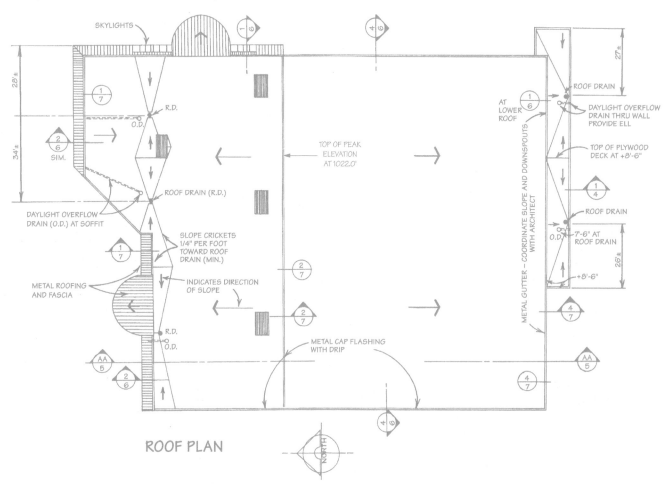

28304-14_F18.EPS

Figure 18 Roof plan.

- Widths and heights of masonry wall configurations
- Fire ratings of various materials, which is measured in terms of hours of resistance
- Sound-barrier placement or materials
- Insulation applications and materials
- In-wall features such as recesses or chases

As with elevations, sections are also drawn for landings, stairways, and backfilled retaining walls. The section drawings detail construction techniques or materials. For instance, a stairway will have details, sections, and elevations with information on tread, landing, and handrail construction. The amount of detail will depend on the complexity of the stairway and on the various building-code specifications.

Building and wall sections show construction features that are expanded on the structural plans. These sections will show the following:

- Relationships of all wall features from the footings through the roof
- Footing and foundation placement in relation to other elevations
- Exterior materials symbols and notations
- Wall heights and floor spacing on multistory buildings
- Framework type and placement

As with elevation drawings, the sections provide an overall view of the proposed structure rather than the detail required for construction. Detail notations or callouts will refer directly to the structural plan sheets. Most callouts will refer to details on roofing beams and trusses, foundations, framing features, or other structural components.

2.2.5 Architectural Drawing Schedules

Schedules are tables that describe and specify the various types and sizes of construction materials used in a building. Door and window schedules (*Figure 19*) and finish schedules (*Figure 20*) are the most commonly included schedules on architectural drawings. For commercial projects, additional schedules are provided for mechanical equipment and controls, plumbing fixtures, lighting fixtures, and any other equipment that needs to be listed separately.

Door, window, and finish schedules are of particular importance to a mason. Door and window types are identified on the various plan and elevation drawings by numbers and/or letters. The door and window schedules list these identifier numbers or letters and describe the corresponding size, type of material, and model number for each different type of door or window used in the structure.

In a finish schedule, each room is identified by name or number. The material and finish for each part of the room (walls, floor, ceiling, base, and trim) are designated, along with any clarifying remarks.

2.3.0 Reading and Interpreting Structural Drawings

Structural drawings (*Figure 21*) provide detailed information on the structural features of the building. This includes information on the load-bearing design and materials, such as masonry, reinforced concrete, steel framing, or oversize timber. The structural drawings include plan views, sections, details, schedules, and notes. They provide information on the size and placement of loadbearing elements. They also show how they are connected to each other and to other parts of the structure.

Typical structural drawings include a foundation plan, floor framing plans, and a roof framing plan (*Figure 22*). The plan view will have sections, details, schedules, and notes located in any available space on the drawing sheet. Each plan view should have a North directional arrow to maintain a consistent orientation. Plan views are typically drawn to the scale of ⅛ or ¼ inch to 1 foot; sections are ½ or ¾ inch to 1 foot; details are 1 or 1½ inches to 1 foot.

Each plan is referenced to a grid or checkerboard identifying the placement of columns and/or footings. The grid is the same on all the plans. The grid also shows dimension lines. In the structural drawings, a callout sequence number or mark will identify and show the placement of columns on the grid. Some project plans do not use a grid, but use a callout sequence marking. In either case, the identification marks will be referenced to schedules or notes. For example, in *Figure 23* the notation B2 may be referencing a footing, pier, or column on the appropriate schedule.

The structural drawings will show the type of framing and loadbearing for the building. For example, if the floor, roof beams, or trusses place their weight directly on the wall materials, the structure has loadbearing walls. This means that the exterior walls are constructed of materials with high compressive strength such as brick, block, or cast concrete. The walls support their own weight as well as that of the various floor and roof elements. The plans for a loadbearing wall will show no structural beams or columns along the wall.

DOOR SCHEDULE

DOOR	WIDTH	HEIGHT	THICK-NESS	MAT'L	TYPE	STORM DOOR	QTY.	THRES-HOLD	REMARKS	MANUFACTURER
2068	2'-0"	6'-8"	1 3/8"	Wood-Ash	Hollow-core	NO	5	None	Oil Stain	LBJ Door Co.
2468	2'-4"	6'-8"	1 3/8"	Wood-Ash	Hollow-core	NO	1	None	Oil Stain	LBJ Door Co.
2668	2'-6"	6'-8"	1"	Wood-Ash	Cafe	NO	1 pr.	None	Oil Stain	LBJ Door Co.
2668	2'-6"	6'-8"	1 3/8"	Wood-Ash	Sliding Pocket	NO	1	None	Oil Stain	LBJ Door Co.
2668	2'-6"	6'-8"	1 3/8"	Wood-Ash	Hollow-core	1 Screen	6	Alum.	Screen door in garage	LBJ Door Co.
2868	2'-8"	6'-8"	1 3/4"	Metal Clad	Fireproof	YES	1	Alum.	Paint	LBJ Door Co.
3068	3'-0"	6'-8"	1 3/4"	Wood-Ash	Solid-core	NO	1	None	Oil Stain	LBJ Door Co.
3668	3'-6"	6'-8"	1 3/4"	Wood-Ash	Solid-core	YES	1	Alum.	Marine Varnish	LBJ Door Co.
6066	6'-0"	6'-6"	1/2"	Glass/Metal	Sliding	YES	1 pr.	Alum.	Sliding Screen	LBJ Door Co.
6068	6'-0"	6'-8"	1 1/4"	Wood-Ash	Bi-Fold	NO	2 sets	None	Oil Stain	LBJ Door Co.
1956	1'-9"	5'-6"	1/2"	Glass/Metal	Sliding Shower door	NO	2 sets	None	Frosted Glass	LBJ Door Co.

WINDOW SCHEDULE

SYMBOL	WIDTH	HEIGHT	MAT'L	TYPE	SCREEN & DOOR	QUANTITY	REMARKS	MANUFACTURER	CATALOG NUMBER
A	3'-8"	3'-0"	ALUM.	DOUBLE HUNG	YES	2	4 LIGHTS, 4 HIGH	LBJ Window Co.	141 PW
B	3'-8"	5'-0"	ALUM.	DOUBLE HUNG	YES	1	4 LIGHTS, 4 HIGH	LBJ Window Co.	145 PW
C	3'-0"	5'-0"	ALUM.	STATIONARY	STORM ONLY	2	SINGLE LIGHTS	H & J Glass Co.	59 PY
D	2'-0"	3'-0"	ALUM.	DOUBLE HUNG	YES	1	4 LIGHTS, 4 HIGH	LBJ Window Co.	142 PW
E	2'-0"	6'-0"	ALUM.	STATIONARY	STORM ONLY	2	20 LIGHTS	H & J Glass Co.	37 TS
F	3'-6"	5'-0"	ALUM.	DOUBLE HUNG	YES	1	16 LIGHTS, 4 HIGH	LBJ Window Co.	143 PW

28304-14_F19.EPS

Figure 19 Examples of a door schedule and a window schedule for architectural drawings.

In reinforced concrete construction, the load-bearing elements usually include reinforced concrete footings, foundations, piers, columns, and pillars. Exterior masonry is typically nonbearing curtain or panel walls. The structural drawings will show the size, type, and placement of reinforcing materials and of various jointing techniques. For instance, sections may provide information for the placement of reinforcing bar or wire-mesh reinforcement, while details may show the types of saddles, chairs, stirrups, or joints to use (*Figure 24*). A reinforcing-bar notation will usually include the bar size and the bar spacing.

High-rise buildings are typically steel frame construction with masonry curtain or panel walls. The framework for the entire structure is formed by bolting or welding various steel elements together. The loads from floors and roofing are transferred to beams and girders, down columns to the footings. When the joints are bolted together, the schedules will designate the number, size, and material requirements for the bolts. *Figure 25* shows the most common shapes for steel frame elements and lists the typical plan designations.

Timber construction is still used in some commercial roofing. The beams usually have 6-inch nominal dimensions. They may be solid or laminated. The drawings and schedules will show manufacturer's designations and notations for connecting hardware. The metal parts, such as strap hangers, brackets, baseplates, and lag screws, are listed on the schedule. They are used to connect the wood to a concrete or steel support.

The standard structural-steel notation gives the type or shape of the beam, the depth of the web, and the weight per foot. For example, the notation "W18 × 77" refers to a wide-flange shape with a nominal depth of 18 inches and a weight of 77

ROOM FINISH SCHEDULE

Column groups: FLOOR (CARPET, CERAMIC TILE, RUBBER TILE, CONCRETE) · CEILING (ACOUSTIC TILE, DRYWALL, PAINT, CERAMIC TILE) · WALL (DRYWALL, PAINT, WALLPAPER, CERAMIC TILE) · BASE (WOOD, RUBBER, CERAMIC TILE, STAIN) · TRIM (WOOD, STAIN, PAINT)

ROOMS	CARPET	CERAMIC TILE	RUBBER TILE	CONCRETE	ACOUSTIC TILE	DRYWALL	PAINT	CERAMIC TILE	DRYWALL	PAINT	WALLPAPER	CERAMIC TILE	WOOD	RUBBER	CERAMIC TILE	STAIN	WOOD	STAIN	PAINT	REMARKS
ENTRY		✓			✓				✓	✓	✓		✓				✓	✓		See owner for all painting
HALL	✓				✓				✓	✓			✓				✓	✓		
BEDROOM 1	✓				✓				✓	✓	✓		✓				✓	✓		See owner for grade of carpet
BEDROOM 2	✓				✓				✓	✓			✓				✓	✓		See owner for grade of carpet
BEDROOM 3	✓				✓				✓	✓			✓				✓	✓		See owner for grade of carpet
BATH 1	✓	✓			✓			✓	✓	✓	✓	✓	✓				✓	✓		Wallpaper 3 walls around vanity
BATH 2		✓			✓			✓	✓	✓	✓	✓			✓		✓	✓		Water-seal tile / Wallpaper w/wall
UTIL + CLOSETS	✓	✓				✓	✓		✓	✓			✓			✓	✓	✓	✓	Use off-white flat latex
KITCHEN		✓			✓				✓	✓				✓			✓	✓		
DINING	✓				✓				✓	✓	✓		✓				✓	✓		
LIVING	✓				✓				✓	✓			✓				✓	✓		See owner for grade of carpet
GARAGE				✓		✓	✓		✓				✓				✓	✓		

28304-14_F20.EPS

Figure 20 Example of a finish schedule.

pounds per linear foot. The length of the beam is found on the plan view or the shop drawings. The fabricator will cut these beams to the specified length, label them, and predrill connection holes.

2.3.1 Foundation Plans and Details

Commercial buildings that carry heavy loads receive a great deal of design attention at the foundation and footing level. Soil sampling, laboratory tests, and engineering analysis determine the foundation type and size. The foundation size is determined by the load to be carried. The type of foundation is determined by a combination of load and soil capacity. Foundations are categorized as shallow, intermediate, or deep.

Shallow foundations are set to a depth just below the frost line or slightly lower to reach soil with adequate bearing capacity. Shallow foundations take these forms:

- A continuous reinforced concrete footing, around the entire building perimeter, carrying wall loads directly
- Isolated reinforced concrete footings located under loadbearing columns
- A concrete slab that is placed in a single operation and can carry wall loads directly, or through columns, or both

Check the Legend

In order to avoid mistakes in reading the drawings, be sure you understand the symbols and abbreviations used on every drawing set. Symbols and abbreviations may vary widely from one drawing set to another.

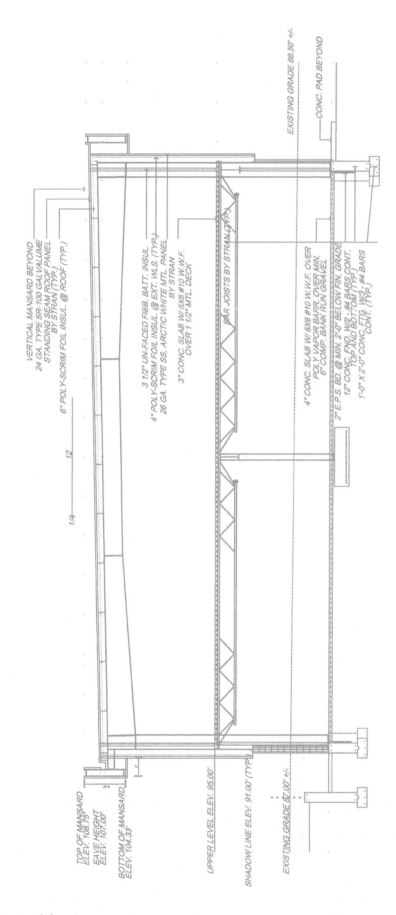

VERTICAL MANSARD BEYOND
24 GA. TYPE SR-100 GALVALUME
STANDING SEAM ROOF PANEL
BY STRAN (TYP.)

6" POLY-SCRIM FOIL INSUL. @ ROOF (TYP.)

3 1/2" UN-FACED FIBB. BATT. INSUL.

4" POLY-SCRIM FOIL INSUL. @ EXT. WLS. (TYP.)
26 GA. TYPE SS, ARCTIC WHITE MTL. PANEL
BY STRAN

3" CONC. SLAB W/ 6X6 #10 W.W.F.
OVER 1 1/2" MTL. DECK

BAR JOISTS BY STRAN (TYP.)

4" CONC. SLAB W/ 6X6 #10 W.W.F. OVER
POLY VAPOR BARR. OVER MIN.
6" COMP. BANK RUN GRAVEL

2" E.P.S. BD. @ MIN. 2'-0" BELOW FIN. GRADE
12" CONC. FND. W/2 - #4 BARS CONT.
TOP AND BOTTOM (TYP.)
1'-0" X 2'-0" CONC. FTG. W/3 - #4 BARS
CONT. (TYP.)

EXISTING GRADE 86.50' +/-

CONC. PAD BEYOND

TOP OF MANSARD
ELEV. 108.75'
EAVE HEIGHT
ELEV. 107.00'
BOTTOM OF MANSARD
ELEV. 104.33'

UPPER LEVEL ELEV. 95.00'

SHADOW LINE ELEV. 91.00' (TYP.)

EXISTING GRADE 87.00' +/-

12
1/4

BUILDING SECTION
SCALE: 1/4" = 1'-0"

Figure 21 Structural drawing.

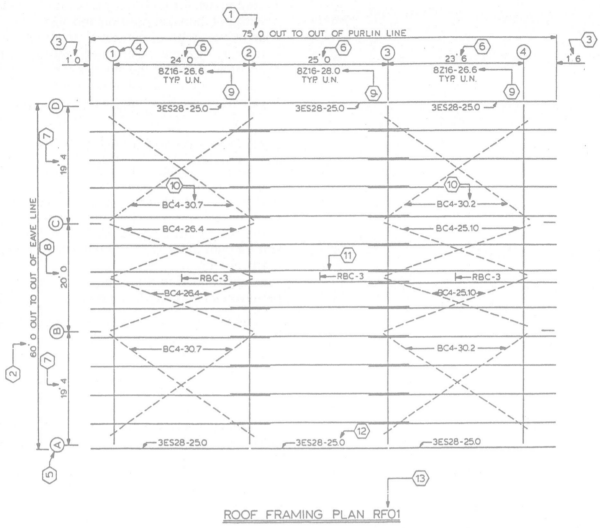

Figure 22 Roof framing plan.

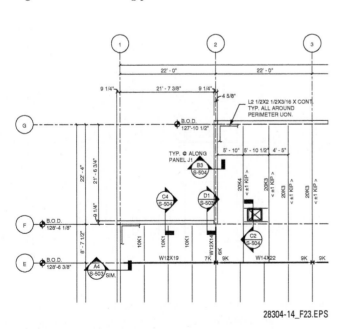

Figure 23 Grid lines for structural plan.

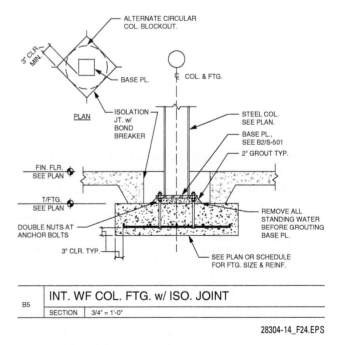

Figure 24 Footing-pier drawing.

NAME	IDENTIFYING SYMBOL	SHAPE
AMERICAN STANDARD BEAMS	S	I
AMERICAN STANDARD CHANNELS	C	[
ANGELS–EQUAL LEGS	L	L
ANGELS–UNEQUAL LEGS	L	L
MISCELLANEOUS CHANNELS	MC	[
MISCELLANEOUS SHAPES	M	I
STRUCTURAL TEES (CUT FROM AM. STD. BEAMS)	ST	T
STRUCTURAL TEES (CUT FROM WIDE-FLANGE SHAPES)	WT	T
WIDE-FLANGE SHAPES	W	I

28304-14_F25.EPS

Figure 25 Structural steel notations.

- Grade beams of reinforced concrete set below grade level and supported by other foundation elements

Intermediate foundations are set to a depth generally not exceeding 15 to 20 feet. Intermediate foundations take these forms:

- *Mat foundations* – large, heavily reinforced concrete mats under the complete building area; sometimes called raft foundations
- *Drilled piers* – concrete or reinforced concrete piers formed by placing concrete in deep holes drilled in the earth; these are designed to carry column loads or grade beams

Deep foundations are set to a depth over 20 to 30 feet. These are used where surface soil is not adequate for building loads. Steel or concrete pilings are driven into the ground until they reach a strong, stable soil layer, or until they generate enough frictional resistance to compensate for the building load. Piles are typically capped with concrete that supports column loads or grade beams.

More detailed information is needed if the foundation goes well below the earth's surface. However, any foundation plan should provide the following information:

- Plan views for footings, piers, and/or columns with notations on position and size
- Schedules with dimensional notations, shapes, reinforcing, and construction requirements
- Sections showing the smaller construction details of footings, columns, connections, and callout marks that refer to the schedules

2.3.2 Framing Plans and Details

The structural engineer draws a framing plan or diagram for the roof and for each floor level that will be framed. On these drawings, the exterior walls or bearing walls are often drawn in lightly while heavier lines represent the framing. The resulting plan looks very much like a graph or diagram, as shown in *Figure 26*.

Looking at this diagram, or at any framing plan, you should see the following:

- Notations identifying beams, joists, and girders by size, shape, and material
- Column, pier, and support locations and their relationships to joists or framing
- Notes or callouts identifying corresponding sections or detail drawings

You may also see details for locations of stairs, recesses, and chimney placements. Details show additional unique framing around these areas. The dimensions on the drawings are center-line dimensions, not actual member dimensions. The member size must be less than the center-line dimensions to allow for construction tolerances.

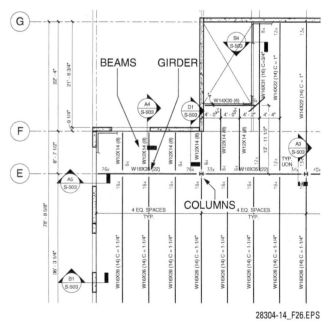

28304-14_F26.EPS

Figure 26 Structural-steel framing diagram (bearing walls not shown).

Coordination Drawings

Coordination drawings are produced by the individual contractors for each trade in order to prevent a conflict in the installation of their materials and equipment. Coordination drawings are produced prior to finalizing shop drawings, cut lists, and other drawings, and before the installation begins. Development of these drawings evolves through a series of review and coordination meetings held by the various contractors.

Some contracts require coordination drawings, while others only recommend it. In the case where one contractor elects to make coordination drawings and another does not, the contractor who made the drawings may be given the installation right-of-way by the presiding authority. As a result, the other contractor may have to bear the expense of removing and reinstalling equipment if the equipment was installed in a space designated for use by the contractor who produced the coordination drawings.

The columns on a framing plan are shown from the top. Lines running between columns are beams. Beams fasten directly to the columns. Joists fasten between beams or between beams and walls. To keep long spans from swaying or twisting in the center, bridging or support members are placed between joists. All of these members will either be referenced to a schedule or to notes directly written on the plans.

Structural drawings include sheets of details, schedules, and notes with the framing plans. They help workers to understand and follow the specifications for the structure. They provide information on the following areas:

- Reinforcing information for all areas where steel rods or wire mesh will be used
- Information for each type of connection made in framing members
- Bearing-plate information detailing the features of all members that will bear directly on other members
- Information for positioning ties, stirrups, or saddles

The details also identify load limits, test strength, fastener types, and uniform specifications, which must be applied where specific information is not given.

There is a great deal of information on the framing plans. It is easier to read such plans by isolating the separate bays or spans between columns. Read the details for that area before moving to other areas of the plan.

2.3.3 Structural Plan Schedules

Schedules are tables that describe and specify the various types and sizes of construction materials used in a building. Wall schedules, column schedules, and ledge schedules are of particular interest to a mason. Always refer to the schedules on struc-

tural plans when preparing to do masonry work on any structural elements.

2.4.0 Reading and Interpreting Shop Drawings

Masons and other trade professionals use shop drawings to fabricate and install components of a construction project. Shop drawings are prepared to provide details on the locations of holes and openings in structural steel members, for example, or to illustrate purchased items and equipment. The design drawing is often placed on the same sheet as the shop drawing.

In masonry work, shop drawings are often prepared for the following types of components:

- Rebar
- Embedded materials
- Post-tensioned locations
- Precast materials
- Doors and windows
- Flashing

Submittal drawings are shop drawings that have been prepared by the fabricator or manufacturer. A submittal drawing that has been approved by the architect or the project engineer is called an approved submittal drawing or approved submittal data. In the event of a conflict, approved submittal drawings typically override the information included in a construction drawing. However, to be safe, you should consult with the architect or project engineer if you find a conflict between an approved submittal drawing and a construction drawing.

2.4.1 Rebar Shop Drawings

Due to the complex reinforcement involved in commercial masonry work, rebar shop drawings are often required as part of the contract documents. These drawings are created by a draftsperson employed by the reinforcing-steel supplier or a third-party detailing service. The drawings typically include a floor plan, wall sections, bending details, schedules, and notes. The floor plan provides layout for each of the vertical reinforcing-bar locations on a given wall and/or indicates wall types for typical spacing requirements.

The wall sections show the lifts that will be used on each wall type and the corresponding vertical rebar lengths for each lift. In addition, the wall section may show the size, quantity, and approximate location of the horizontal bars. Details show specific configurations of reinforcement at wall corners, jambs, intersections, or any other specially reinforced sections. Bending details show the exact dimensions for any bent bars, if applicable. Schedules correspond with structural schedules such as those for masonry lintels and columns. Notes clarify any typical information or information that is not found in the other views. This can include typical spacings, lap lengths, and standard horizontal bar lengths.

2.4.2 Embed Shop Drawings

Masonry loadbearing structures use plates and anchors embedded in mortar and grout to connect the masonry walls to structural members such as beams and girders. Because these items are often referred to collectively as embeds, the drawings that illustrate their dimensions and placement are called embed shop drawings. Due to the many variables involved, embed shop drawings are often provided by the structural-steel supplier for the project. These drawings typically include framing plans, sections, and details.

The framing plan is often referred to as the placement drawing. It includes the layout or location of each embed used on the project. Embeds should never be used interchangeably on a project. The thickness and size of metal plate, and the size, quantity, and configuration of anchors, vary depending on the part of the structure for which they are to be used. On the framing plan, each type of embed is marked with a specific designation that corresponds to the fabrication drawing for that embed. The fabrication drawing will show exactly how each particular embed is to be manufactured. Typically, the framing plan also indicates the height of each embed. Sections and details will be used to clarify and communicate any additional information required for the proper layout and installation of embeds.

2.4.3 Precast Shop Drawings

Projects that involve the use of custom-manufactured and custom-molded precast concrete units require precast shop drawings to ensure accuracy in their manufacture and installation. Precast concrete units are made to the exact sizes, finishes, and configurations needed for each wall on a project. These drawings typically include a plan view and unit details. The plan view will designate where each of the individual units on a project will be placed. The unit details will provide the exact dimensions, shape, color, and finish of each unit to be manufactured for the order.

2.4.4 Door and Window Shop Drawings

On large commercial projects, door and/or window shop drawings may be required. Much like the other types of shop drawings discussed in this section, door and window shop drawings include a plan view along with details and schedules related to the openings. The plan view will identify the locations where each different door or window will be located. The details and schedules provide additional information regarding materials, configurations, types, and any relevant installation notes.

2.4.5 Flashing Shop Drawings

Shop drawings are specialized drawings that show how to fabricate and install components of a construction project. One type of shop drawing that may be created after an engineer designs the structure is a flashing shop drawing. A flashing shop drawing shows the sizes of all flashing, along with locations of all seams, hems, holes, and openings in the flashing. These shop drawings also provide notes specifying how the flashing components are to be made and to fit together.

Flashing shop drawings are commonly used to generate material takeoffs for the flashing (*Figure 27*). Flashing dimensions are taken from the shop drawings and entered into a spreadsheet for automatic calculation of material lengths. An estimate, which includes labor charges, can then be generated from the material takeoff.

2.5.0 Understanding and Applying Emerging Technologies

For centuries, architectural plans have been used to pictorially describe buildings and structures before they are actually built. In the past, draftspersons would draw these plans by hand. Today, most construction drawings are generated electronically, and masons can use apps to render them on mobile devices such as tablets and even smartphones. The variety of digital and online tools available to masons grows every day. This section serves as a broad overview of the technologies that have become well established in the trade. By the time you read this, newer digital and online technologies may be widely used as well.

Modern construction drawings are prepared using architectural software called computer-aided design (CAD). The CAD operator, working with architects and engineers, creates the drawings on the computer. The drawings can then be printed or plotted onto paper.

Some CAD software is capable of performing building information modeling (BIM). BIM is a digital representation of a structure and its characteristics, which masons can use to prepare construction drawings, identify design conflicts

Kopf Suites, Dubuque, IA — Flashing Estimate, January 15, 2015

	A	B	C	D	E	F	G	H	I	J	K	L	M	N
1														
2	TotalFlash®		18 inch			Length:	552.45	LF	without waste first floor only					
3	Additional Alternate					Length:	577.95	LF	per upper floor perimeter detail not provided					
4	TotalFlash®		12 inch			Length:	596.30	LF	without waste					
5			•	565.44	LF-window heads									
6			•	30.86	LF-door heads									
7														
8		Opening			Flashing Length		Location		Quantity	Material			Total	
9	Windows:	4.00	LF	5.34	LF		heads		12	12 inch material			64.08	LF
10		4.50	LF	5.94	LF				78	12 inch material			463.32	LF
11		5.00	LF	6.34	LF		heads		6	12 inch material			38.04	LF
12												Total:	565.44	
13	Doors:	3.50	LF	4.84	LF		heads		3	12 inch material			14.52	LF
14		15.00	LF	16.34	LF		heads		1	12 inch material			16.34	LF
15								Total:	100			Total:	30.86	
16														
17	TotalFlash® Components:													
18	End Dams			108	pieces									
19	Inside Corner Boots			15	pieces									
20	Outside Corner Boots			19	pieces									
21	Inside Stainless Steel Corners			15	pieces									
22	Outside Stainless Steel Corners			19	pieces									
23														
24														
25	The estimate assumes:													
26	Exterior wall section was not provided, a quantity is provided for each floor above first floor of 577.95 LF													
27	Window dimensions were taken from the window schedule detail 2 A1.1 that conflicted with detail 6 on A 2.0.													
28	Drawings indicated .67 LF additional flashing per side of each opening.													
29	Mortar Net Solutions will custom-make window lintel flashings at no additional charge to the customer.													
30	TotalFlash® is installed above grade in all applications.													
31	Flashing was not typically shown at storefront window and door heads.													
32	Custom length material will be marked on boxes.													
33	Lengths greater than 5 feet will be made in two pieces.													
34	Site drawings were not reviewed for this takeoff.													
35	Estimate was provided free of charge as a service to the customer.													
36	Mortar Net Solutions does not assume liability for accuracy.													
37	Contractor must perform a takeoff to verify product quantities.													
38														

28304-14_F27.EPS

Figure 27 Sample material takeoff for flashing.

and problems, and map the complete life cycle of the structure. BIM allows designers to prepare digital models of an entire structure and all its components (*Figure 28*). BIM allows people to virtually "walk" or even "fly" through a building before it has been built, in order to see how it will look when it has been completed. This allows designers to identify problems and conflicts with various components in time to correct them before construction has begun or is completed. BIM also allows designers to estimate the costs of constructing the building and even the cost of facilities maintenance over time. BIM files can be shared electronically by email or they can be posted online for download.

BIM can also be used to prepare coordination drawings, which are used to prevent conflicts during construction, and as-built drawings, which show the project as completed, including all changes made along the way. BIM has not taken the place of construction drawings, but it does offer a new and improved way to look at the systems within a building. Standards for BIM are being developed in the United States, Canada, and Europe. Other terms for BIM include virtual building environment (VBE) and virtual design and construction (VDC).

In addition to BIM, masons can choose from a variety of smartphone apps (*Figure 29*) and calculators that can assist them with measurements and materials estimating, including:

- Computing the number of block or brick needed by entering wall area or length and width dimensions
- Converting between weight and volume for standard construction materials
- Performing dimensional math and estimating material volumes and cost

- Calculating and converting dimensions, plotting right-angle conversions, and finding area and volume
- Determining measures such as board feet, cost per unit, and even the angles of an equal-sided polygon
- Calculating stair dimensions or the length and weight of rebar

On-screen estimators, such as those developed by Tradesmen's Software, Inc., have the ability to turn digital construction drawings into 3-D renderings that automatically calculate the amount of materials required to complete the structure (*Figure 30*). This allows masons to spot mistakes before they happen, saving the company time and money from the start of a project.

On-screen estimators are growing in popularity among the masonry trade because of their speed and ease of use. It takes just a few seconds for the software to estimate the number of masonry units and the amount of mortar or grout for a wall, pier, or other masonry structure. Material estimates are automatically updated whenever a dimensional change is made in the plans. It can even calculate labor rates, mortar yields, and lay rates based on information entered into the system. The information can also be exported into BIM files and used to create 3-D proposal drawings for clients and contractors.

28304-14_F28.EPS

Figure 28 A BIM rendering of building components in three dimensions.

28304-14_F29.EPS

Figure 29 Estimating calculator for masons.

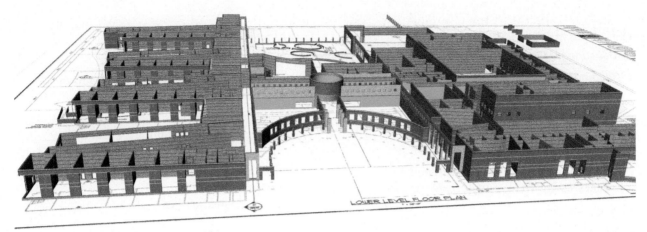

28304-14_F30.EPS

Figure 30 A 3-D rendering of a building in an on-screen estimator.

Green Building XML

Green Building XML (gbXML) is a type of BIM that focuses on the design and operation of green buildings. Designers can use gbXML to simulate energy consumption in a building to help improve its energy efficiency. The gbXML system is only one of several new simulation technologies being developed to help designers design more energy-efficient buildings in the future.

Additional Resources

Measuring, Marking, and Layout. Newtown, CT: Taunton Press.

Reading Architectural Plans: For Residential and Commercial Construction. 1998. Ernest R. Weidhaas. Upper Saddle River, NJ: Prentice Hall.

2.0.0 Section Review

1. Drawings that show enlargements of special features drawn to a larger scale are called _____.

   a. site plans
   b. elevation views
   c. section views
   d. detail views

2. Each of the following is usually provided in a wall section *except* _____.

   a. insulation applications and materials
   b. fire ratings of materials
   c. horizontal and vertical joint reinforcement
   d. sound-barrier placement or materials

3. In a framing plan, columns are shown _____.

   a. from the top
   b. from the bottom
   c. as solid circles
   d. as dashed circles

4. Framing plans are often referred to as _____.

   a. shop drawings
   b. isometric drawings
   c. placement drawings
   d. building and wall sections

5. VDC stands for _____.

   a. virtual design and construction
   b. validated drawing collection
   c. vital document control
   d. virtual design calculator

NCCER – *Masonry Level Three* 28304-14

SECTION THREE

3.0.0 WRITTEN SPECIFICATIONS FOR COMMERCIAL DRAWINGS

Objective

Explain the purpose of written specifications.
  a. Describe how specifications are written.
  b. Explain the format of specifications.

Performance Task

Locate the section of a set of specifications that shows the type of mortar to be used.

The written specifications for a building or project are the written descriptions of work and duties required of the owner, architect, and consulting engineer. Together with the working drawings, these specifications form the basis of the contract requirements for the construction of the building or project. Those who use the construction drawings and specifications must always be alert to discrepancies between the working drawings and the written specifications. These are some situations where discrepancies may occur:

- Architects or engineers use standard or prototype specifications and attempt to apply them without any modification to specific working drawings.
- Previously prepared standard drawings are changed or amended by reference in the specifications only and the drawings themselves are not changed.
- Items are duplicated in both the drawings and specifications, but an item is subsequently amended in one and overlooked in the other contract document.

In such instances, the person in charge of the project has the responsibility to ascertain whether the drawings or the specifications take precedence. Such questions must be resolved, preferably before the work begins, to avoid added costs to the owner, architect/engineer, or contractor.

3.1.0 How Specifications Are Written

Writing accurate and complete specifications for building construction is a serious responsibility for

those who design the buildings because the specifications combined with the working drawings govern practically all important decisions made during the construction span of every project. Compiling and writing these specifications is not a simple task, even for those who have had considerable experience in preparing such documents.

A set of written specifications for a single project will usually contain thousands of products, parts, and components, and the methods of installing them, all of which must be covered in either the drawings and/or specifications. No one can memorize all of the necessary items required to accurately describe the various areas of construction. One must rely on reference materials such as manufacturer's data, catalogs, checklists, and, most of all, a high-quality master specification.

3.1.1 Special and General Conditions Sections

The special and general conditions sections of the specifications cover the nontechnical aspects of the contractual agreements. Special conditions cover topics such as safety and temporary construction. General conditions cover the following points of information:

- Contract terms
- Responsibilities for examining the construction site
- Types and limits of insurance
- Permits and payments of fees
- Use and installation of utilities
- Supervision of construction
- Other pertinent items

The general conditions section is the area of the construction contract where misunderstandings often occur. Therefore, these conditions are usually much more explicit on large, complicated construction projects. Note that residential specifications often do not spell out general conditions and are basically material specifications only.

3.1.2 Technical Aspects Section

The technical aspects section includes information on materials that are specified by standard numbers and by standard national testing organizations such as the American Society for Testing and Materials (ASTM). The technical aspects section of specifications can be of three types:

- *Outline specifications* – These specifications list the materials to be used in order of the basic parts of the job, such as foundation, floors, and walls.

- *Fill-in specifications* – This is a standard form filled in with pertinent information. It is typically used on smaller jobs.
- *Complete specifications* – For ease of use, most specifications written for large construction jobs are organized in the format called the *MasterFormat™*.

3.2.0 Format of Specifications

For convenience in writing, speed in estimating, and ease of reference, the most suitable organization of the specifications is a series of sections that deals with the construction requirements, products, and activities, and that is easily understandable by the different trades. Those people who use the specifications must be able to find all information needed without spending too much time looking for it.

The most commonly used specification format in North America is *MasterFormat™*. This standard was developed jointly by the Construction Specifications Institute (CSI) and Construction Specifications Canada (CSC). Prior to 2004, the organization of construction specifications and supplier's catalogs was based on a standard with 16 sections, known as divisions. The divisions and their subsections were individually identified by a five-digit numbering system. The first two digits represented the division number and the next three individual numbers represented successively lower levels of breakdown.

In 2004, the *MasterFormat™* standard underwent a major change. What had been 16 divisions was expanded to four major groupings and 49 divisions with some divisions reserved for future expansion. The first 14 divisions of *MasterFormat™* 2012 (*Figure 31*) are essentially the same as the old format. Subjects under the old division 15 (Mechanical) have been relocated to new divisions 22 and 23. The basic subjects under old division 16 (Electrical) have been relocated to new divisions 26 and 27.

In addition, the numbering system for the new *MasterFormat™* organization was changed to six digits preceding the decimal point, to allow for more subsections in each division. In the new numbering system, the first two digits represent the division number. The next two digits represent subsections of the division, and the two remaining digits represent the third-level sub-subsection numbers. The fourth level, if required, is a decimal point and number added to the end of the last two digits. Masonry specifications are typically found under Division 04 (see *Figure 31*). Masons also refer frequently to Division 03, *Concrete*, and Division 07, *Thermal and Moisture Protection*.

DIVISIONS NUMBERS AND TITLES

PROCUREMENT AND CONTRACTING REQUIREMENTS GROUP

Division 00   Procurement and Contracting Requirements

SPECIFICATIONS GROUP

GENERAL REQUIREMENTS SUBGROUP

Division 01   General Requirements

FACILITY CONSTRUCTION SUBGROUP

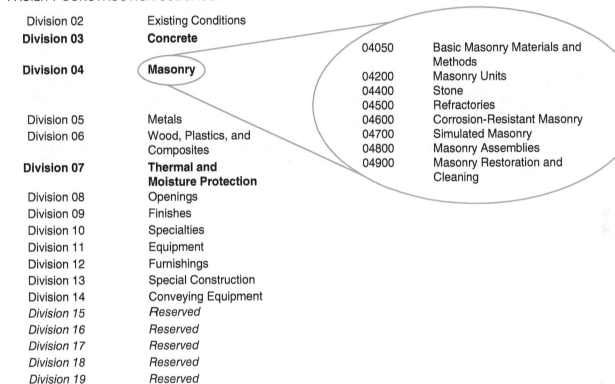

Division 02	Existing Conditions	
Division 03	**Concrete**	
Division 04	**Masonry**	
		04050 Basic Masonry Materials and Methods
		04200 Masonry Units
		04400 Stone
		04500 Refractories
		04600 Corrosion-Resistant Masonry
Division 05	Metals	04700 Simulated Masonry
Division 06	Wood, Plastics, and Composites	04800 Masonry Assemblies
		04900 Masonry Restoration and Cleaning
Division 07	**Thermal and Moisture Protection**	
Division 08	Openings	
Division 09	Finishes	
Division 10	Specialties	
Division 11	Equipment	
Division 12	Furnishings	
Division 13	Special Construction	
Division 14	Conveying Equipment	
Division 15	*Reserved*	
Division 16	*Reserved*	
Division 17	*Reserved*	
Division 18	*Reserved*	
Division 19	*Reserved*	

28304-14_F31.EPS

Figure 31 2012 *MasterFormat*™.

Additional Resources

Architectural Drawing and Light Construction. 2009. Philip A. Grau III, Edward J. Muller, and James G. Fausett. Upper Saddle River, NJ: Prentice Hall.

MasterFormat™, Latest Edition. Alexandria, VA: The Construction Specifications Institute (CSI) and Construction Specifications Canada (CSC).

Plan Reading & Material Takeoff. Kingston, MA: R.S. Means Company.

3.0.0 Section Review

1. Each of the following is a type of technical aspects section that can appear in specifications *except* _____.

   a. digest
   b. fill-in
   c. outline
   d. complete

2. The numbering system in *MasterFormat*™ contains _____.

   a. six digits following the decimal point
   b. four digits following the decimal point
   c. six digits preceding the decimal point
   d. four digits preceding the decimal point

SUMMARY

In order to build commercial structures, a mason must be able to understand and interpret commercial plans. The basic principles, such as scaling and dimensioning, apply to both residential and commercial drawings. However, commercial drawings are much more detailed. They can contain additional structural, mechanical, and electrical systems.

A typical commercial plan set includes architectural, structural, shop, and MEP (mechanical, electrical, and plumbing) drawings. Architectural drawings include the site plan, floor plans, wall sections, door and window details, and schedules. Structural drawings include framing plans, support details, schedules, and notes. Mechanical drawings include HVAC plans. Plumbing drawings and mechanical drawings are similar in format. They provide a plan view and isometric view of the fresh-water system and the waste-water system. Electrical drawings usually include a lighting plan.

When reading commercial plans, follow a step-by-step process to avoid confusion and make sure you see all the details involved. Begin by reading the project specifications to pick up details not found on the drawings. Quickly review all the drawings in order to get a general impression of the shape, size, and appearance of the structure. Begin correlating the floor plans with the exterior elevations. Look at the wall sections to determine the wall types and construction details.

Review the structural plans to determine the foundation requirements and the type of structural system. Carefully consider all items that may require special construction procedures. Review all interior elevations. Try to get a clear picture of what the interior of the building will look like. Review the finish schedule and the mechanical and electrical plans. In particular, look for work items to be performed by masons or features that will affect the work of masons.

Specifications provide written instructions for the owner, architect, and engineer. A set of written specifications for a large commercial project will usually contain thousands of products, parts, components, and the methods of installing them. A standard format has been widely adopted for commercial construction throughout North America. Manufacturers and suppliers key their catalogs and data to this standard format.

The standard format makes understanding the specifications much easier. However, discrepancies can arise if the standard specifications are not modified to match the drawings of a particular project. All discrepancies must be resolved by the person in charge before work begins.

1. The number of types of drawings that make up a typical commercial drawing set is _____.
   a. three
   b. five
   c. six
   d. ten

2. Symbols used in a large commercial drawing set are listed in a key block or _____.
   a. graphic index
   b. table
   c. legend
   d. code panel

3. Except in unusual circumstances, a type of structural support that is *not* used in commercial construction is _____.
   a. steel framing
   b. precast concrete elements
   c. cast-in-place concrete
   d. wooden framing

4. Specialized drawings that show how to fabricate and install components are described as _____.
   a. shop drawings
   b. technical drawings
   c. detail drawings
   d. instructional drawings

28304-14_RQ01.EPS
Figure 1

5. The HVAC symbol shown in Review Question *Figure 1* is a _____.
   a. wall or space heater
   b. cold-air return duct
   c. warm-air supply duct
   d. convector

6. Isometric drawings used in plumbing drawings are called _____.
   a. connection views
   b. riser diagrams
   c. runner plans
   d. piping schematics

7. Locations and dimensions of easements are shown on the _____.
   a. mechanical drawings
   b. floor plans
   c. site plans
   d. section views

8. The drawing scale used for floor plans of commercial structures is typically _____.
   a. ⅛ inch per foot
   b. ¼ inch per foot
   c. ½ inch per foot
   d. 1 inch per foot

9. The main drawing of the entire commercial drawing set is the _____.
   a. floor plan
   b. elevation view
   c. foundation plan
   d. site plan

10. Drawings that appear to be a cut through a wall or other structural feature are called _____.
    a. cutaways
    b. detail views
    c. slices
    d. section views

11. On most commercial floor plans, a legend or chart is often included to specify _____.
    a. various interior wall types
    b. flooring materials
    c. partition locations
    d. door and window openings

12. Ceiling plans that keep features in proper relation to the floor plan are described as _____.
    a. mirrored
    b. scaled
    c. reflected
    d. correlated

13. Tables describing and specifying types and sizes of materials to be used in a building are referred to as _____.
    a. punch lists
    b. schedules
    c. specifications
    d. materials logs

14. Timber roof beams used in commercial construction usually have a nominal dimension of _____.
    a. 6 inches
    b. 8 inches
    c. 10 inches
    d. 12 inches

15. In North America, the most commonly used specification format is _____.
    a. MCAAformat®
    b. AutoCAD/BIM
    c. *MasterFormat*™
    d. BOCA

Trade Terms Quiz

Fill in the blank with the correct term that you learned from your study of this module.

1. A drawing that shows proposed plantings and other landscape features is called a(n) _____.

2. A(n) _____ is a drawing that shows the overall shape of the building site. Also called a site plan.

3. A large steel or wooden beam supporting a building, usually around the perimeter, is called a(n) _____.

4. A(n) _____ is a point established by the surveyor on or close to the building site and used as a reference for determining elevations during the construction of a building.

5. A physical structure that marks the location of a survey point is called a(n) _____.

6. _____ are the set of construction drawings that consists of mechanical, electrical, and plumbing drawings.

7. A construction drawing that shows a project as it was completed, including all changes incorporated into the design during the construction process, is called a(n) _____.

8. A(n) _____ is a three-dimensional drawing in which the object is tilted so that three faces are equally inclined to the picture plane.

9. A type of isometric drawing that depicts the layout, components, and connections of a piping system is called a(n) _____.

10. A(n) _____ is the recorded legal boundary of a piece of property.

11. A person who provides detailed shop drawings for the fabrication of components and who fabricates them in a shop for later installation at the job site is called a(n) _____.

12. The _____ is the distance from the property line to the front of the building.

13. A drawing that represents a view looking down on an object is called a(n) _____.

14. A(n) _____ is a loadbearing horizontal framing element supported by walls or columns and girders.

15. A horizontal member of wood or steel supported by beams and holding up the planks of floors or the lathes of ceilings is called a(n) _____.

16. A(n) _____ is a drawing that shows the details of the ceiling as though the ceiling were reflected by a mirror on the floor.

17. A marking or identifying tag describing parts of a drawing on detail drawings, schedules, or other drawings is called a(n) _____.

18. A(n) _____ is a legal right-of-way provision on another person's property.

19. A drawing giving a view from the front or side of a structure is called a(n) _____.

Trade Terms

As-built drawing	Civil drawing	Front setback	Landscape drawing	Property line
Beam	Easement	Girder	MEP drawings	Reflected ceiling
Benchmark	Elevation view	Isometric drawing	Monument	plan
Callout	Fabricator	Joist	Plan view	Riser diagram

Trade Terms Introduced in This Module

As-built drawing: A construction drawing that shows a project as it was completed, including all changes incorporated into the design during the construction process.

Beam: Loadbearing horizontal framing element supported by walls or columns and girders.

Benchmark: A point established by the surveyor on or close to the building site and used as a reference for determining elevations during the construction of a building.

Callout: Marking or identifying tag describing parts of a drawing on detail drawings, schedules, or other drawings.

Civil drawing: A drawing that shows the overall shape of the building site. Also called a site plan.

Easement: A legal right-of-way provision on another person's property (for example, the right of a neighbor to build a road or a public utility to install water and gas lines on the property). A property owner cannot build on an area where an easement has been identified.

Elevation view: A drawing giving a view from the front or side of a structure.

Fabricator: A person who provides detailed shop drawings for the fabrication of components and who fabricates them in a shop for later installation at the job site.

Front setback: The distance from the property line to the front of the building.

Girder: Large steel or wooden beam supporting a building, usually around the perimeter.

Isometric drawing: A three-dimensional drawing in which the object is tilted so that three faces are equally inclined to the picture plane.

Joist: Horizontal member of wood or steel supported by beams and holding up the planks of floors or the lathes of ceilings. Joists are laid edgewise to form the floor support.

Landscape drawing: A drawing that shows proposed plantings and other landscape features.

MEP drawings: The set of construction drawings that consists of mechanical, electrical, and plumbing drawings.

Monument: A physical structure that marks the location of a survey point.

Plan view: A drawing that represents a view looking down on an object.

Property line: The recorded legal boundary of a piece of property.

Reflected ceiling plan: A drawing that shows the details of the ceiling as though the ceiling were reflected by a mirror on the floor.

Riser diagram: A type of isometric drawing that depicts the layout, components, and connections of a piping system.

Additional Resources

This module presents thorough resources for task training. The following resource material is suggested for further study.

Architectural Drawing and Light Construction. 2009. Philip A. Grau III, Edward J. Muller, and James G. Fausett. Upper Saddle River, NJ: Prentice Hall.

Architectural Graphic Standards. 1998. The American Institute of Architects. New York: John Wiley & Sons, Inc.

A Manual of Construction Documentation: An Illustrated Guide to Preparing Construction Drawings. 1989. Glenn E. Wiggins. New York: Whitney Library of Design.

Masonry Design and Detailing for Architects, Engineers and Contractors, Sixth Edition. 2012. Christine Beall. New York: McGraw-Hill.

MasterFormat™, Latest Edition. Alexandria, VA: The Construction Specifications Institute (CSI) and Construction Specifications Canada (CSC).

Measuring, Marking, and Layout. Newtown, CT: Taunton Press.

Plan Reading & Material Takeoff. Kingston, MA: R.S. Means Company.

Reading Architectural Plans: For Residential and Commercial Construction. 1998. Ernest R. Weidhaas. Upper Saddle River, NJ: Prentice Hall.

Figure Credits

Section Review Answers

Answer	Section Reference	Objective
Section One		
1. b	1.1.0	1a
2. d	1.2.3	1b
Section Two		
1. d	2.1.4	2a
2. c	2.2.4	2b
3. a	2.3.2	2c
4. c	2.4.2	2d
5. a	2.5.0	2e
Section Three		
1. a	3.1.2	3a
2. c	3.2.0	3b

NCCER CURRICULA — USER UPDATE

NCCER makes every effort to keep its textbooks up-to-date and free of technical errors. We appreciate your help in this process. If you find an error, a typographical mistake, or an inaccuracy in NCCER's curricula, please fill out this form (or a photocopy), or complete the online form at **www.nccer.org/olf**. Be sure to include the exact module ID number, page number, a detailed description, and your recommended correction. Your input will be brought to the attention of the Authoring Team. Thank you for your assistance.

Instructors – If you have an idea for improving this textbook, or have found that additional materials were necessary to teach this module effectively, please let us know so that we may present your suggestions to the Authoring Team.

NCCER Product Development and Revision

13614 Progress Blvd., Alachua, FL 32615

Email: curriculum@nccer.org
Online: www.nccer.org/olf

❏ Trainee Guide   ❏ Lesson Plans   ❏ Exam   ❏ PowerPoints   Other _____

Craft / Level: _____   Copyright Date: _____

Module ID Number / Title: _____

Section Number(s): _____

Description: _____

_____

_____

_____

_____

Recommended Correction: _____

_____

_____

_____

_____

Your Name: _____

Address: _____

_____

Email: _____   Phone: _____

28305-14

Estimating

Estimating material quantities is an important part of the planning and scheduling of a construction project. Whether the project is a 20-story high-rise or a bungalow, the estimating process is basically the same. Accurate material estimates require good math and plan-reading skills, as well as access to estimating tables. This module introduces trainees to the skills and techniques required to estimate masonry units, mortar, grout, and accessories for many types of masonry construction.

Module Five

Trainees with successful module completions may be eligible for credentialing through the NCCER Registry. To learn more, go to **www.nccer.org** or contact us at **1.888.622.3720**. Our website has information on the latest product releases and training, as well as online versions of our *Cornerstone* magazine and Pearson's product catalog.

Your feedback is welcome. You may email your comments to **curriculum@nccer.org**, send general comments and inquiries to **info@nccer.org**, or fill in the User Update form at the back of this module.

This information is general in nature and intended for training purposes only. Actual performance of activities described in this manual requires compliance with all applicable operating, service, maintenance, and safety procedures under the direction of qualified personnel. References in this manual to patented or proprietary devices do not constitute a recommendation of their use.

Objectives

When you have completed this module, you will be able to do the following:

1. Explain how to estimate block, mortar, and grout.
   a. Describe how to use the coursing method for block.
   b. Describe the square-foot method for block.
   c. Explain how to estimate openings and lintels.
   d. Explain how to estimate mortar for single-wythe walls.
   e. Explain how to estimate mortar for multiwythe walls.
   f. Explain how to estimate grout.
2. Explain how to estimate brick and mortar.
   a. Explain the coursing method for brick.
   b. Explain the square-foot method for brick.
   c. Describe how to allow for openings in an estimate.
   d. Explain how to estimate mortar for brick.
3. Describe how to estimate accessory items.
   a. Explain how to estimate joint reinforcement.
   b. Explain how to estimate structural reinforcement.
   c. Explain how to estimate masonry ties.
   d. Explain how to estimate other masonry units.
   e. Explain how to estimate other masonry accessories.

Performance Tasks

Under the supervision of your instructor, you should be able to do the following:

1. Estimate the amounts of block, mortar, and grout required for a hypothetical backing wall, using plans provided by the instructor.
2. Estimate the amounts of brick and mortar required for a hypothetical veneer wall, using plans provided by the instructor.
3. Estimate the amounts of rebar and ties required for hypothetical walls, using plans provided by the instructor.

Trade Terms

Square-foot method
Takeoff

Industry-Recognized Credentials

If you're training through an NCCER-accredited sponsor, you may be eligible for credentials from NCCER's Registry. The ID number for this module is 28305-14. Note that this module may have been used in other NCCER curricula and may apply to other level completions. Contact NCCER's Registry at 888.622.3720 or go to **www.nccer.org** for more information.

Code Note

Codes vary among jurisdictions. Because of the variations in code, consult the applicable code whenever regulations are in question. Referring to an incorrect set of codes can cause as much trouble as failing to reference codes altogether. Obtain, review, and familiarize yourself with your local adopted code.

Contents ——————————

Topics to be presented in this module include:

Figures and Tables

1.0.0 ESTIMATING BLOCK

Objective

Explain how to estimate block, mortar, and grout.

a. Describe how to use the coursing method for block.
b. Describe the square-foot method for block.
c. Explain how to estimate openings and lintels.
d. Explain how to estimate mortar for single-wythe walls.
e. Explain how to estimate mortar for multiwythe walls.
f. Explain how to estimate grout.

Performance Task

Estimate the amounts of block, mortar, and grout required for a hypothetical backing wall, using plans provided by the instructor.

Trade Terms

Square-foot method: A method of estimating materials by calculating the area, in square feet, of a structural unit.

Takeoff: The process of measuring and counting individual items from a set of plans in order to estimate material quantities and associated items for construction projects.

Making estimates of the required quantities of masonry materials involves assembling the following pieces of preliminary information:

- A list of the different structural elements to be built, such as walls, patios, walkways, stairs, and foundations
- The size of each item to be built
- The type of wall and the type of bond
- The size and type of masonry units needed
- The size of the mortar joints
- The number of openings
- The type of mortar to be used
- The type and spacing of any metalwork and other accessories

The first step in estimating, then, is to find out what needs to be built. This information comes from a close study of the working drawings, schedules, and specifications. Review those documents, the schedules, and the specifications to find out how the structural items are to be built and make a list of the structural items to be estimated. It is helpful to use colored pencils to check off items on the drawings as they are listed. Use one color as items are taken off onto the worksheet; use a second color when checking the worksheet against the drawings.

The next step is to estimate the materials needed to build each of the structural items on the initial list. Use a takeoff worksheet like the one shown in *Figure 1* to list the details for each structural item. The takeoff is the process of measuring and counting individual items from a set of plans in order to estimate material quantities and associated items for construction projects. For example, if the north wall and the south wall are identical, the calculations need only be done once, then copied for the other. However, the plans must be checked very closely to make sure that the two items are identical.

A full-page version of *Figure 1* is available in the *Appendix*. Use a copy of that worksheet or use your company's estimating form as a guide to ensure that all the materials, elements, and procedures have been included in the estimate. Once the materials have been calculated for each structural item, combine the calculations into a master worksheet by material type. This type of worksheet, shown in *Figure 2*, is useful for summarizing information. It is also useful for a total check and review. For example, something may be wrong if one quantity of unit ties has been estimated for a wall and a very different amount appears for a similar wall. A full-page version of *Figure 2* is also available in the *Appendix*.

Once all of the estimates have been completed, the information can be combined in a recapitulation sheet (recap), as shown in *Figure 3*. This serves as a summary for the entire project. A full-page version of *Figure 3* is also available in the *Appendix*.

The two most common mistakes in estimating material quantities are omitting all or part of an item, and making arithmetic errors. Always recheck the drawings against the worksheets to make sure all the required items have been taken off. Always recheck calculations before accepting the final results. Review the summary worksheet against individual item worksheets to make sure all the data has been transferred without errors.

Different masonry units are each estimated in a variety of ways. The amount of block can be estimated by either the coursing method or the square-foot method. Each method has some advantages, depending on the type and size of the structure. In most cases, masons use the method they find easiest.

REF.	DESCRIPTION	DIMENSIONS				EXTENSION	QUANTITY	UNIT	TOTAL		REMARKS
		NO.	LENGTH	WIDTH	HEIGHT				QUANTITY	UNIT	

Worksheet header:

Takeoff By: _____
Checked By: _____

WORKSHEET   PAGE #

DATE _____
SHEET ____ of _____

PROJECT _____
ARCHITECT _____

28305-14_F01.EPS

Figure 1 Quantity takeoff sheet.

1.1.0 Using the Coursing Method for Block

Use of the coursing method for estimating requires knowing the values for the following items:

- The size of the masonry unit
- The total linear feet of wall
- The number of masonry units needed for a single course of the length of the wall
- The number of courses needed for the height of the wall

For example, say that you must calculate the number of standard 8 × 8 × 16-inch concrete block needed to build the single-wythe garden wall shown in *Figure 4*. As shown, the wall is 8 feet high with one 4-foot-wide gate, enclosing a 20-foot by 10-foot garden.

Step 1 Determine the total linear feet around the wall by adding the length of all four walls together and subtracting the opening.

$$L = 20 + 10 + 10 + 20 - 4 = 56 \text{ feet}$$

Step 2 This figure includes the corners. To avoid counting them twice, subtract the width of the wall for each corner. In this case, the width of the wall is the width of one standard block, or 8 inches.

$$L = \text{linear feet} - (4 \times \text{width of corners})$$
$$= 56 - (4 \times 0.66 \text{ feet})$$
$$= 56 - 2.64 = 53.36 \text{ feet}$$

In some cases, this step can be omitted and the extra block included as part of the waste estimate.

Building the Pentagon

The largest office building in the world is not the Sears Tower, but the Pentagon. Although it is only 77 feet tall, it has 6.5 million square feet of floor space. (The Sears Tower, at 1,470 feet, only has 4.4 million square feet of space.) The five-story building has five concentric rings with five sides each. There are 17.5 miles of corridors.

The Pentagon has reinforced concrete walls built over wood framing. Construction started in 1941 when steel and other metals were scarce. Over 410,000 cubic yards of concrete went into the building, using 680,000 tons of sand and gravel dredged from the adjacent Potomac River. There are 4,900 fixtures in the 284 restrooms. There are 691 drinking fountains, 7,754 windows, and 16,250 light fixtures.

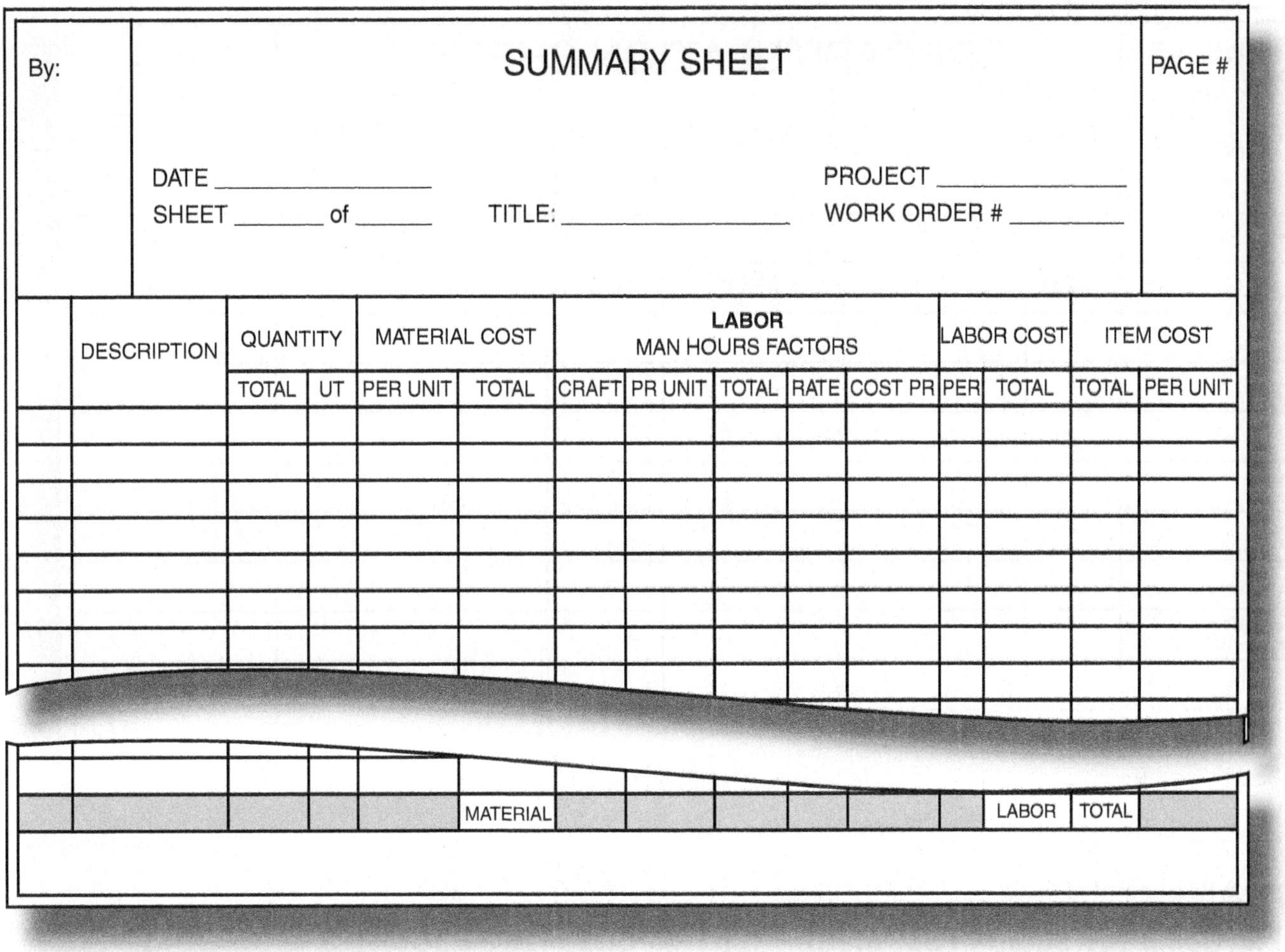

By:		SUMMARY SHEET											PAGE #	

DATE _____   PROJECT _____

SHEET _____ of _____   TITLE: _____   WORK ORDER # _____

	DESCRIPTION	QUANTITY		MATERIAL COST		LABOR MAN HOURS FACTORS					LABOR COST		ITEM COST	
		TOTAL	UT	PER UNIT	TOTAL	CRAFT	PR UNIT	TOTAL	RATE	COST PR	PER	TOTAL	TOTAL	PER UNIT
				MATERIAL							LABOR	TOTAL		

28305-14_F02.EPS

Figure 2 Summary sheet.

Step 3 To determine the number of block needed to lay one course, multiply the linear feet by 0.75.

$$Bc = 53.36 \times 0.75$$

$$= 40.02 \text{ or 41 block to lay one course}$$

Step 4 To determine the number of courses, multiply the height by 1.5.

$$Bh = \text{wall height} \times 1.5$$

$$= 8 \text{ feet} \times 1.5$$

$$= 12 \text{ courses high}$$

Step 5 To find the total number of block, multiply the number of block in one course by the number of courses.

$$Bt = Bc \times Bh$$

$$= 41 \times 12 = 492 \text{ block}$$

Step 6 Add an appropriate percentage for breakage and waste. Use 6 percent for this example.

$$= 492.00 \times 1.06 = 521.52 \text{ or 522 total block}$$

Step 7 Recheck the accuracy of all figures.

This method works well for simple structural elements without complicated features.

Alternate Coursing Method for Block

The number of block needed to lay one course can also be determined by dividing the linear feet by the 16-inch nominal length of the block, or 1.33 feet. This method can be substituted in Step 3 of the instructions in this section.

Listed By: ____	RECAPITULATION SHEET					Page ____ of ____

PROJECT _____ DATE _____

ARCHITECT_____ S/SF _____

PAGE REF.	ITEM	MATERIAL	LABOR	SUB	EQUIPMENT	TOTAL

28305-14_F03.EPS

Figure 3 Recapitulation sheet.

1.2.0 Using the Square-Foot Method for Block

Because standard block is so large and the conversion factor is a constant, most block is estimated by the square-foot method. This is simple, because one 8-inch by 16-inch block has a face of 128 square inches. With 144 square inches in a square foot, it takes 1⅛, or 1.125, block masonry units to fill a square foot. *Figure 5* shows the measurements for a standard block.

Consider building a single-wythe wall 24 feet long and 8 feet high with standard 8 × 8 × 16-inch block.

Step 1 Determine the number of square feet in the wall.

$$Sf = l \times w$$

$$= 24 \times 8 = 192 \text{ square feet}$$

Step 2 Calculate the number of block required for the total square feet.

$$B = 192 \times 1.125$$

$$B = 216 \text{ block}$$

Step 3 Add a percentage for waste and breakage. According to experience, 6 percent is reasonable.

$$TB = 216 \times 1.06$$

$$= 228.96 \text{ or } 229 \text{ block (total required)}$$

Step 4 Recheck all calculations for accuracy.

This method gives a slightly smaller figure than the course method. However, it is considered more accurate for items with architectural features, and generally is a reliable method for bidding a job or ordering materials.

1.3.0 Estimating Openings and Lintels

If a wall has openings, calculate the area of the openings in square feet, then deduct that amount from the total square feet of the wall area. The steps for estimating the amounts for block are exactly the same as for brick. Remember, the adjustment for breakage is slightly higher where there are many openings.

If the openings in the wall are windows or doors, check the drawings to find if the openings are large enough to require lintels. The lintels

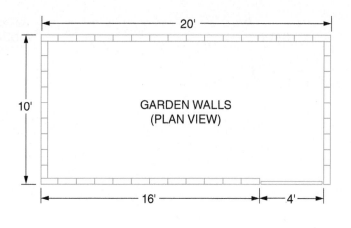

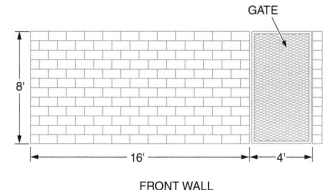

FRONT WALL

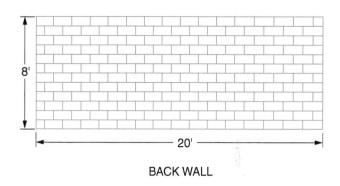

BACK WALL

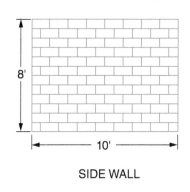

SIDE WALL

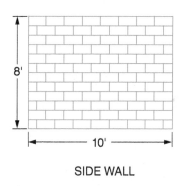

SIDE WALL

28305-14_F04.EPS

Figure 4 Plans for a garden wall.

may be of steel, masonry, or concrete. As an example, refer to the garden wall shown in *Figure 6*. Assume the garden wall will have two windows that are 2 feet, 8 inches wide × 3 feet tall on the front and back 20-foot walls. Also, assume that the 4-foot opening for the gate will have a lintel over it.

Standard lintel block must extend a minimum of one-half block length, or 8 inches, on each side of an opening. The actual length is determined by the specifications. To figure the amount of lintel block needed, determine the width of the openings in inches. Add 1 foot 4 inches (8 inches on each side) for the overlap needed for each opening. Divide this figure by 16 to get the number of block.

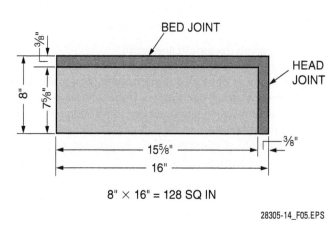

8" × 16" = 128 SQ IN

28305-14_F05.EPS

Figure 5 Measurements for a standard block.

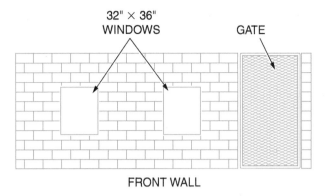

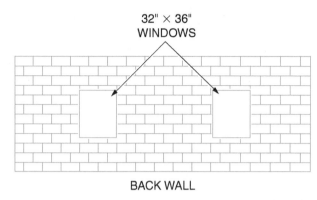

28305-14_F06.EPS

Figure 6 Garden wall with windows.

Step 1 Add the width of the openings and their overlaps.

W = width of openings + overlap

= 32 + 32 + 32 + 32 + 48 + (5 × 16)

= 176 + 80 = 256 inches

= 21 feet 4 inches

Step 2 Divide by 16 inches to get the number of block.

B = w ÷ 16 = 256 ÷ 16

= 16 lintel block

Step 3 Recheck the accuracy of all figures.

Be sure to add the lintels to the worksheets.

1.4.0 Estimating Mortar for Single-Wythe Block Walls

Mortar for block walls can be estimated by rule-of-thumb or by looking up the conversion figures in a standard chart used for estimating purposes.

1.4.1 The Rule-of-Thumb Method

The rule-of-thumb method for estimating masonry cement for block is as follows:

- 6 to 8 bags of masonry cement for 1 ton of sand
- 1 bag of masonry cement for 30 standard block (8 × 8 × 16)

As with brick, the rule-of-thumb estimate is good for most mortar types. The process is to estimate the total number of block, then divide by

Calculators for Estimating

Masons use specialized calculators to estimate materials. In addition to the standard calculator functions, these calculators can perform mathematical operations on dimensional measurements. Dimensional measurements can be easily converted into other English or metric units.

These calculators include specialized functions designed to perform many typical calculations used in estimating:

- Area, given the length and width
- Volume, given the length, width, and height
- The number of block or brick needed for a given area
- Cost estimates given block or brick estimates and per-unit cost
- The volume of concrete needed for footings, walls, or curbs, given the cross-sectional area
- The weight of materials given the volume, or the volume for a given weight
- The square-up, or diagonal, length of a rectangle, given the length and width

28305-14_SA01.EPS

30 to get the number of bags of masonry cement. Use the 24-foot by 8-foot wall with 229 block, discussed earlier for this example.

Step 1 Recall the rule for the number of block that can be laid per bag of masonry cement.

$$\text{Block per bag} = 30$$

Step 2 Divide the total number of block by 30 to get the number of bags of masonry cement, then round to the nearest whole number.

$$\text{Bags} = \text{number of block} \div \text{block masonry units per bag}$$

$$= 229 \div 30$$

$$= 7.63$$

$$= 8$$

Step 3 Find the amount of sand needed per bag of masonry cement. The amount of sand may vary according to the relative humidity in your area of the country. Unless experience suggests using another value, use the base value given here.

$$\text{Sand} = \text{weight of sand per 1,000 brick} \div \text{number of bags of masonry cement per 1,000 brick}$$

$$= 2,000 \text{ pounds (1 ton)} \div 8$$

$$= 250 \text{ pounds of sand per bag of masonry cement}$$

Step 4 Find the amount of sand needed for the total number of bags of masonry cement. Assume 8 bags per ton of sand for this calculation.

$$\text{Sand} = \text{number of bags} \times \text{weight per bag plus waste (15 percent)}$$

$$= 8 \times 250 = 2,000 \text{ pounds}$$

$$= 2,000 \text{ pounds} \times 1.15 \text{ (waste)}$$

$$= 2,300 \text{ pounds or } 1.15 \text{ tons}$$

Step 5 Recheck the accuracy of all figures.

1.4.2 The Table Method

The table method of estimating mortar for block is based on an estimate of the amount of mortar per 100 square feet of surface or per 100 block. *Table 1* gives these quantities for different sizes of concrete block.

The calculations for this method require knowing the number and the size of the block. For the 24-foot by 8-foot wall with 229 block of $8 \times 8 \times 16$-inch size discussed, the calculation is as follows:

Step 1 Refer to *Table 1*. For the type of block used, every 100 square feet of surface area requires 112.5 units. For every 112.5 units, the table specifies 8.5 cubic feet of mortar.

Square-Foot Method

The square-foot method is not limited to structures made of block. Before estimating the number of masonry units needed, the number of units per square foot or 100-square-foot area must be known. Industry associations or masonry manufacturers provide tables that give the number of units per square foot or 100-square-foot area.

Unit type	Nominal height and length of units in inches	Number of units per 100 sq ft
Block	8×16	112.5
Block	8×12	150.0
Block	5×12	221.0
Block	4×16	225.0
Modular concrete brick	$2\frac{1}{4} \times 8$	675.0
Jumbo concrete brick	4×8	450.0
Double concrete brick	5×8	340.0
Roman concrete brick	2×12	600.0
Roman concrete brick	2×16	450.0

28305-14_SA02.EPS

Table 1 Material Quantities for Single-Wythe CMU Walls

Nominal Wall Thickness, Inches		Nominal Size (w × h × l) of Concrete Masonry Units, Inches	Material Quantities for 100 Sq Ft Wall Area*		
			Number of Units	Mortar, Cu Ft	Mortar for 100 Units, Cu Ft†
a	4	4 × 4 × 16	225	13.5	6.0
b	6	6 × 4 × 16	225	13.5	6.0
c	8	8 × 4 × 16	225	13.5	6.0
d	4	4 × 8 × 16	112.5	8.5	7.5
e	6	6 × 8 × 16	112.5	8.5	7.5
f	8	8 × 8 × 16	112.5	8.5	8.0
g	10	10 × 8 × 16	112.5	8.5	9.0
h	12	12 × 8 × 16	112.5	8.5	10.0

* Based on 3/8-inch joints
† With face-shell mortar bedding. Mortar quantities include 10% allowance for waste.

28305-14_T01.EPS

Step 2 To calculate the amount of mortar required, divide the total number of block in the wall by the unit value; then, multiply by the volume of mortar per unit.

Mortar = (229 ÷ 112.5) × 8.5 cubic feet

= 2.04 × 8.5 = 17.34 cubic feet

Step 3 Recheck the accuracy of all figures.

Look at the last column in *Table 1*. For 100 units, the table calls for 8 cubic feet of mortar. This includes a 10 percent allowance for waste. For 229 units with face shell mortar bedding, this would come to 2.29 × 8, or 18.32 cubic feet of mortar. This is a little more than the result of the previous calculation with the unit volume of 8.5 cubic feet per 112.5 units.

1.5.0 Estimating Mortar for Multiwythe Walls

Estimating mortar for multiwythe block and brick walls takes a few more steps than estimating for single-wythe walls. The amount of mortar and other material will depend on the configuration of the wall, as well as the block and brick selected. *Figure 7* shows six different wall configurations. All walls pictured are made using both block and brick. Walls A, B, and C use 4-inch-wide block; walls D, E, and F use 8-inch-wide block. These cross sections should help clarify how the block and brick are constructed to form a solid brick composite wall.

Table 2 shows material quantities per 100 square feet for various types of composite walls. The wall type identifications A through F in the second col-

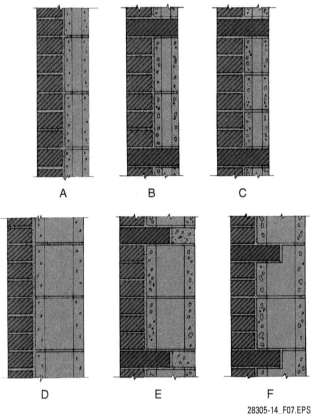
28305-14_F07.EPS

Figure 7 Cross sections of multiwythe wall configurations.

umn refer to the type of bonding shown in *Figure 7*. The third column shows the number of block for each size per 100 square feet. The number of brick and the cubic feet of mortar are given in the last two columns. Mortar quantities in reference tables are typically based on ⅜-inch joints with face-shell bedding, and include an allowance for waste.

NCCER – *Masonry Level Three* 28305-14

Table 2 Materials for Composite Walls per 100 Square Feet

Wall thickness, (inches)	Type of Bonding	Number and Size of Block		# of Brick	Mortar, (cu ft)*
		Stretchers	Headers		
8	A—metal ties	112.5 – 4 × 8 × 16	—	675	20.0
	B—7th course headers	97 – 4 × 8 × 16	—	770	12.2
	C—7th course headers	197 – 4 × 5 × 12	—	770	13.1
12	D—metal ties	112.5 – 8 × 8 × 16	—	675	20.0
	E—7th course headers	97 – 8 × 8 × 16	—	868	13.5
	F—course headers	57 – 8 × 8 × 16	57 – 8 × 8 × 16	788	13.6

* Mortar quantities are based on ⅜-in mortar joints with face-shell bedding for the block; mortar quantities include allowance for waste. All unit sizes are normal.

28305-14_T02.EPS

1.6.0 Estimating Grout

Grouting of block walls is a common practice for structures such as foundations and retaining walls. Grout is typically ordered in cubic yards. A general rule of thumb is that the volume of grout will equal half the volume of the wall itself. To illustrate, the following calculations for a 1-foot-deep, reinforced masonry wall 48 feet long and 12 feet high include the requirement for grout:

Step 1 Determine the volume of the wall.

$$Vw = \text{wall area} \times \text{wall thickness}$$

$$= (48 \text{ feet} \times 12 \text{ feet}) \times$$
$$(8 \text{ inches} \div 12 \text{ inches per foot})$$

$$= (576) \times (0.67)$$

$$= 385.92 \text{ or } 386 \text{ cubic feet}$$

Step 2 Determine the required volume of grout.

$$Vg = \text{wall volume} \times 0.5$$

$$= 386 \times 0.5$$

$$= 193 \text{ cubic feet, or } 7.15 \text{ cubic yards of grout required}$$

Step 3 Recheck the accuracy of all figures.

There are more sophisticated ways of calculating grout, and tables are available for calculating this material, as well. *Table 3* shows the volume of grout in two-wythe grouted concrete brick walls. *Table 4* shows the volume of grout in grouted concrete block walls. Each of these tables uses the unit-value method of providing factors that can be used to determine the amount of grout required to fill the voids in the two types of CMU walls.

For the two-wythe brick wall, the variable is the width of the void between the two wythes. For the concrete block wall, the variables are the wall thickness and the spacing of the cores in the block. Both tables include a 3 percent allowance for waste.

Table 3 Grout in Two-Wythe Concrete Brick Walls

Width of grout space (inches)	Grout, cu yd, for 100 sq ft wall area	Wall area for 1 cu yd of grout
2.0	0.64	154
2.5	0.79	126
3.0	0.96	105
3.5	1.11	89
4.0	1.27	79
4.5	1.43	70
5.0	1.59	63
5.5	1.75	57
6.0	1.91	53
6.5	2.06	49
7.0	2.22	45

28305-14_T03.EPS

Table 4 Grout in Concrete Block Walls

Wall thickness, (inches)	Spacing of grouted cores, (inches)	Grout, cu yd, for 100 sq ft wall area	Wall area (sq ft), for 1 cu yd of grout
6	All cores grouted	0.79	126
	16	0.40	250
	24	0.28	357
	32	0.22	450
	40	0.19	526
	48	0.17	588
8	All cores grouted	1.26	79
	16	0.74	135
	24	0.58	173
	32	0.49	204
	40	0.44	228
	48	0.39	257
12	All cores grouted	1.99	50
	16	1.18	85
	24	0.91	110
	32	0.76	132
	40	0.70	143
	48	0.64	156

28305-14_T04.EPS

Additional Resources

Basics for Builders: Plan Reading & Material Takeoff. 1994. Wayne J. DelPico. Kingston, MA: R.S. Means Company, Inc.

Concrete Masonry Handbook, Fifth Edition. W. C. Panerese, S. K. Kosmatka, and F. A. Randall, Jr. Skokie, IL: Portland Cement Association.

1.0.0 Section Review

1. When using the coursing method to determine the number of block needed to lay one course, multiply the linear feet of the wall by _____.

   a. 1.33
   b. 1.125
   c. 0.83
   d. 0.75

2. The surface area of the face of a standard block is _____.

   a. 218 square inches
   b. 144 square inches
   c. 128 square inches
   d. 64 square inches

3. The minimum that a standard lintel block must extend on each side of an opening is _____.

   a. 2 inches
   b. 4 inches
   c. 6 inches
   d. 8 inches

4. The rule of thumb for the number of standard block per bag of masonry cement is _____.

   a. 15
   b. 30
   c. 45
   d. 60

5. When estimating mortar for multiwythe walls, keep in mind that mortar quantities in reference tables are typically based on _____.

   a. 8-inch brick
   b. the weight of the mortar bags
   c. the number of masonry units
   d. ⅜-inch joints

6. When estimating grout, the variable for two-wythe brick walls is the _____.

   a. width of the void between the two wythes
   b. thickness of the wider wythe minus the thickness of the narrower wythe
   c. height of the lift of the two wythes divided by 2
   d. sum of the spacing of the cores in the two wythes

2.0.0 ESTIMATING BRICK AND MORTAR

Objective

Explain how to estimate brick and mortar.
  a. Explain the coursing method for brick.
  b. Explain the square-foot method for brick.
  c. Describe how to allow for openings in an estimate.
  d. Explain how to estimate mortar for brick.

Performance Task

Estimate the amounts of brick and mortar required for a hypothetical veneer wall, using plans provided by the instructor.

As with block, masons can calculate the amount of brick by using either the coursing method or the square-foot method. The conversion factors for brick are different, because brick is smaller than block; however, the estimating steps are basically the same. The examples in the following sections are similar to those used in the section *Estimating Block*. These examples assume the use of nominal 8-inch brick.

2.1.0 Using the Coursing Method for Brick

To estimate brick using the coursing method, the following information is required:

- The size of the brick being used
- The total linear feet of the wall
- The number of masonry units needed for a single course of wall length
- The number of courses needed for the height of the wall

For example, you need to calculate the number of engineer modular brick to build the single-wythe garden wall shown in *Figure 4*. Remember, the wall is 8 feet high with one 4-foot-wide gate enclosing a 20-foot by 10-foot garden.

The first step for this method requires finding the size of the brick. *Table 5* gives the nominal and specific dimensions of various types of modular brick. The second row identifies the information needed for an engineer modular brick. With this information, begin calculating the size of the wall area and the number of brick required.

Step 1 Determine the total linear feet around the wall by adding the length of all four walls together.

$$L = 20 + 10 + 10 + 16 \text{ feet}$$

$$= 56 \text{ feet}$$

Step 2 This number includes the brick units in each corner. To avoid counting them twice, subtract the width of the wall for each corner. In this case, the width of the wall is the width of one engineer modular brick, or 4 inches.

$$L = \text{linear feet} - (\text{width of corners})$$

$$= 56 \text{ feet} - (4 + 4 + 4 + 4 \text{ inches})$$

$$= 56 \text{ feet} - 16 \text{ inches}$$

$$= 56 \text{ feet} - 1.3 \text{ feet}$$

$$= 54.7 \text{ feet}$$

Step 3 To determine the number of engineer modular brick needed to lay one course, multiply the linear feet by 1.5. For other brick sizes, consult the appropriate brick sizing table.

$$Bc = 54.7 \times 1.5 = 82.05 =$$
$$83 \text{ brick to lay one course}$$

Step 4 To determine the number of courses, divide the height by the nominal height of the brick, or use the last column in *Table 5*.

$$Bh = \text{wall height} \div \text{brick height}$$

$$= 8 \text{ feet} \div 0.27 \text{ decimal feet}$$
(or 3.2 inches converted to decimal feet)

$$= 29.6, \text{ or } 30 \text{ courses high}$$

Or, using the engineer-modular coursing value from *Table 5*:

$$Bh = (\text{wall height} \div \text{Table 5 value}) \times$$
$$\text{course count}$$

$$= (8 \text{ feet} \div 1 \text{ foot 4 inches}) \text{ per 5 courses}$$

$$= (96 \text{ inches} \div 16) \times 5 = 6 \times 5 =$$
$$30 \text{ courses high}$$

Table 5 Sizes of Modular Brick

Unit Designation	Nominal Designation Inches			Joint Thickness Inches	Specified Designation Inches			# of Courses In 16"
	w	h	l		w	h	l	
Modular	4	$2^2/3$	8	$3/8$	$3^5/8$	$2^1/4$	$7^5/8$	6
				$1/2$	$3^1/2$	$2^1/4$	$7^1/2$	
Engineer Modular	4	$3^1/5$	8	$3/8$	$3^5/8$	$2^3/4$	$7^5/8$	5
				$1/2$	$3^1/2$	$2^{13}/16$	$7^1/2$	
Closure Modular	4	4	8	$3/8$	$3^5/8$	$3^5/8$	$7^5/8$	4
				$1/2$	$3^1/2$	$3^1/2$	$7^1/2$	
Roman	4	2	12	$3/8$	$3^5/8$	$1^5/8$	$11^5/8$	8
				$1/2$	$3^1/2$	$1^1/2$	$11^1/2$	
Norman	4	$2^2/3$	12	$3/8$	$3^5/8$	$2^1/4$	$11^5/8$	6
				$1/2$	$3^1/2$	$2^1/4$	$11^1/2$	
Engineer Norman	4	$3^1/5$	12	$3/8$	$3^5/8$	$2^3/4$	$11^5/8$	5
				$1/2$	$3^1/2$	$2^{13}/16$	$11^1/2$	
Utility	4	4	12	$3/8$	$3^5/8$	$3^5/8$	$11^5/8$	4
				$1/2$	$3^1/2$	$3^1/2$	$11^1/2$	

28305-14_T05 .EPS

Step 5 To find the total number of brick, multiply the number of brick in one course by the number of courses.

$$Bt = Bc \times Bh$$

$$= 83 \times 30 = 2,490 \text{ brick}$$

Step 6 Add a percentage for waste and breakage. For this example, use 5 percent.

$$2,490.00 \times 1.05 = 2614.50 = 2,615.00 \text{ total brick}$$

Step 7 Recheck the accuracy of all figures.

This method works best for straightforward structural items without intricate bond patterns or openings.

Alternate Coursing Method for Brick

The number of brick needed to lay one course can also be determined by first converting the linear feet to linear inches, then dividing the linear inches by the nominal length of a brick, or 8 inches in the case of an engineer modular brick. This method can be substituted in Steps 1–3 of the instructions in this section.

2.2.0 Using the Square-Foot Method for Brick

Today, most brick masonry units are based on the 4-inch modular system. This simplifies estimating with the square-foot method. To use this method, the following information is needed:

- The area of each wall in square feet
- The number of brick per square foot
- The number of brick eliminated by openings
- The percentage for breakage and cutting

The breakage figure can range from 3 to 15 percent for block and brick, and 10 to 25 percent for mortar.

Other factors in estimating brick are the bond pattern, the wall type, and the number of specialized brick types required, such as headers and face brick. If a multiwythe wall is specified, the mortar estimates must be increased to allow for full collar joints. A collar joint is the mortar joint between the wythes of masonry in a multiwythe wall system.

Most charts or tables make allowances for such variations, but be careful how you apply the values. These correction factors are applied after the initial figures are estimated. Standard charts and tables can be used to look up additional calculations for patterned bonds and determine the quantity of mortar materials.

To understand the square-foot estimating procedure, follow the steps in the example given below. This estimate is for building one wall of a structure, a 4-inch-thick wall with ⅜-inch joints, using standard modular brick. The wall is to be built 24 feet long and 8 feet high in a running bond pattern with full headers every sixth course.

The following calculations use *Table 6*, which gives the number of modular brick and the cubic feet of mortar for a single-wythe wall in running or stack bond. The figures in this table do not allow for any waste.

Step 1 Determine the total wall area of the construction.

24 feet (specified length) × 8 feet (specified height) =

192 square feet (wall area)

Step 2 Look at *Table 6* to determine the number of brick needed for each 100 square feet of wall. Since the specifications are for 4-inch × 2⅔-inch × 8-inch brick, use the value on the first line in the second column.

675 brick masonry units per 100 square feet, or

6.75 brick masonry units per square foot

Step 3 Determine the number of brick required.

192 square feet × 6.75 brick masonry units per square foot

= 1,296 total number of brick masonry units

Step 4 Add the correction factor for bond patterns; the example has full headers every sixth course. Checking *Table 7*, it is easy to see that using headers every sixth course adds ⅙ to the amount of brick needed. These correction factors work only with brick that are twice as long as they are wide.

1,296.00 brick × 0.1667 (⅙ adjustment)

= 216.04 = 217 additional brick

1,296 + 217 = 1,513 total number of brick

Step 5 Add an appropriate percentage for breakage and waste. Use 6 percent for this example.

1,513.00 × 1.06 = 1,603.78 or 1,604 total brick

Step 6 Recheck the accuracy of all figures.

2.3.0 Allowing for Openings

When estimating for openings, a calculated number of brick should be deducted from the total number of brick estimated for the structure without openings. Do this by subtracting the total square feet of openings from the total wall area. To illustrate this procedure, consider the wall in the previous example with the addition of one rectangular opening 3 feet wide and 6 feet high.

Table 6 Brick and Mortar Equivalent Table

Modular Brick								
Nominal Size of Brick (In)			Number of Bricks per 100 Sq Ft	Cubic Feet of Mortar				
				Per 100 Sq Ft		Per 1,000 Brick		
T	H	L		3/8" Joints	1/2" Joints	3/8" Joints	1/2" Joints
4 ×	2⅔ ×	8	675	5.5	7.0	8.1	10.3
4 ×	3⅕ ×	8	563	4.8	6.1	8.6	10.9
4 ×	4 ×	8	450	4.2	5.3	9.2	11.7
4 ×	5⅓ ×	8	338	3.5	4.4	10.2	12.9
4 ×	2 ×	12	600	6.5	8.2	10.8	13.7
4 ×	2⅔ ×	12	450	5.1	6.5	11.3	14.4
4 ×	3⅕ ×	12	375	4.4	5.6	11.7	14.9
4 ×	4 ×	12	300	3.7	4.8	12.3	15.7
4 ×	5⅓ ×	12	225	3.0	3.9	13.4	17.1
6 ×	2⅔ ×	12	450	7.9	10.2	17.5	22.6
6 ×	3⅕ ×	12	375	6.8	8.8	18.1	23.4
6 ×	4 ×	12	300	5.6	7.4	19.1	24.7

NOTE: Equivalents are for single-wythe wall in running or stack bond.

28305-14_T06.EPS

Table 7 Corrections for Bond Patterns

Bond	Correction Factor
Full headers every 5th course only	1/5
Full headers every 6th course only	1/6
Full headers every 7th course only	1/7
English bond (full headers every 2nd course)	1/2
Flemish bond (alternate full headers and stretchers every course)	1/3
Flemish headers every 6th course	1/18
Flemish cross bond (Flemish headers every 2nd course)	1/6
Double-stretcher garden-wall bond	1/5
Triple-stretcher garden-wall bond	1/7

28305-14_T07.EPS

Step 1   Determine the total wall area without any openings.

24 feet × 8 feet = 192 square feet

Step 2   Determine the square feet of wall area in each opening.

3 feet × 6 feet = 18 square feet

Step 3   Deduct the square feet of openings from the total square feet.

192 square feet − 18 square feet
= 174 square feet

Step 4   Recheck the accuracy of all figures.

Continue the estimating process as before, using Steps 2 through 6. If there are numerous small openings in the masonry structure, increase the percentage for waste to account for the units that must be cut around the openings.

2.4.0 Estimating Mortar for Brick

Quantities of mortar can be estimated by two different methods. The first method relies on past experience and trial and error from previous work. This is typically called the rule-of-thumb method. The second method uses the more direct approach of applying standard values to specific dimensions of the structure.

2.4.1 The Rule-of-Thumb Method

The rule-of-thumb method uses equivalent values for the materials used to make masonry mortar. This method uses the following base values:

- Eight bags of masonry cement per 1,000 brick
- One ton of sand per 1,000 brick

This method works for Type N or O mortar made with masonry cement. Use the earlier example of the solid 24-foot by 8-foot wall that needs 1,604 brick, and assume it will use type N mortar. The calculation is simplified using the following steps:

Step 1   Determine the number of brick that can be laid per bag of masonry cement.

1,000 brick ÷ 8 bags = 125 brick per bag

Step 2   Divide the total number of brick by 125 to find the number of bags of cement, rounding to the nearest whole number of bags.

Bags = number of brick ÷
brick masonry units per bag

= 1,604 ÷ 125 = 12.83 or
13 bags of masonry cement

It's Not Rocket Science

In September 1999, National Aeronautics and Space Administration (NASA) engineers directed the Mars Climate Orbiter to execute a burn that would place it into orbit around Mars. The spacecraft passed behind Mars and was never heard from again.

The problem stemmed from some spacecraft commands being sent in English units instead of being converted to metric. This caused the spacecraft to miss its intended altitude and enter the Martian atmosphere too low. It was destroyed by atmospheric stress and friction at the lower altitude.

Always double-check your math. Be sure to check that you are using the correct units.

Step 3 From the calculation for block, you know the amount of sand needed per bag of masonry cement is 250 pounds. The next step is to find the amount of sand needed for the total number of bags of masonry cement. The amount of sand may vary according to the relative humidity in the area of the country. Unless experience suggests using another value, use the base value given here.

Sand = weight of sand per 1,000 brick ÷ number of bags of masonry cement per 1,000 brick

= 2,000 pounds (1 ton) ÷ 8

= 250 pounds of sand per bag of masonry cement

Step 4 Find the amount of sand for the estimated number of bags.

Sand = number of bags of masonry cement × weight of sand per bag, plus waste (15 percent)

= 13 × 250 = 3,250 pounds

= 3,250 × 1.15 (waste)

= 3,737.5 pounds

= 3,735.5 lbs ÷ 2,000 pounds per ton

= 1.87, or 2 tons

Step 5 Recheck the accuracy of all figures.

> **NOTE**
> When ordering material by the ton, always round up to the next ton, no matter how small a fraction of a ton over the previous ton your measurement ends up being.

Always add a waste value to the amount of sand. Because sand is delivered in bulk and spread on the ground, there will always be some amount of waste. If more than 3 tons of sand are needed, allow ½ ton for waste. If about 1½ tons of sand are needed, allow at least ¼ ton for waste.

A rule of thumb can be established for calculating mixtures of portland cement–lime–sand mortar, but it can be more complicated because the percentage of ingredients varies considerably with the mortar type. This can be seen by comparing the mix proportions by volume requirements in *Table 8* for the various mortar types. *Table 9* is a useful reference for estimating the typical cubic-foot quantities of common sizes of bags of preblended mortar, portland cement, and bulk mortar mix.

2.4.2 The Table Method

The table method of estimating relies on standard volumes of materials required to meet mix proportions for one unit of mortar. In this case, the volume is figured in cubic feet. This is the information on the left half of *Table 8*. To illustrate this calculation, use the same structure from the previous example: the 24-foot by 8-foot wall using modular brick sized 4-inches × 2⅔-inches × 8-inches with ⅜-inch joints. From the previous calculation, we know there are a total of 1,604 brick required to build the surface area outlined by the wall dimensions.

As indicated in *Table 6*, the wall will need 5.5 cubic feet of mortar per 100 square feet of wall, or 8.1 cubic feet of mortar per 1,000 brick.

- For 192 square feet, this will be 1.92 × 5.5, or 10.6 cubic feet of mortar. This figure does not include any waste calculations.
- For 1,603 brick, this will be 1.603 × 8.1 or 12.98 cubic feet of mortar. This figure is larger because it includes the waste calculation added to the brick.

Use the proportions in *Table 8* to calculate the amount of masonry cement, or portland cement and lime, plus sand. Locate the column for quantities by volume for type N mortar. Using 12.98 cubic feet of mortar as the total requirement, the calculation for the individual materials is as follows:

- *Portland cement (Type N)* – 12.98 × 0.167 = 2.16766 rounded to 2.17 cubic feet
- *Lime* – 12.98 × 0.167 = 2.16766 rounded to 2.17 cubic feet
- *Sand* – 12.98 × 1.00 = 12.98 cubic feet

> **NOTE**
> The volume of mortar is always roughly equal to the volume of sand in the mix.

Table 8 Material Quantities per Cubic Foot of Mortar

Material	Quantities by Volume				Quantities by Weight			
	Mortar Type and Proportions by Volume				Mortar Type and Proportions by Volume			
	M 1:1/4:3	S 1:1/2:4-1/2	N 1:1:6	O 1:2:9	M 1:1/4:3	S 1:1/2:4-1/2	N 1:1:6	O 1:2:9
Cement	0.333	0.222	0.167	0.111	31.33	20.89	15.67	10.44
Lime	0.083	0.111	0.167	0.222	3.33	4.44	6.67	8.89
Sand	1.000	1.000	1.000	1.000	80.00	80.00	80.00	80.00

28305-14_T08.EPS

Table 9 Cubic Foot Equivalents of Common Sizes of Mortar Bags

Type of Bag	Cubic Foot Equivalent
Preblended 80-pound bag	1.7 cu ft
Masonry cement bag	4.2 cu ft
Bulk bag	50 cu ft

28305-14_T09.EPS

Table 10 Mortar Correction for Collar Joints

Joint Size (In)	Cu Ft of Mortar per 100 Sq Ft of Wall
1/4	2.08
3/8	3.13
1/2	4.17

28305-14_T10.EPS

2.4.3 Estimating for Collar Joints

A collar joint is the mortar joint between the wythes of masonry in a multiwythe wall system. Increase the amount of mortar if the collar joints are to be filled. *Table 10* provides values for the additional amount of mortar for collar joints.

The amount depends on the joint size. The table shows the cubic foot amount per 100 square feet of wall. The amount is calculated and then added to the total amount of mortar from the previous calculation. This chart does not include any waste calculation.

Exact Estimates

The construction of the Great Wall of China began between the seventh and eighth centuries BC and was completed during the Ming Dynasty in the 1300s. With an average height of 10 meters and a width of 5 meters, the wall runs up and down along the mountain ridges and valleys for 6,700 kilometers (4,163 miles).

One of the engineers building a fort on the Great Wall of China ordered exactly as many brick masonry units as he needed. Other engineers said that he should add some overage for waste or the possibility that he had not estimated correctly. The designer added one brick to the order. When the fort was finished, there was one brick left over. It was placed above the main gate as if to say, "I told you so."; at least that is the legend of the one loose brick above the gate.

Additional Resources

Bricklaying: Block and Brick Masonry. 1988. Brick Industry Association. Orlando, FL: Harcourt Brace & Company.

Pocket Guide to Brick Construction. 1990. Reston, VA: Brick Industry Association.

2.0.0 Section Review

1. When using the coursing method with engineer modular brick, the number of brick needed to lay one course can be found by _____.

   a. dividing the linear feet by 1.33
   b. multiplying the linear feet by 1.3.3
   c. dividing the linear feet by 1.5
   d. multiplying the linear feet by 1.5

2. When using the square-foot method for brick, the step before adding an appropriate percentage for breakage and waste is _____.

   a. adding the correction factor for bond patterns
   b. finding the amount of sand per bag of masonry cement
   c. determining the number of brick required
   d. calculating the number of brick needed for each 100 square feet of wall

3. Increase the percentage for waste when the masonry structure has _____.

   a. few large openings
   b. numerous small openings
   c. horizontal and/or vertical joint reinforcement
   d. control or expansion joints

4. A mortar joint between the wythes of masonry in a multiwythe wall system is called a(n) _____.

   a. expansion joint
   b. collar joint
   c. airspace
   d. lift

3.0.0 ESTIMATING ACCESSORY ITEMS

Objective

Describe how to estimate accessory items.
  a. Explain how to estimate joint reinforcement.
  b. Explain how to estimate structural reinforcement.
  c. Explain how to estimate masonry ties.
  d. Explain how to estimate other masonry units.
  e. Explain how to estimate other masonry accessories.

Performance Task

Estimate the amounts of rebar and ties required for hypothetical walls, using plans provided by the instructor.

Estimating wall ties and reinforcement involves different types of calculations, but the basic approach is the same. Block and brick are easy to estimate. They are available in standard sizes, so the number of units in a square foot is easy to determine. Most other masonry items are also estimated based on the square footage of the wall, but different conversion factors are used.

3.1.0 Estimating Continuous Joint Reinforcement

Continuous joint reinforcement is usually placed every second or third course for block, with placement every second course or 16 inches on center being the most common. Occasionally, joint reinforcement is placed 8 inches on center. Joint re-

Check Your Math

Electronic aids such as calculators, smartphone apps, and online estimators increase the accuracy and speed of the estimating process. However, it is still important to double-check all calculations. Calculators and computer programs can still produce incorrect estimates if the information entered is incorrect. As computer programmers say, "garbage in, garbage out."

inforcement for brick is commonly placed every sixth course, or as the plans specify.

Joint reinforcement is estimated using a linear-foot or square-foot method. The linear-foot method is easier to use if the job has odd or unusual spacing requirements. For standard applications of joint reinforcement, the square-foot method is easier.

3.1.1 The Linear-Foot Method

The linear-foot method requires determining the length of the wall and the number of times joint reinforcement is required in the wall. The joint-reinforcement estimate is simply the product of these two items. To illustrate, consider a 1-foot-thick perimeter block wall 40 feet × 20 feet × 8 feet high, as shown in *Figure 8*.

As shown in the section drawing, the specification requires joint reinforcement to be placed in the bottom two courses, then every 16 inches on center, and then in the top two courses.

Step 1 Determine the length of the wall, minus the depth of the corners.

$$Lw = \text{length of perimeter} - \text{depth of 4 corners}$$

$$Lw = [(40 \text{ feet} + 20 \text{ feet}) \times 2] - (4 \times 1 \text{ foot})$$

$$= 120 - 4 = 116 \text{ linear feet}$$

Step 2 Determine the number of times joint reinforcement is required. From the wall section, count the locations of required joint reinforcement (there are seven in *Figure 8*). If there is no wall section, make a sketch and mark all reinforcement locations on it.

Step 3 Determine the total length of reinforcement.

$$Lr = \text{wall length} \times \text{number of lengths required}$$

$$= 116 \times 7 = 812 \text{ linear feet}$$

Step 4 Add a percentage for lapping and waste. Joint reinforcement is normally lapped 6 inches. This adds 5 percent to the length needed (6-inch lap divided by the 10-foot length). Add another 5 percent for general waste.

$$TL = (\text{length from Step 3}) \times (1 + \text{waste percentage} + \text{lap percentage})$$

$$= 812 \times 1.10 = 893.2 = 894 \text{ linear feet}$$

Step 5 Check all calculations for accuracy.

The process is the same for brick walls. The critical items are the length of the wall and the number of times the joint reinforcement is required. Follow the same steps listed above.

3.1.2 *The Square-Foot Method*

The square-foot method requires determining the area of the wall in square feet and using a conversion factor. This conversion factor is normally based on wall area, but can also be based on the number of block. *Table 11* is a conversion table for block joint reinforcement.

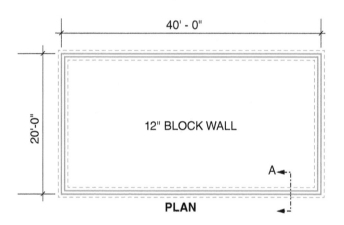

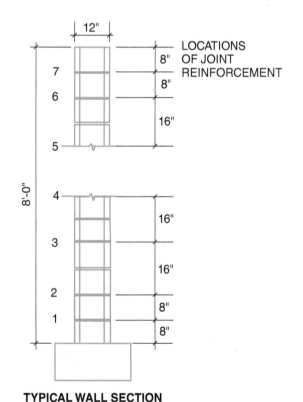

TYPICAL WALL SECTION

28305-14_F08.EPS

Figure 8 Block-wall plan and section.

To illustrate this process, consider the previous example, in which joint reinforcement is required every 16 inches on center.

Step 1 Determine the wall area.

$$A = \text{wall length} \times \text{height}$$

$$= 116 \times 8 = 928 \text{ square feet}$$

Step 2 Obtain the conversion factor. A spacing of 16 inches on center requires that 1 linear foot of reinforcement be placed for every 1.33 square feet of wall area.

$$12 \text{ inches length} \times 16 \text{ inches height}$$
$$= 192 \text{ square inches}$$

$$192 \div 144 \text{ inches per square foot}$$
$$= 1.33 \text{ square feet}$$

This results in a conversion factor of 0.75 linear feet per square foot of wall area (1 ÷ 1.33 = 0.75). This can be found in a reference table similar to *Table 11*, if one is available.

Step 3 Determine the linear feet of reinforcement.

$$L = \text{wall area} \times \text{conversion factor}$$

$$= 928 \times 0.75 = 696 \text{ linear feet}$$

Step 4 Add a percentage for lapping and waste.

$$LT = \text{total linear feet} \times$$
$$(1 + 0.05 \text{ waste percentage}$$
$$+ 0.05 \text{ lap percentage})$$

$$= 696 \times 1.10 = 765.60 = 766 \text{ linear feet}$$

Step 5 Check all calculations for accuracy.

The most dangerous type of error in estimating continuous joint reinforcement is not a calculation error, but a material specification error. If the wrong material is ordered, it cannot be used at all.

Table 11 Block Joint Reinforcement Conversion Table

Spacing of Joint Reinforcement, Inches	Number of Courses between Reinforcement	Reinforcement (Linear Feet)	
		Wall Area (Sq Ft)	Block
8	1	1.50	1.33
16	2	0.75	0.67
24	3	0.50	0.44

28305-14_T11.EPS

Joint Reinforcement

Prefabricated reinforcement for embedment in the horizontal mortar joints of masonry can be truss or tab design. They are usually manufactured in 10-foot lengths. The wire conforms to ASTM A82, *Standard Specification for Steel Wire, Plain, for Concrete Reinforcement*, for cold drawn steel wire.

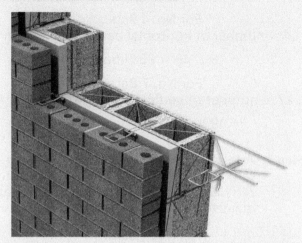

28305-14_SA03.EPS

Be sure to check the accuracy on these items:

- The width required: 3, 4, 6, 8, 10, 12, 13, 14, or 16 inches
- The style required: truss type or ladder type
- The weight required: standard or extra heavy
- The number of longitudinal wires: 2, 3, or 4
- The finish required: uncoated, zinc, or stainless steel

Joint reinforcement is shipped in bundles of 500 feet. Each bundle contains two 250-linear-foot packages, each with twenty-five 10-foot sections. Each package weighs about 50 pounds, for easy handling. Special-order joint reinforcement can be supplied in different packaging.

3.2.0 Estimating Structural Reinforcement

Structural reinforcement for masonry walls is normally estimated by the ton. The reinforcement is first estimated in linear feet and then converted to weight, using a chart similar to the one shown in *Table 12*.

3.2.1 Calculating Rebar Quantities

Consider a 1-foot-deep, reinforced masonry wall 48 feet long and 12 feet high. The specifications call for a fully grouted wall with two No. 4 bars placed horizontally, 16 inches on center, and No. 7 vertical bars spaced 3 feet on center.

Step 1 Determine the number of bars required. Remember to add 1 bar to the total of vertical bars to account for the starting location.

Horizontal:

$$N_h = \text{wall height} \div \text{vertical spacing}$$

$$= (12 \text{ feet} \times 12 \text{ inches per foot}) \div 16\text{-inch spacing}$$

= 9 locations of reinforcing and two bars are required at each location

$$= 9 \times 2 = 18 \text{ horizontal bars}$$

Vertical: $N_v = \text{wall length} \div \text{horizontal spacing}$

$$= (48 \text{ feet} \times 12 \text{ inches per foot}) \div 32\text{-inch spacing}$$

$$= 576 \div 32 + 1 \text{ (for the starting location)}$$
$$= 19 \text{ locations for vertical bars}$$

Table 12 Steel Rebar Weights by Size

Bar Size	Approximate Bar Diameter, Inches	Weight, Lb per Linear Foot	Cross-Sectional Area (Sq In/Ft)
#3	⅜	0.376	0.11
#4	½	0.668	0.20
#5	⅝	1.043	0.31
#6	¾	1.502	0.44
#7	⅞	2.044	0.60
#8	1	2.670	0.79
#9	1⅛	3.400	1.00
#10	1¼	4.303	1.27
#11	1⅜	5.313	1.56
#14	1¾	7.650	2.25
#18	2¼	13.600	4.00

28305-14_T12.EPS

Step 2 Determine the length of all reinforcing.

For No. 4 Bars:
L4 = number of horizontal bars × wall length

= 18 × 48 = 864 linear feet

For No. 7 Bars:
L7 = number of vertical bars × wall height

= 19 × 12 = 228 linear feet

Step 3 Add an amount for lapping. Rebar typically is lapped 48 bar diameters. Refer to *Table 12* to determine the appropriate bar diameters.

For No. 4 bars:
½ inch × 48 = 24 inches = 2 feet

L4 = 864 + 2 = 866 linear feet

For No. 7 bars:
⅞ inch × 48 = 42 inches = 3.5 feet

L7 = 228 + 3.5 = 231.5 = 232 linear feet

Step 4 Check all figures for accuracy.

Other considerations with structural reinforcements can include:

- The strength requirement of the steel
- Coating requirements of the steel surface
- Straight- or bent-bar requirements
- Lapping requirements (5 percent allowance for end laps)
- Special support requirements for the steel

3.2.2 Calculating Accessory Items

Specifications may call for wire baskets or braces to support rebar in place. The number of baskets, also called rebar positioners, can be calculated using a linear-foot method. This requires finding the number of linear feet of rebar to be run through the baskets, as well as the specified distance between the baskets. These steps calculate the number of baskets needed to support the horizontal rebar in the wall in the previous example.

Step 1 Check the specification, which indicates that the horizontal bars are run doubled; the baskets are specified at 1½ feet apart.

Step 2 Determine the running feet of the bars. According to Step 3 in the previous example:

L4 (horizontal bars) = 866 linear feet

Smartphone Apps for Construction

Although the craft of masonry has been around for centuries, not everything in masonry is old. Smartphones are becoming an increasingly popular form of communication, and also offer a great deal of versatility for craftworkers. Smartphone cameras can be used to document on-the-job activities or potential safety violations. Best practices can be communicated to crewmembers using video clips. Construction calculators can be downloaded, providing craftworkers with the same (or even greater) capabilities than a handheld calculator.

The smartphone version of Calculated Industries, Inc.'s popular Construction Master® Pro, for example, lets craftworkers calculate and convert dimensions, plot right-angle conversions, find area and volume, and determine measures such as board feet, cost per unit, and even the angles of an equal-sided polygon. A separate app (see the figure), allows masons to calculate stair dimensions, the length and weight of rebar, and the number of brick loads and mortar bags needed for a job.

28305-14_SA04.EPS

Since the bars are laid double for this wall, divide the linear feet by 2 to get the running feet; then divide again by 1.5 to get the number of baskets.

$$B = \text{running feet} \div \text{basket distance}$$

$$= (866 \div 2) \div 1.5 = 289 \text{ baskets}$$

Note that the figure used does not include waste for the rebar or the baskets. Do not include the waste figure for the rebar because it should not affect the basket count. However, there may be defects in the baskets, so add 5 percent for waste to the number of baskets.

$$289 \times 1.05 = 303.45$$
$$= 304 \text{ baskets (including waste)}$$

This procedure is also useful for calculating accessory items other than baskets.

3.3.0 Estimating Masonry Ties

The required number of masonry ties is estimated using the square-foot method. Typically, the number of ties required for a square foot of wall area according to the tie spacing will need to be determined. *Table 13* gives the number of ties according to spacing standards.

The next item to determine is the square feet of wall area. To illustrate, consider a cavity wall measuring 64 feet wide by 8 feet tall, with wall ties spaced 2 feet on center horizontally and 8 inches on center vertically. Determine the number of ties required.

Step 1 Determine the wall area.
$$A = \text{wall length} \times \text{height}$$
$$= 64 \times 8 = 512 \text{ square feet}$$

Step 2 Determine area of wall covered by one tie.
$$T = \text{horizontal spacing} \times \text{vertical spacing}$$
$$= 24 \times 8 = 192 \text{ square inches}$$

Reinforcing Bars

Reinforcing bar, or rebar, is usually delivered in bundles. The standard lengths for rebar are 20 feet, 40 feet, and 60 feet. However, rebar can be cut and sold in other lengths, as specified by customers.

Table 13 Masonry Ties per Square Foot

Vertical Tie Spacing (In)	Number of Ties Required Horizontal Tie Spacing (In)		
	8	16	32
8	2.25	1.13	0.56
12	1.50	0.75	0.38
16	1.13	0.56	0.28
18	1.00	0.50	0.25
24	0.75	0.38	0.19
32	0.56	0.28	0.14
36	0.50	0.25	0.13

28305-14_T13.EPS

Step 3 Determine the number of ties per square foot, according to the spacing specifications.

$$Nt = \text{number of square inches in a square foot} \div \text{area covered by one tie}$$

$$= 144 \div 192 = 0.75 \text{ ties per square foot}$$

Step 4 Determine the number of ties required.
$$N = \text{wall area} \times \text{ties per square foot}$$
$$= 512 \times 0.75 = 384 \text{ ties}$$

Step 5 Check all calculations for accuracy.

Ties are available in a variety of styles, sizes, and weights. Be sure to carefully check the drawings and specifications before ordering, to determine:

- *The tie style* – Rectangular, Z, strap, flange, or triangular
- *The tie size* – To fit the wall size
- *The weight required* – Standard, heavy, extra heavy, etc.
- *The finish required* – Uncoated, galvanized, or stainless steel

Masonry ties are available in boxes of 100, 250, 500, or 1,000, depending on the particular type of tie.

3.4.0 Estimating Other Masonry Units

Most other masonry units are estimated by the square foot. The area of the wall surface is determined and converted to masonry units using the appropriate conversion factor. *Table 14* provides conversion factors for other types of masonry units.

Remember that these conversion factors can be easily derived by a two-step process:

- Calculating the nominal area of the masonry face
- Dividing the nominal area into 144 square inches (1 square foot)

By using the nominal dimensions, all conversion factors allow for mortar joints.

To illustrate the use of *Table 14*, figure the number of 8-inch glass-block units required for a panel measuring 40 feet by 8 feet.

Step 1   Determine the total square feet of wall area.

$$A = length \times height$$

$$= 40 \times 8 = 320 \text{ square feet}$$

Step 2   Multiply this figure by the conversion factor for 8-inch glass block.

$$A_g = \text{square feet wall area} \times \text{conversion factor}$$

$$= 320 \times 2.25 = 720 \text{ units}$$

Step 3   Check all calculations for accuracy.

3.5.0 Estimating Additional Items

Many miscellaneous items not previously mentioned must also be estimated when planning any masonry work. Miscellaneous items are often more difficult to estimate than the actual masonry units. The miscellaneous items may not be

Table 14 Conversion Factors for Odd Units

Type of Masonry Unit	Nominal Face Dimension, Inches	Number of Units per Square Foot
Structural Glazed 6T	5⅓ × 12	2.25
Structural Glazed 8W	8 × 16	1.125
Spectra-Glaze®	8 × 16	1.125
Glass Block–6 inch	6 × 6	4.00
Glass Block–8 inch	8 × 8	2.25
Glass Block–12 inch	12 × 12	1.00

28305-14_T14.EPS

shown on the drawings. They may be noted with only a small note or may be referred to only in the specifications. In the worst case, the specifications may simply state that all masonry work is to be constructed in compliance with all local building codes. This requires the mason to know all aspects of the local codes and to construct the masonry accordingly, including specifying, estimating, and supplying all necessary anchors, joints, and reinforcement.

Additional items that may need to be ordered are flashing, control- and expansion-joint filler, water stops, caulking, and insulation. Most of these items are estimated in terms of linear feet, but some require a square-foot or cubic-foot estimate. Flashing, for instance, is calculated in linear feet, while batt or board insulation is calculated in square feet. For these items, there are no rules of thumb or general rules to follow; use the customary unit of measure for that item.

Square Iron Rebar

Ernest Ramsone wrote the book on reinforced concrete in 1912. He developed a system for using twisted square iron rods to reinforce concrete. He found that by using twisted square bars he could create much greater tensile strength in the surrounding concrete than others could with smooth round rods. He patented his system in 1884 and popularized the use of reinforced concrete as a building material. Many of the buildings he designed are significant landmarks today.

Additional Resources

Building Block Walls: A Basic Guide, Latest Edition. Herndon, VA: National Concrete Masonry Association.

Principles of Brick Masonry. Reston, VA: Brick Industry Association.

3.0.0 Section Review

1. Joint reinforcement is normally lapped _____.

   a. 10 inches
   b. 8 inches
   c. 6 inches
   d. 4 inches

2. Structural reinforcement for masonry walls is normally estimated by the _____.

   a. linear foot
   b. square foot
   c. pound
   d. ton

3. Masonry ties and joint reinforcement are estimated by the _____.

   a. linear foot
   b. square foot
   c. pound
   d. ton

4. Conversion factors for other types of masonry units can be derived by calculating the nominal area of the masonry face and then _____.

   a. subtracting 144 square inches from the nominal area
   b. adding 144 square inches to the nominal area
   c. multiplying the nominal area by 144 square inches
   d. dividing the nominal area into 144 square inches

5. Batt insulation is calculated in _____.

   a. rolls
   b. pounds
   c. square feet
   d. linear feet

SUMMARY

Estimating material quantities is an important part of the planning and scheduling of a construction project. Accurate estimates require good math and plan-reading skills, as well as a set of estimating tools. The first step in estimating is to find out what needs to be built. Plans and drawings provide this guidance. Worksheets can be drawn up to help keep the information organized. The two most common mistakes in estimating material quantities are omitting all or part of an item and making an arithmetic error. Always have the work checked before finalizing the estimate.

The coursing method and the square-foot method can be used to perform block and brick estimates. Estimating mortar for block and brick walls can be done by using a rule of thumb or by referring to standard tables that use unit values for different types of materials. This requires determining a square-foot or cubic-foot value and using a conversion factor.

Joint reinforcement can be calculated using either a linear- or a square-foot method. Structural reinforcement such as rebar is typically estimated first by linear feet, and then converted into tons. Structural accessory items can be estimated by the linear foot. Masons use the square-foot method to estimate masonry ties and most other masonry units. Conversion factors for odd units can be easily derived by calculating the nominal area of the masonry face, and then dividing the nominal area into 144 square inches. When estimating other items such as flashing, control- and expansion-joint filler, water stops, caulking, and insulation, keep in mind that not all of these items may be shown on the drawings. Use the customary unit of measure for these items.

With time and practice, a mason can become skilled at estimating the masonry units, mortar, grout, and accessories required to build any masonry installation.

1. Once all the estimates for the masonry project have been completed, they can be used to fill out a _____.
   a. master list
   b. summary report
   c. recapitulation sheet
   d. project grid

2. Block quantities required for a project can be calculated by using either the square-foot method or the _____.
   a. volume method
   b. linear-foot method
   c. tabular method
   d. coursing method

3. The nominal dimensions of a standard concrete block are _____.
   a. 4 × 8 × 12
   b. 4 × 4 × 12
   c. 8 × 8 × 16
   d. 4 × 8 × 16

4. To determine the number of block needed for a wall, multiply the square footage by the constant _____.
   a. 0.95
   b. 1.125
   c. 1.150
   d. 1.333

5. Finding the number of lintel blocks needed for an opening is done by adding 1 foot 4 inches to the opening width, then dividing by _____.
   a. 8
   b. 12
   c. 16
   d. 24

6. Eight bags of masonry cement and 1 ton of sand will produce enough mortar for _____.
   a. 240 block
   b. 360 block
   c. 480 block
   d. 720 block

7. Tables used in estimating mortar needs for block express the quantities needed in _____.
   a. cubic feet
   b. square yards
   c. square feet
   d. cubic yards

8. Grout is commonly used for block structures such as retaining walls and _____.
   a. cisterns
   b. partition walls
   c. screens
   d. foundations

9. The amount of grout needed for a block wall with a volume of 596 cubic feet would be _____.
   a. 10 cubic yards
   b. 11 cubic yards
   c. 12 cubic yards
   d. 13 cubic yards

10. When estimating brick requirements by the square-foot method, apply a correction factor based on the _____.
    a. size of the masonry unit
    b. mortar-joint thickness
    c. bond pattern
    d. wall height

11. If a wall has many small openings, the waste percentage must be increased because _____.
    a. breakage of brick is more likely
    b. the work will take longer
    c. specialized brick shapes are needed
    d. a number of brick will need to be cut

12. Mixing mortar for a structure that will use 450 brick will require 4 bags of masonry cement and _____.
    a. 500 lbs of sand
    b. 750 lbs of sand
    c. 1,000 lbs of sand
    d. 1,250 lbs of sand

13. The amount of additional mortar needed to fill collar joints in multiwythe walls is calculated based on the _____.

   a. joint size
   b. wythe thickness
   c. type of mortar
   d. wall volume

14. For brick walls, joint reinforcement is generally placed every _____.

   a. third course
   b. fourth course
   c. fifth course
   d. sixth course

15. If a job has unusual reinforcement spacing, the easiest estimating method to use is the _____.

   a. rule-of-thumb method
   b. linear-foot method
   c. square-foot method
   d. table method

16. Assuming a conversion factor of 0.75 linear feet per square foot of wall area, the amount of reinforcement needed for a 1,250-square-foot wall would be _____.

   a. 93.8 linear feet
   b. 930.8 linear feet
   c. 938 linear feet
   d. 9,380 linear feet

17. Continuous joint reinforcement is available in various widths and two styles: truss and _____.

   a. interlocked
   b. ladder
   c. diamond
   d. overlapping

18. For ease of handling, joint reinforcement is provided in packages weighing approximately _____.

   a. 20 pounds
   b. 30 pounds
   c. 40 pounds
   d. 50 pounds

19. When estimating rebar, the usual allowance for laps is _____.

   a. 60 bar diameters
   b. 48 bar diameters
   c. 36 bar diameters
   d. 24 bar diameters

20. Depending on the type, masonry ties are furnished in boxes of up to _____.

   a. 1,000 pieces
   b. 500 pieces
   c. 250 pieces
   d. 100 pieces

Trade Terms Quiz

Fill in the blank with the correct term that you learned from your study of this module.

1. The process of measuring and counting individual items from a set of plans in order to estimate material quantities and associated items for construction projects is called the _____.

2. The _____ is a method of estimating materials by calculating the area (square feet) of a structural unit.

Trade Terms

Square-foot method
Takeoff

Trade Terms Introduced in This Module

Square-foot method: A method of estimating materials by calculating the area, in square feet, of a structural unit.

Takeoff: The process of measuring and counting individual items from a set of plans in order to estimate material quantities and associated items for construction projects.

ESTIMATING FORMS

REF.	DESCRIPTION	DIMENSIONS				EXTENSION	QUANTITY	UNIT	TOTAL		REMARKS
		NO.	LENGTH	WIDTH	HEIGHT				QUANTITY	UNIT	

Takeoff By:

Checked By:

WORKSHEET

PAGE #

DATE _____

SHEET _____ of _____

PROJECT _____

ARCHITECT _____

28305-14_A01.EPS

Figure A-1 Quantity takeoff sheet.

SUMMARY SHEET

By: PAGE #

DATE _____

SHEET _____ of _____   TITLE: _____

PROJECT _____

WORK ORDER # _____

DESCRIPTION	QUANTITY		MATERIAL COST		LABOR MAN HOURS FACTORS					LABOR COST		ITEM COST	
	TOTAL	UT	PER UNIT	TOTAL	CRAFT	PR UNIT	TOTAL	RATE	COST PR	PER	TOTAL	TOTAL	PER UNIT
				MATERIAL							LABOR	TOTAL	

Figure A-2 Summary sheet.

28305-14_A02.EPS

Listed By: ____		RECAPITULATION SHEET				Page ____ of ____
Checked By: ____		PROJECT _____ DATE _____ ARCHITECT_____ S/SF _____				

PAGE REF.	ITEM	MATERIAL	LABOR	SUB	EQUIPMENT	TOTAL

28305-14_A03.EPS

Figure A-3 Recapitulation sheet.

Additional Resources

This module presents thorough resources for task training. The following resource material is suggested for further study.

Basics for Builders: Plan Reading & Material Takeoff. 1994. Wayne J. DelPico. Kingston, MA: R.S. Means Company, Inc.

Bricklaying: Block and Brick Masonry. 1988. Brick Industry Association. Orlando, FL: Harcourt Brace & Company.

Building Block Walls: A Basic Guide, Latest Edition. Herndon, VA: National Concrete Masonry Association.

Concrete Masonry Handbook, Fifth edition. W. C. Panerese, S. K. Kosmatka, and F. A. Randall, Jr. Skokie, IL: Portland Cement Association.

Pocket Guide to Brick Construction. 1990. Reston, VA: Brick Industry Association.

Principles of Brick Masonry. Reston, VA: Brick Industry Association.

Figure Credits

Section Review Answers

Answer	Section Reference	Objective
Section One		
1. d	1.1.0	1a
2. c	1.2.0	1b
3. d	1.3.0	1c
4. b	1.4.1	1d
5. d	1.5.0	1e
6. a	1.6.0	1f
Section Two		
1. d	2.1.0	2a
2. a	2.2.0	2b
3. b	2.3.0	2c
4. b	2.4.3	2d
Section Three		
1. c	3.1.1	3a
2. d	3.2.0	3b
3. b	3.3.0	3c
4. d	3.4.0	3d
5. c	3.5.0	3e

NCCER CURRICULA — USER UPDATE

NCCER makes every effort to keep its textbooks up-to-date and free of technical errors. We appreciate your help in this process. If you find an error, a typographical mistake, or an inaccuracy in NCCER's curricula, please fill out this form (or a photocopy), or complete the online form at **www.nccer.org/olf**. Be sure to include the exact module ID number, page number, a detailed description, and your recommended correction. Your input will be brought to the attention of the Authoring Team. Thank you for your assistance.

Instructors – If you have an idea for improving this textbook, or have found that additional materials were necessary to teach this module effectively, please let us know so that we may present your suggestions to the Authoring Team.

NCCER Product Development and Revision
13614 Progress Blvd., Alachua, FL 32615

Email: curriculum@nccer.org
Online: www.nccer.org/olf

❏ Trainee Guide   ❏ Lesson Plans   ❏ Exam   ❏ PowerPoints   Other _____

Craft / Level: _____   Copyright Date: _____

Module ID Number / Title: _____

Section Number(s): _____

Description: _____

_____

_____

_____

_____

Recommended Correction: _____

_____

_____

_____

_____

Your Name: _____

Address: _____

_____

Email: _____   Phone: _____

28306-14

Site Layout—Distance Measurement and Leveling

Masons need to know how to perform site layout, which involves measuring distances and establishing level. Depending on the size of the project, the lead mason may do many of the same layout tasks at the job site as a field engineer. On very large construction jobs, the mason may work with the field engineers. On smaller projects, the mason may be responsible for the layout of the entire project. This module focuses on the principles, equipment, and basic methods used to perform the site layout tasks of distance measurement and differential leveling.

Module Six

Trainees with successful module completions may be eligible for credentialing through the NCCER National Registry. To learn more, go to www.nccer.org or contact us at 1.888.622.3720. Our website has information on the latest product releases and training, as well as online versions of our *Cornerstone* magazine and Pearson's product catalog.

Your feedback is welcome. You may email your comments to curriculum@nccer.org, send general comments and inquiries to info@nccer.org, or fill in the User Update form at the back of this module.

This information is general in nature and intended for training purposes only. Actual performance of activities described in this manual requires compliance with all applicable operating, service, maintenance, and safety procedures under the direction of qualified personnel. References in this manual to patented or proprietary devices do not constitute a recommendation of their use.

Objectives

When you have completed this module, you will be able to do the following:

1. Describe the elements of site plans and topographic maps.
   a. List characteristics of contour lines.
2. Describe layout control points.
   a. Explain how to convert between distance measurement systems.
   b. Identify the types of control points.
   c. Explain how to place control points and other markers.
   d. Describe how to communicate information on control points and other markers.
   e. Discuss how control markers are color-coded.
3. Identify distance measurement tools and equipment.
   a. Explain how to use tapes.
   b. Explain how to use range poles.
   c. Explain how to use plumb bobs and gammon reels.
   d. Explain how to use hand sight levels.
4. Describe how to make distance measurements.
   a. Explain how to estimate distances by pacing.
   b. Describe how to measure distances electronically.
5. Identify differential-leveling tools and equipment.
   a. Identify leveling instruments.
   b. Describe the use of tripods.
   c. Describe the use of leveling rods.
   d. Explain how to set up and adjust leveling instruments.
   e. Explain how to test the calibration of leveling instruments.
6. Explain the basics of differential leveling.
   a. Define differential-leveling terminology.
   b. Explain the differential-leveling procedure.
   c. Explain how field notes are recorded and used.
7. Identify leveling applications.
   a. Explain how to transfer elevations up a structure.
   b. Explain profile, cross-section, and grid leveling.
8. Describe how to lay out building corners.
   a. Explain how to construct batter boards.
   b. Describe how to use the 3-4-5 rule.

Performance Tasks

Under the supervision of your instructor, you should be able to do the following:

1. Interpret a construction site plan and relate the man-made and topographic features and other project information to the layout and topography of the actual site.
2. Convert measurements stated in feet and inches to equivalent decimal measurements stated in feet, tenths, and hundredths, and vice versa.
3. Set up, adjust, and field-test a leveling instrument.
4. Use a builder's level and leveling rod to determine site and building elevations.
5. Use differential-leveling and distance-measurement procedures to transfer elevations up a structure.
6. Check and/or establish 90-degree angles using the 3-4-5 rule.

Trade Terms

Backsight (BS)	Height of instrument (HI)
Benchmark	Leveling rod
Control point	Offset
Crosshairs	Parallax
Cut	Station
Differential leveling	Tape
Earthwork	Taping
Field notes	Temporary benchmark
Fill	Transit level
Foresight (FS)	Turning point (TP)

Industry-Recognized Credentials

If you're training through an NCCER-accredited sponsor, you may be eligible for credentials from NCCER's Registry. The ID number for this module is 28306-14. Note that this module may have been used in other NCCER curricula and may apply to other level completions. Contact NCCER's Registry at 888.622.3720 or go to **www.nccer.org** for more information.

Code Note

Codes vary among jurisdictions. Because of the variations in code, consult the applicable code whenever regulations are in question. Referring to an incorrect set of codes can cause as much trouble as failing to reference codes altogether. Obtain, review, and familiarize yourself with your local adopted code.

Contents

Topics to be presented in this module include:

Contents (continued)

Figures

SECTION ONE

1.0.0 SITE PLANS AND TOPOGRAPHIC MAPS

Objective

Describe the elements of site plans and topographic maps.
  a. List characteristics of contour lines.

Performance Task

Interpret a construction site plan and relate the man-made and topographic features and other project information to the layout and topography of the actual site.

Trade Terms

Benchmark: A reference point established by the surveyor on or close to the property, usually at one corner of the lot.

Control point: A horizontal or vertical point established in the field to serve as part of a known framework for all points on the site.

Site layout involves the extensive use of site plans, also called plot plans. These plans show existing man-made and topographic features and other relevant project information pertaining to the job site, viewed as if looking down from above. Man-made features include roads, sidewalks, utilities, and buildings. Topographic features include trees, streams, springs, and contours. Project information includes the building outline, general utility information, proposed sidewalks, parking areas, roads, landscape information, control points or property corners, the direction and length of property lines or control lines, proposed contours, and any other information that will convey what is to be constructed or changed on the site. *Figure 1* shows a typical site plan.

Figure 2 shows an example of a topographic map. It depicts the natural and man-made features of a place or region, showing their relative positions and elevations.

Like other drawings, site plans are drawn to scale. The engineering scale used depends on the size of the project. A project covering a large area typically will have a small scale, such as 1 inch = 100 feet, while a project on a small site might have a large scale, such as 1 inch = 10 feet.

Normally, the dimensions shown on site plans are stated in feet and tenths of a foot. However, some site plans state the dimensions in feet, inches, and fractions of an inch. Dimensions to the property lines are shown to establish code requirements. Frequently, building codes require that nothing be built on certain portions of the land. For example, local building codes may have a front setback requirement that dictates the minimum distance that must be maintained between the street and the front of a structure. Normally, side yards have a minimum width specified from the property line to allow for access to rear yards and to reduce the possibility of fire spreading to adjacent buildings.

A property owner cannot build on an area where an easement has been identified. Examples of typical easements are the right of a neighbor to build a road; a public utility to install water, gas, or electric lines on the property; or an area set aside for drainage of groundwater.

Site plans show finish grades (also called elevations) for the site, based on data provided by a surveyor or engineer. It is necessary to know these elevations for grading the lot and for construction of the structure. Finish grades are typically shown for all four corners of the lot as well as other points within the lot. Finish grades are also shown for the corners of the structure and relevant points within the building.

All the finish grade references shown are keyed to a single reference point. This is called the benchmark, also called the job datum, monument, or primary control point. This is established by the surveyor on or close to the property, usually at one corner of the lot. At the site, this point may be marked by a plugged pipe driven into the ground, a brass marker, or a wood stake. The location of the benchmark is shown on the plot plan with a grade figure next to it. It is also typically included on the title sheet of the set of plans. This grade figure may indicate the actual elevation relative to sea level, or it may be an arbitrary elevation, such as 100 feet or 500 feet. All other grade points shown on the site plan are relative to the benchmark.

A site plan usually shows the finish floor elevation of the building. This is the first floor of the building relative to the job-site benchmark. For example, if the benchmark is labeled 100 feet and the finish floor elevation indicated on the plan is marked 105 feet, the finish floor elevation is 5 feet above the benchmark. During construction, many important measurements are taken from the finish floor elevation point.

The content and layout of site plans are covered in the Masonry modules, *Residential Plans and Drawing Interpretation* and *Commercial Drawings*. You may wish to review this information at this

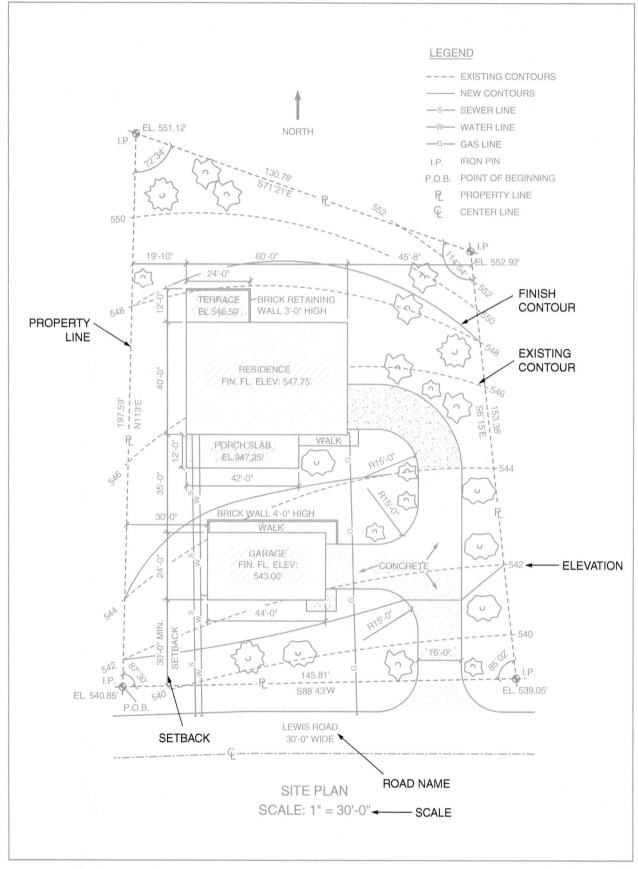

Figure 1 A typical site plan.

28306-14_F01.EPS

CONTOUR LINES

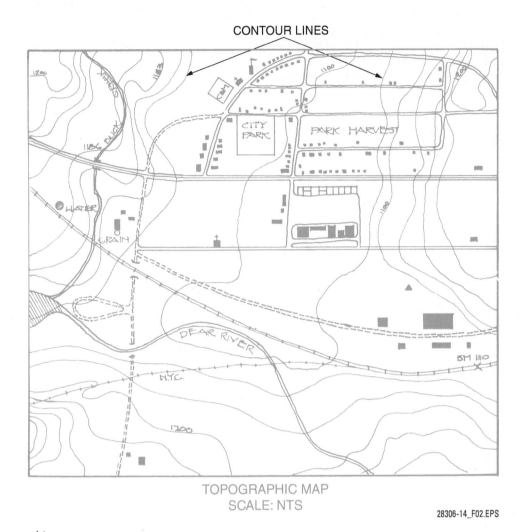

TOPOGRAPHIC MAP
SCALE: NTS

28306-14_F02.EPS

Figure 2 Topographic map.

time, as these modules provide important information about contour lines shown on site plans and topographic maps.

1.1.0 Understanding the Characteristics of Contour Lines

The topography of a job site or other area can be represented by contour lines. Contour lines show changes in the elevation and contour of the land. The lines may be dashed or solid. Generally, dashed lines are used to show the natural or existing grade, and solid lines show the finish grade to be achieved during construction.

Each contour line across the plot of land represents a line of constant elevation relative to some point, such as sea level or a local feature. *Figure 3* shows an example of a contour map for a hill. As shown, contour lines are drawn in uniform elevation intervals called contour intervals. Commonly used intervals are 1 foot, 2 feet, 5 feet, and 10 feet.

On some plans and surveys, every fifth contour line is drawn using a heavier-weight line and is

Surveying in American History

The Mason-Dixon line is a boundary between Pennsylvania and Maryland surveyed by Charles Mason and Jeremiah Dixon from 1763 to 1767. It was one of the earliest survey projects in North America. The survey was ordered by the Court Chancery in England to settle the disputes between the heirs of Lord Baltimore and William Penn.

In 1779, it was extended to present-day West Virginia. In the 19th century, it was continued along the course of the Ohio River. Before the Civil War, it designated the boundary between slave and free states. Today, it is still used on occasion to mark the line of separation between the North and the South.

labeled with its elevation to help the user more easily determine the contour. The elevation is marked above the contour line, or the line is interrupted for it. This method of drawing contour lines is called indexing contours.

As shown in *Figure 3*, contour lines form a closed loop within the map. A person starting out at any point on the contour and following its path will eventually return to the starting point. A contour may close on a site plan or map, or it may be discontinued at any two points at the borders of the plan or map. Examples of this are shown on the topographic survey map in *Figure 4*. Such points mark the ends of the contour on the map, but the contour does not end at these points. The contour is continued on a plan or map of the adjacent land.

Some rules for interpreting contours include the following:

- Contour lines do not cross.
- Contour lines crossing a stream form a pattern that appears to point upstream (refer to *Figure 3*, in which the stream starts near the highest elevation).
- The horizontal distance between contour lines represents the degree of slope. Closely spaced contour lines represent steep ground and widely spaced contour lines represent nearly level ground with a gradual slope. Uniform spacing indicates a uniform slope.

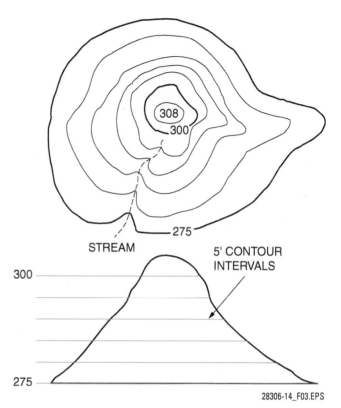

Figure 3 Contour map of a hill.

- Contour lines are at right angles to the slope. Therefore, water flow is perpendicular to contour lines.
- Straight contour lines parallel to each other represent man-made features such as terracing.

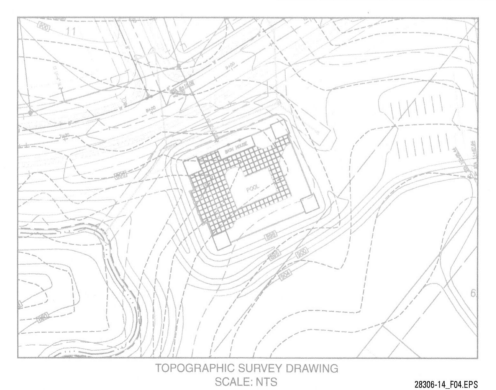

TOPOGRAPHIC SURVEY DRAWING
SCALE: NTS

28306-14_F04.EPS

Figure 4 Topographic survey drawing.

Additional Resources

Architectural Drawing and Light Construction. 2009. Philip A. Grau III, Edward J. Muller, and James G. Fausett. Upper Saddle River, NJ: Prentice Hall.

Architectural Graphic Standards. 1998. The American Institute of Architects. New York: John Wiley & Sons, Inc.

Plan Reading & Material Takeoff. Kingston, MA: R.S. Means Company.

Reading Architectural Plans: For Residential and Commercial Construction. 2002. Ernest R. Weidhaas. Upper Saddle River, NJ: Prentice Hall.

1.0.0 Section Review

1. The method of drawing contour lines in which every fifth contour line is drawn using a heavier-weight line and is labeled with its elevation is called _____.

    a. tabulating contours
    b. pointer contours
    c. indicia contours
    d. indexing contours

2.0.0 SITE LAYOUT CONTROL POINTS

Objective

Describe layout control points.
  a. Explain how to convert between distance measurement systems.
  b. Identify the types of control points.
  c. Explain how to place control points and other markers.
  d. Describe how to communicate information on control points and other markers.
  e. Discuss how control markers are color-coded.

Performance Task

Convert measurements stated in feet and inches to equivalent decimal measurements stated in feet, tenths, and hundredths, and vice versa.

Trade Terms

Cut: To remove soil or rock on site to achieve a required elevation.

Field notes: A permanent record of field measurement data and related information.

Fill: Adding soil or rock on site to achieve a required elevation.

Offset: To position a stake at a specified distance and direction from the control point to allow that area to be worked in without disturbing the stake. Offset stakes include the distance from, and direction to, the control point.

Tape: A measuring tape, usually made of fiberglass, cloth, or stainless steel.

Site plans show the locations of property corners and the direction and length of property lines or control lines. In most states, registered surveyors are required to perform any layout work that establishes legal property lines or boundaries. This is because the surveyor assumes the liability for any mistake in the surveying work.

The surveyor is legally responsible if the building ends up on the wrong property or at some location that violates setback requirements or other regulations. Because of the tremendous liability involved, the mason should never make any layout measurements that relate to property lines or boundaries. This is a task for the professional land surveyor.

2.1.0 Converting between Distance Measurement Systems

Construction project drawings can show dimensions in feet and inches, in decimals of a foot, or both. For example, the dimensions of structures are usually shown in feet and inches. Land measurements and ground elevations are typically shown in decimals of a foot. For this reason, it is often necessary to convert between the two measurement systems.

Conversion tables are available in many trade-related reference books that can be used for this purpose. However, you should be familiar with the methods used to make the conversions mathematically, in case conversion tables are not readily available.

2.1.1 Converting Feet and Inches to Decimals of a Foot

To convert values given in feet and inches into their equivalent values in decimals of a foot, use the following procedure. For this example, convert 45 feet 4⅜ inches to decimals of a foot.

Step 1 Convert the inch-fraction ⅜ to a decimal. This is done by dividing the numerator of the fraction (top number) by the denominator of the fraction (bottom number). In this example, ⅜ inch = 3 ÷ 8 = 0.375 inch.

Step 2 Add 0.375 inches to 4 inches to obtain 4.375 inches.

Step 3 Divide 4.375 inches by 12 to obtain 0.3646 feet, which rounds to 0.36 feet.

Step 4 Add 0.36 feet to 45 feet to obtain 45.36 feet.

2.1.2 Converting Decimals of a Foot to Feet and Inches

To convert values given in decimals of a foot into equivalent feet and inches, use the following procedure. For this example, convert 45.3646 feet to feet and inches (convert to the nearest eighth inch).

Step 1 Subtract 45 feet from 45.3646 feet to get 0.3646 feet.

Step 2 Convert 0.3646 feet to inches by multiplying 0.3646 inches by 12, to get 4.3752 inches.

Step 3 Subtract 4 inches from 4.3752 inches, to get 0.3752 inches.

Step 4 Convert 0.3752 inches into eighths of an inch by multiplying 0.3752 by 8 = 3.0016 eighths or, when rounded off, ⅜ inch. Therefore, 45.3646 feet equals 45 feet 4⅜ inches.

2.2.0 Identifying Types of Control Points

Site layout involves establishing a network of control points on a site that serve as a common reference for all construction. The exact locations of these control points are marked at the site and recorded in the field notes as they are made. Annotating control-point location reference data in the field notes is important for two reasons. First, it makes it possible to locate a point should it become covered up or otherwise hidden. Second, it makes it possible to re-establish a point accurately if the marker is damaged or removed.

There are three basic categories of control points:

- Primary control points
- Secondary control points
- Building-layout or working control points

Primary control points (*Figure 5*) are permanent points used as the basis for locating secondary control points and other points on the site. Primary control points are located where they are accessible and protected from damage for the duration of the job. They can be located and marked on many kinds of permanent and immovable objects, such as fire hydrants and power poles.

When no suitable permanent objects are available for use as primary control-point markers, iron stakes driven into the ground or concrete monuments dug and poured into the ground can be used. If a poured concrete monument is used, it must be dug a foot deeper than the frost line to prevent freezing and thawing from moving it. It must also have a distinct high point. This can be a rounded brass cap or rebar that sticks up out of the top of the concrete.

Primary control point markers are typically established by a registered surveyor. They are commonly referred to as monuments or benchmarks.

Secondary control points (*Figure 6*) are additional control points located within the job site to aid in the construction of the individual structures on the site. Secondary control points may be marked by a hub stake surrounded by protective laths, posts, or fencing. The hub stake is typically a

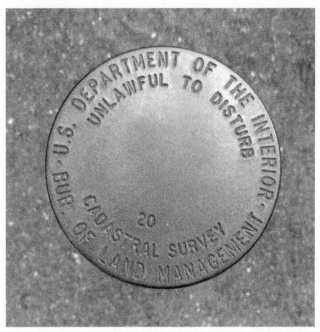

28306-14_F05.EPS

Figure 5 Primary control-point marker.

1½-inch-square piece of wood pointed on one end. Its length is normally determined by the hardness of the ground it must be driven into, with lengths between 8 inches and 12 inches being typical. The hub stake is driven into the ground until flush or nearly so. A surveyor's tack, with a depression in the center of the head, is driven into the top of the hub stake to locate the exact point.

Building-layout or working control points are usually located with reference to the secondary control points. They are the points from which

28306-14_F06.EPS

Figure 6 Secondary control-point marker.

Error

Errors have a cumulative effect. The more measurements you make, the larger the error can be upon completion of the measurements. Take your time and repeat a measurement one or two times to make sure you are correct before moving on to the next measurement.

---

actual measurements for construction are taken. Building-layout points are used to locate the corners of buildings and building lines. They usually are marked with a hub and a related marker stake (*Figure 7*). The marker stake is typically a ¾-inch × 1½-inch piece of wood that varies in length, with 24–36 inches being typical. In addition to serving as hub markers, these stakes are also used to mark line or grade and other information for center lines, offset lines, or slope stakes.

2.3.0 Placing Control Points and Other Markers

The placement of the numerous on-site control-point markers depends on the nature of the job, the terrain, other work in progress, the sequence of work, and many other factors. Accuracy in site layout work requires that benchmarks, control points, and other important markers be referenced in a way that ensures they can be easily located at a later date. These points should be referenced to permanent objects in the surrounding area.

Guidelines for establishing such references are summarized as follows:

- Establish several (three or more) definable permanent or semipermanent references for each point. Some widely used reference objects include the following:

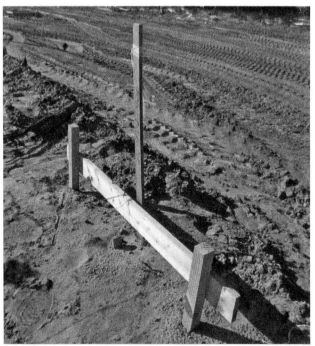

28306-14_F07.EPS

Figure 7 Working control-point marker.

- A bonnet bolt located on a nearby fire hydrant
  - Power or telephone poles
  - Sidewalks
  - Trees
  - Fences
  - Building corners
  - Signposts
- To locate a point more accurately, reference it in all directions (north, south, east, west) from the point, rather than in the same direction. This is because when arcs are swung from the reference points, there is a very small, yet distinct area where they intercept.
- If possible, stay within one tape length of the point being used as a reference.
- Draw a clear, complete sketch in the field notes.

The placement and positioning of stakes at the construction site must be done properly. Some guidelines for performing this task include the following:

- Face the stakes so that they can be read from the direction of use.
- Offset the stakes as required for their protection.
- Set the stakes within tolerances.
- Place the stakes solidly in the ground.
- Place the stakes and laths so that they are plumb.
- Center the hubs and stakes.

2.4.0 Communicating Information on Control Markers and Other Markers

Hubs, stakes, and laths must be legibly and accurately marked to correctly communicate location, elevation, and other pertinent construction information. Some guidelines for marking stakes are as follows:

- Use a permanent marker.
- Print neatly. Start from the top of the stake and work toward the bottom. Use all capital letters, being careful not to crowd your words or numbers.
- Avoid the use of abbreviations or use standard abbreviations. The *Appendix* lists some of the common abbreviations used for layout tasks.
- All sides of a stake can be marked with information; however, the main information should always be marked in the direction of use.
- Mark only pertinent information. Too much or too little information can be confusing.

- Mark the following types of information on a stake, as applicable:
  - Alignment information
  - Center line
  - Cut or fill data
  - Elevation data
  - General information description
  - Grade information
  - Offset
  - Reference information
  - Slope
  - Other specific information

In addition to marking stakes, it is often necessary to mark reference lines and points on walls, curbs, foundations, etc. When doing so, the same general guidelines should be followed that were given for marking information on stakes. In addition, any such reference lines must be marked straight, level, and/or plumb. *Figure 8* shows an example of a reference line indicating a height of 1 mile above sea level.

Good practice dictates that a control point should always be placed so that three other control points are visible from it at all times. This way, if the control point is covered up or destroyed, it can be relocated easily. Two tapes can be stretched from the other points to relocate the control point.

28306-14_F08.EPS

Figure 8 The "Mile High Row" at Coors Field in Denver, CO, marking 5,280 feet above sea level, is a famous example of a reference line.

2.5.0 Color-Coding Control Markers

Control points and other markers are often identified by color-coding. Color coding allows for easy identification of control points or markers, as well as providing information about their purpose. Color coding can be done by applying paint to monuments, hubs, or other field markers; applying ribbons on stakes; or attaching flags on wire markers.

Color coding of field markers is not standardized; however, many construction or field-engineering organizations have established their own color-coding systems. When performing layout work at a site, ask your supervisor what color-coding scheme to use, and follow it.

Hubs

Hubs are commonly made by sawing 2 × 4s in half. Sometimes it is necessary to drive a wooden hub into hard-packed ground. This task is easier if you use a pilot hole. Drive a tempered steel pin, called a gad, into the ground first to start a pilot hole. Remove the pin and drive the hub into the ground.

NOTE

In some states, mandatory color-coding schemes are used. A contractor can be fired for failing to use the correct color. Always check your local laws.

Additional Resources

Construction Surveying and Layout: A Step-by-Step Engineering Methods Manual. 2002. Wesley G. Crawford. West Lafayette, IN: Creative Construction Publishing.

Measuring, Marking, and Layout. Newtown, CT: Taunton Press.

Surveying Principles and Applications. 2008. Barry F. Kavanagh. Upper Saddle River, NJ: Prentice Hall.

Surveying with Construction Applications. 2009. Barry F. Kavanagh. Upper Saddle River, NJ: Prentice Hall.

2.0.0 Section Review

1. To convert values given in decimals of a foot into equivalent feet and inches, subtract the whole-number feet and divide the fractional foot by _____.

   a. 4
   b. 8
   c. 12
   d. 16

2. Control points located within the job site to aid in the construction of the individual structures on the site and typically marked by a hub stake are called _____.

   a. building-layout points
   b. primary control points
   c. working control points
   d. secondary control points

3. When establishing references for control points, whenever possible the control point should be placed no more than _____.

   a. one tape length from the reference point
   b. two tape lengths from the reference point
   c. three tape lengths from the reference point
   d. four tape lengths from the reference point

4. Each of the following types of information should be marked on a stake, as applicable, *except* _____.

   a. slope
   b. general information description
   c. latitude and longitude
   d. center line

5. The color coding of control markers is _____.

   a. not standardized
   b. standardized
   c. established by the authority having jurisdiction (AHJ)
   d. referenced in the local applicable code

NCCER – *Masonry Level Three* 28306-14

3.0.0 DISTANCE MEASUREMENT TOOLS AND EQUIPMENT

Objective

Identify distance measurement tools and equipment.
a. Explain how to use tapes.
b. Explain how to use range poles.
c. Explain how to use plumb bobs and gammon reels.
d. Explain how to use hand sight levels.

Trade Terms

Crosshairs: A set of lines, typically horizontal and vertical, placed in a telescope used for sighting purposes.

Leveling rod: A vertical measuring device that consists of two or more movable sections with graduated markings.

Taping: The process of making horizontal and vertical distance measurements.

Site layout involves making horizontal and vertical distance measurements by a process commonly called taping. Taping is also sometimes called chaining, which is a reference to past surveying practices when a metal chain was used to measure distances instead of a measuring tape. Today, the terms *taping* and *chaining* are used interchangeably in the trade when referring to the measurement of distances using a steel tape or other type of tape. The term *taping* will be used throughout the module.

The task of taping involves two people working together and communicating with each other. The major items of equipment (*see Figure 9*) needed to perform taping typically include the following:

• 100-foot or longer steel tape
• Range poles
• Plumb bobs and gammon reels
• Hand sight levels
• Taping pins
• Accessories and stakes

3.1.0 Using Tapes

Tapes can be made of cloth, fiberglass, or steel. However, fiberglass tapes are the most widely used for precision measuring tasks. Steel tapes are made in a variety of graduated lengths, with 100 feet being common. Typically, the tape is mounted on a reel for ease of handling and storage.

Tapes are made with graduations in feet and inches or feet and tenths of a foot. Metric versions are also available. Some steel tapes are nylon coated to increase their durability. The ends of steel tapes are equipped with heavy loops that provide a place to attach leather thongs, also called tension handles. These allow the user to tighten or tension the tape firmly. The physical construction of a tape requires that it be handled and used properly.

Some guidelines for the proper care and handling of tapes are as follows:

• Keep the tape on the reel and rolled up when not in use.
• Dry a tape that is wet. Once dry, wipe it with an oiled cloth for added protection against rust.
• Do not allow vehicles to run over the tape.
• Remove all loops in a tape immediately. This is important to prevent kinks from deforming the tape.
• When making measurements, the proper amount of tension must be applied to get accurate results. The tension is specified by the tape manufacturer. Pulling too hard will eventually stretch and permanently elongate the tape. If a tension spring is not available, try to pull as consistently as possible.

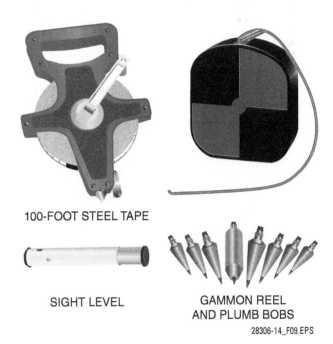

100-FOOT STEEL TAPE

SIGHT LEVEL   GAMMON REEL AND PLUMB BOBS

28306-14_F09.EPS

Figure 9 Common taping equipment.

3.2.0 Using Range Poles

Range poles are used to help maintain alignment for taping and to mark measurement points at the site so that they are more visible. Range poles are made of wood, fiberglass, or metal, and come in various lengths. Typically, they are painted with 1-inch-wide red or orange stripes alternating with white stripes to make them highly visible. Range poles that come in sections are also available, which can be connected together to obtain increased length.

Some guidelines for the proper care and handling of range poles are as follows:

- Clean dirt and mud from the poles and pole tips after each use.
- Store poles in a protective case when not in use.
- Repaint poles when necessary.

3.3.0 Using Plumb Bobs and Gammon Reels

When suspended vertically from a string, a plumb bob creates a plumb vertical line that can be used to position yourself or an instrument directly over a reference point. Plumb bobs are usually made of brass and have a replaceable tip. Depending on the model, they can weigh between 8 and 24 ounces, with 16 ounces being typical.

The gammon reel is used to store the plumb-bob string. The string automatically retracts into the reel to help prevent the string from becoming tangled, broken, or muddy. The case of the gammon reel is colored so that it can be used as a target for sighting purposes.

Some guidelines for the proper care and handling of plumb bobs and gammon reels are as follows:

- Keep plumb bobs and gammon reels clean.
- Never use a plumb bob as a hammer.
- Do not use a plumb-bob tip to mark hard surfaces because this will damage the tip.

- Clean the plumb-bob string before allowing it to be retracted into the gammon reel. This prevents dirt from getting inside the gammon reel where it can damage the retracting mechanism.
- Check the string for wear and knots, and replace when necessary.

3.4.0 Using Hand Sight Levels

When taping distances or making layout measurements, a hand sight level can be used to determine your position on line and/or the correct horizontal position of the tape needed to plumb measurement points. The hand sight level is a short, handheld telescope with a bubble level built into it. The bubble level is visible when sighting through the telescope. When the bubble is centered, the crosshairs of the scope fall on some object that is at about the same elevation as the eye of the user.

For leveling tasks, a hand sight level can be used to help determine where to set up the leveling instrument so that its line of sight will intercept the leveling rod. This is done by using a hand level to sight on a benchmark or other object to see if the height is above or below eye level. Some guidelines for the proper care and handling of a sight level are as follows:

- Do not drop the level.
- Wipe the lens with a clean cloth as needed.
- Keep the level in its protective case when it is not being used.
- Check its calibration frequently.

Use of Tapes

How would you rank steel, fiberglass, and cloth tapes in terms of their usefulness in making precise measurements?

Additional Resources

Basic Surveying Technology, Latest Edition. Stillwater, OK: The Mid-America Vocational Curriculum Consortium, Inc.

Code Check, Latest Edition. Newtown, CT: Taunton Press.

Measuring, Marking, and Layout. Newtown, CT: Taunton Press.

Principles and Practices of Commercial Construction. 2008. Cameron K. Andres and Ronald C. Smith. Upper Saddle River, NJ: Prentice Hall.

Surveying Principles and Applications. 2008. Barry F. Kavanagh. Upper Saddle River, NJ: Prentice Hall.

Surveying with Construction Applications. 2009. Barry F. Kavanagh. Upper Saddle River, NJ: Prentice Hall.

3.0.0 Section Review

1. The most common material for tapes used for precision measuring tasks is _____.
   - a. fiberglass
   - b. nylon
   - c. cloth
   - d. steel

2. To make them highly visible, range poles are typically painted with _____.
   - a. diagonal 2-inch orange stripes
   - b. alternating 1-inch-wide red or orange and yellow stripes
   - c. alternating 1-inch-wide red or orange and white stripes
   - d. a yellow-and-white checkerboard pattern

3. Store the plumb bob string _____.
   - a. in a box
   - b. on a gammon reel
   - c. around the plumb bob
   - d. using range poles

4. To allow the operator to establish level when in use, hand sight levels are fitted with _____.
   - a. transits
   - b. plumb bobs
   - c. laser levels
   - d. bubble levels

4.0.0 Distance Measurements

Objective

Describe how to make distance measurements.
  a. Explain how to estimate distances by pacing.
  b. Describe how to measure distances electronically.

Distance can be estimated using one of two common methods. The first method is pacing, which involves taking a measurement based on a person's average stride. The second method is using electronic distance-measuring instruments (EDMIs). Both methods have their uses, and it's important for masons to be skilled at both methods so that they can use the appropriate method for a given set of circumstances. The following sections describe both methods in detail.

4.1.0 Estimating Distances by Pacing

The ability to pace a distance with reasonable accuracy can be very helpful. Pacing can be used to check measurements that have been made by others or to estimate an unknown distance. An average pace length can be determined by walking a known distance that has been previously measured accurately with a steel tape and dividing that length by the number of paces taken.

When pacing the distance, walk naturally with a consistent pace length. Some people count each step as a pace. Others only count full strides or two steps as a pace. The count does not matter as long as it is consistent. Also, for the last pace in the measurement, which is normally less than a full pace, record to the nearest ½ or even ¼ pace, if possible.

The procedures for finding your average pace length and determining an unknown distance by pacing are briefly outlined here:

Step 1 Use a tape to lay out a level distance of 100 feet.

Step 2 Starting at the beginning point, walk naturally to pace the 100-foot distance. Record the number of paces required to travel the distance.

Step 3 Repeat Step 2 a minimum of four more times and record the number of paces required for each time.

Step 4 Calculate your average number of paces per 100 feet. For example, assume the total number of paces for the five trips equals 199 paces (40.5 + 39 + 40 + 39.5 + 40 = 199). In this case, the average number of paces per 100 feet equals 39.8 paces (199 total paces ÷ 5 trips).

Step 5 Calculate your length of pace in feet by dividing the average number of paces into the distance traveled. For this example, the average pace length is 2.51 feet (100 feet ÷ 39.8 paces).

Once the average length of your pace is known, other distances can be determined by pacing the distance and calculating its length by multiplying your pace length by the number of paces needed to travel the distance. For example, if it takes 60 paces to travel a distance and your average pace length is 2.51 feet, then the distance is approximately 151 feet (60 paces × 2.51 feet = 150.6 feet, rounded up to 151 feet).

4.2.0 Performing Electronic Distance Measurements

Electronic distance measurement (EDM) is a widely used technology. It provides a fast and extremely accurate method for making long distance measurements. It can measure over obstacles such as lakes, ravines, and roadways.

EDM involves the use of a EDMI. There are two classes of EDMIs: electro-optical instruments (*Figure 10*) and microwave instruments. The difference is in the wavelength of the distance-measurement signal that is transmitted by the device. Most site layout work performed by masons is done using the electro-optical type; therefore, the remainder of this discussion will focus on this type.

The basic electro-optical measurement system consists of an EDMI and a reflector. The EDMI is set up at one end of the line to be measured and the reflector at the other end (*Figure 11*). The reflector consists of one or more prisms mounted on a tripod or range pole. The number of prisms used is determined by the length of the distance to be measured. The longer the distance, the more prisms are used. Depending on the instrument's design, the EDMI transmits either a modulated, visible, low-power laser light signal or an invisible infrared light signal.

During a distance measurement, this directional signal is aimed at the reflector. When the signal strikes the reflector's prism(s), it is reflected back to the EDMI in the same direction from which it came. However, a phase shift is imparted to this reflected signal relative to the phase of the transmitted signal. Within the EDMI, the time

28306-14_F10.EPS

Figure 10 Typical electro-optical prism used for electronic distance measuring.

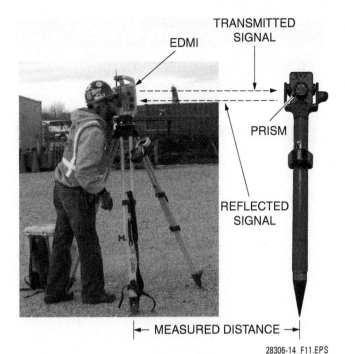

28306-14_F11.EPS

Figure 11 Simplified electronic distance-measuring system.

and phase relationships between the transmitted and received reflected signals are compared and the differences are processed electronically to produce the resultant distance measurement. Many EDMIs can also provide an electronic record of the work done.

Because of the wide variety of EDMIs and prisms that are available, they should be set up, aligned, and operated according to the manufacturer's instructions. This procedure may involve entering several pieces of data into the instrument before or during the measurement.

It should be emphasized that the common use of EDMIs to make distance measurements does not eliminate the need to make distance measurements by taping. Because of the time required to set up the EDMI and prisms for a measurement, it is common practice to make shorter distance measurements by taping because it is often quicker and more convenient.

Additional Resources

Architectural Drawing and Light Construction. 2009. Philip A. Grau III, Edward J. Muller, and James G. Fausett. Upper Saddle River, NJ: Prentice Hall.

Construction Surveying and Layout: A Step-by-Step Engineering Methods Manual. 2002. Wesley G. Crawford. West Lafayette, IN: Creative Construction Publishing.

Measuring, Marking, and Layout. Newtown, CT: Taunton Press.

Principles and Practices of Commercial Construction. 2008. Cameron K. Andres and Ronald C. Smith. Upper Saddle River, NJ: Prentice Hall.

Surveying Principles and Applications. 2008. Barry F. Kavanagh. Upper Saddle River, NJ: Prentice Hall.

Surveying with Construction Applications. 2009. Barry F. Kavanagh. Upper Saddle River, NJ: Prentice Hall.

4.0.0 Section Review

1. When estimating distance by pacing, the last pace in a measurement is normally _____.

   a. less than a full pace
   b. more than a full pace
   c. measured with a steel tape
   d. the same as a full pace

2. The two classes of EDMIs are _____.

   a. short and long range
   b. electro-optical and microwave
   c. electro-optical and laser
   d. laser and microwave

5.0.0 DIFFERENTIAL-LEVELING TOOLS AND EQUIPMENT

Objective

Identify differential-leveling tools and equipment.

a. Identify leveling instruments.
b. Describe the use of tripods.
c. Describe the use of leveling rods.
d. Explain how to set up and adjust leveling instruments.
e. Explain how to test the calibration of leveling instruments.

Performance Tasks

Set up, adjust, and field-test a leveling instrument.

Use a builder's level and leveling rod to determine site and building elevations.

Trade Terms

Differential leveling: A method of leveling used to determine the difference in elevation between two points.

Parallax: The apparent movement of the crosshairs in a surveying instrument caused by movement of the eyes.

Transit level: An optical instrument used in surveying.

Differential leveling is the process used to determine or establish elevations such as those needed for setting slope stakes, grade stakes, footings, anchor bolts, slabs, decks, and sidewalks. This section describes the equipment used to perform differential-leveling tasks.

5.1.0 Using Leveling Instruments

A wide variety of leveling instruments can be used to perform leveling and other on-site layout tasks. The procedural data presented in this module relating to differential leveling emphasizes the use of conventional leveling instruments, such as the builder's level.

5.1.1 Builder's Level

The builder's level (*Figure 12*) is an instrument used to check and establish grades and elevations in order to set up level points over long distances. It consists of a telescope, a bubble spirit level (leveling vial) mounted parallel with the telescope, and a leveling head mounted on a circular base with a horizontal circle scale graduated in degrees. The telescope can be rotated 360 degrees for measuring horizontal angles. The telescope can be tilted slightly for sighting purposes. Builder's levels are mounted on a tripod when in use.

Depending on the model, builder's levels are made with telescope powers ranging from 12 power (12x) to 32 power (32x). The 20 power (20x) is the most common. The power of a telescope determines how much closer an object will appear when viewed through the telescope.

There are two types of leveling-head systems used in builder's levels: a four-screw system and a three-screw system. The advantage of the three-screw system is that it allows the instrument to be leveled more quickly. Four-screw systems are more common in older models.

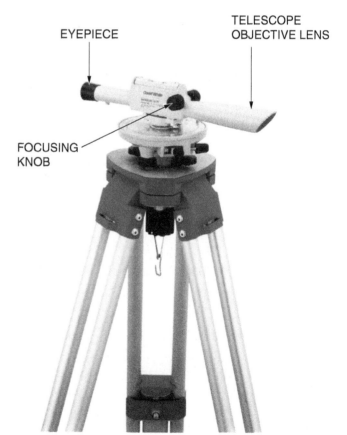

EYEPIECE

TELESCOPE OBJECTIVE LENS

FOCUSING KNOB

28306-14_F12.EPS

Figure 12 Typical builder's level.

5.1.2 Transit

The transit (*Figure 13*) is also commonly called a transit level because it can be used for similar purposes as a level. Its telescope can be tilted vertically up and down. This feature enables it to be used to make vertical angle measurements and perform other operations that are not possible with the builder's level.

The transit has a vertical scale used to measure angles. This scale is typically graduated from 0 to 45 degrees in two directions and moves with the up-and-down motion of the telescope. The telescope is set horizontally when used as a level. Like the builder's level, the transit is mounted on a tripod when in use.

5.1.3 Automatic Leveling Instruments

Automatic levels (*Figure 14*) are used to perform the same measurements and operations as described for the builder's level. These instruments have a built-in compensator mechanism that works to automatically maintain a true-level line of sight. Compensator instruments still have to be leveled within the range of the compensator by three screws located on the base. Automatic leveling instruments must be kept upright and should never be carried over your shoulder. They contain a prism that can be damaged if the level is handled carelessly.

> **CAUTION**
>
> Levels, transits, and other optical instruments are sensitive devices. They provide accurate readings only if in good condition, properly calibrated, and used according to the procedures recommended by the manufacturer.

5.1.4 Laser Leveling Instruments

Laser-beam levels (*Figure 15*) can be used to perform all of the tasks that can be performed with a conventional leveling instrument. The laser level does not depend on the human eye. Instead, it emits a high-intensity light beam, which is detected by an electronic sensor (target) at distances up to 1,000 feet.

Both fixed and rotating laser models are available. Rotating models enable one person, instead of two, to perform any layout operation. When a rotating laser is operated in the sweep mode, the head rotates through 360 degrees, allowing the laser beam to sweep multiple sensors placed at different locations.

> **WARNING!**
>
> Some laser units emit a very powerful and highly focused beam. Direct eye exposure to the laser beam can seriously injure the eyes or even cause blindness. This can also happen if the laser beam is reflected into the operator's eyes from a bright object such as a piece of shiny metal or a mirror. Eye damage occurs most often when the laser beam is stationary and a high-powered laser is used.

> **CAUTION**
>
> You must be trained and possess a certification card before you can operate a laser-beam instrument. Government regulations also require that manufacturers provide a warning in their literature regarding the hazards associated with the use of laser instruments.

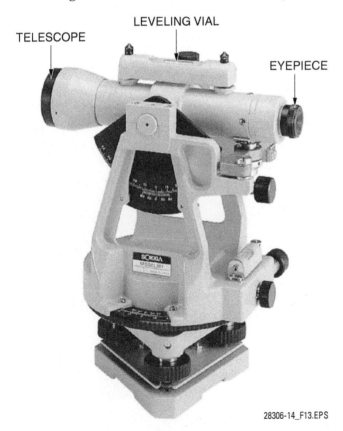

TELESCOPE
LEVELING VIAL
EYEPIECE

28306-14_F13.EPS

Figure 13 Transit or transit level.

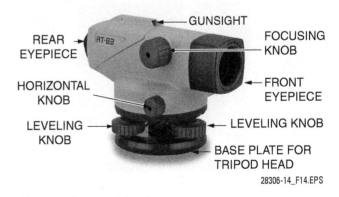

GUNSIGHT
REAR EYEPIECE
FOCUSING KNOB
HORIZONTAL KNOB
FRONT EYEPIECE
LEVELING KNOB
LEVELING KNOB
BASE PLATE FOR TRIPOD HEAD

28306-14_F14.EPS

Figure 14 Automatic level.

5.1.5 Total Station Instruments

The total station (*Figure 16*) is an electronic instrument widely used for site surveying and layout. The instrument is called a total station because it combines the functions of a theodolite, an EDMI, and an internal computer (electronic data collector) into a single instrument that can be used to make both distance and angular measurements. Total stations are commonly used to measure horizontal and vertical angles, measure slope distances, compute the horizontal and vertical components of these distances, and determine the coordinates of the observed points.

Total stations are microprocessor controlled and typically have a large built-in memory capability that is used to store data for thousands of layout points and/or data records. Under software control, two-way data flow between the to-tal station and a remote location is possible via a standard communications interface such as cable, wireless, or Bluetooth®. This capability enables the data collected and stored in the instrument to be recalled and downloaded to a local or remote printer for hard copy or to a computer for calculations. It also enables data that has been processed at a remote location to be sent to the instrument for field use. Because of the wide variety of total stations available, they should be set up, aligned, and operated in accordance with the manufacturer's instructions.

Modern total stations are battery-powered, automatic instruments. They measure the slope distance from the instrument to the reflector, along with the vertical and horizontal angles. The unit's microprocessor then computes the horizontal and vertical components of the slope distance. Using the computed components of the slope distance and the azimuth of the line, the microprocessor determines the north-south and east-west components of the line and the coordinates of the new point. These coordinates are then stored in memory. These instruments

LASER-BEAM LEVEL

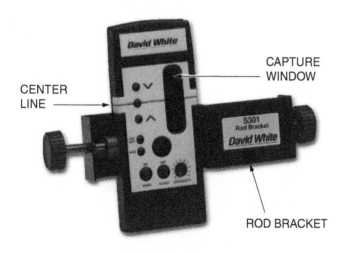

LASER-BEAM DETECTOR

28306-14_F15.EPS

Figure 15 Typical laser-beam level and detector.

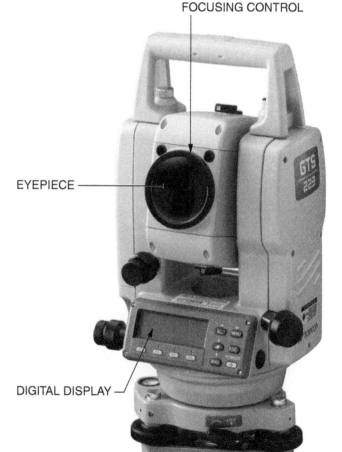

28306-14_F16.EPS

Figure 16 Total station.

18   NCCER – *Masonry Level Three* 28306-14

typically use infrared as the carrier signal for distance measurements.

The distance accuracy of a total station normally depends on the quality of the instrument. Total stations used for construction site layout work can generally measure distances ranging between 2,000–3,600 feet when using a single prism. Manufacturers state the accuracy of their instruments in terms of a constant and scalar instrument error. For example, the distance accuracy may be stated as ± (5 mm +5 parts per million, or ppm). The constant error part (5 mm) does not change regardless of whether a long or short distance is being measured. The scalar error part (+5 ppm) is proportional to the distance being measured. This means that the standard deviation of a single measurement with this instrument is a combination of 5 mm (0.016 foot) and 5 ppm, which varies depending on the distance being measured. For example, in a measurement of 1,000 feet, the error from the scalar part is 0.005 foot. The constant and scalar parts are added to determine the error for a single measurement. For the example of 1,000 feet, the combined error is ±0.021 foot (0.016 foot + 0.005 foot).

Modern total stations have a built-in memory known as a data collector. It is capable of storing data for thousands of layout points and/or data records. The amount of memory in a unit is normally determined by the price of the instrument, with more expensive units having larger memory capabilities. Most total stations can be interfaced with electronic field books and external data collectors to store data for thousands of points or to lay out previously calculated information.

Some of the latest technology improvements for total stations are as follows:

- Units that allow high-accuracy distance measurements to be made with or without the use of prisms or reflective sheet targets
- Wireless control of a total station that allows one person to control the total station operation from a remote handheld controller at the target point itself, or from any other desired location
- Real-time positioning via total stations that are integrated into a global positioning system (GPS)

5.1.6 Care and Handling of Leveling Instruments

Leveling instruments should always be maintained and handled in accordance with the manufacturer's instructions.

General guidelines for the proper use, care, and handling of leveling instruments are as follows:

- Only use an instrument if you know how to operate it.

- Keep the instrument in its closed carrying case when not in use.
- Handle the instrument by its base when removing it from the case or attaching it to the tripod.
- Never force any parts of the instrument. All moving parts should turn freely and easily by hand.
- Keep the instrument clean and free of dust and dirt. Clean the objective and eyepiece lenses using a soft brush or lens tissue. Rubbing with a cloth may scratch the lens coating and impair the view. Clean the instrument with a soft, nonabrasive cloth and mild detergent.
- Do not disassemble the instrument.
- Keep the equipment as dry as possible. If it gets wet, dry it before returning it to its case. It may be necessary to leave it out of its case overnight to dry.
- When moving the instrument over a long distance, by foot or by vehicle, remove it from the tripod and place it in its protective case.
- When moving a tripod-mounted instrument, handle it with care. Carry it only in an upright position. Never carry it over your shoulder or in a horizontal position.
- Periodically have the instrument cleaned, lubricated, checked, and adjusted by a qualified instrument-repair facility or by the manufacturer.

5.2.0 Using Tripods

Levels are mounted on a tripod (*Figure 17*). A tripod consists of a head for attaching the instrument, wooden or metal legs, and metal leg points with foot pads to help force the leg points into the ground. Wing nuts located under the tripod head lock the legs in position. Some tripods have fixed-length legs, while others have adjustable extension legs that help when setting them up on sloping or uneven ground.

Depending on the tripod, one of two types of fastening arrangements is used to fasten the instrument to the tripod head. If the tripod head is threaded, the base of the instrument is screwed directly onto it. If it has a cup assembly, a threaded mounting stud at the base of the instrument is screwed into the cup assembly.

Tripods are often thrown into the backs of vehicles, left out on the ground, exposed to snow and rain, and seldom cleaned. Such misuse can result in damage to or instability of the tripod that can contribute to measurement errors. Guidelines for the use and proper care and handling of tripods are as follows:

- When setting up the tripod, position the tripod legs properly. The legs should have about a 3-foot spread, positioned so that the top of the tripod head is horizontal.
- If the tripod's legs are adjustable, make sure that the leg levers are securely tightened.
- If setting up on dirt, make sure that the tripod points are well into the ground. Apply your full weight to each leg to prevent settlement.
- When setting up on a smooth floor or paved surface, secure the points of the legs by attaching chains between the legs or putting a brick or similar object in front of each leg.
- Attach the instrument to the tripod securely. Do not overtighten the attaching hardware.
- Frequently lubricate the joints and adjustable legs of the tripod using an appropriate lubricant.
- When not in use, protect the head of the tripod from damage.

- When transporting a tripod in a vehicle, never pile other materials on top of the tripod. Make sure to protect it from damage that can be caused by shifting equipment or materials.
- Keep the tripod clean and dry.
- When not in use, store the tripod in its protective case.

5.3.0 Using Leveling Rods

Two people are required when a conventional leveling instrument is used; the first operates the instrument and the second holds a vertical measuring device, called a Leveling rod (*Figure 18*), in the area where the grade or elevation is being checked. Leveling rods are made in many sizes, shapes, and colors. They can be made of wood, fiberglass, metal, or a combination of these materials. Leveling rods consist of two or more movable sections, allowing the rod to be adjusted to different lengths. Telescoping rods are also available.

Many styles of leveling rods are given geographic names, such as Philadelphia rods, Chicago rods, San Francisco rods, and Florida rods. The Philadelphia rod is a two-section rod with scales on both the front and back, which can be extended to about 13 feet. The Chicago and San Francisco rods consist of three sliding sections, with the Chi-

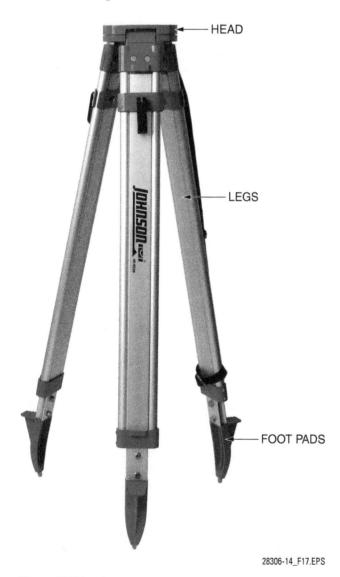

HEAD

LEGS

FOOT PADS

28306-14_F17.EPS

Figure 17 Tripod.

STANDARD ROD

TELESCOPING ROD

28306-14_F18.EPS

Figure 18 Leveling rods.

cago rod being 12 feet long and the San Francisco rod available in several lengths. The Florida rod is a 10-foot long rod graduated with alternating 0.10-foot-wide red and white stripes.

There are two types of leveling rods: direct-reading architect's rods and engineer's rods. An architect's rod is graduated in feet, inches, and eighths of an inch (*Figure 19*). As shown, each line and space on an architect's rod is ⅛ inch wide. An engineer's rod is marked in feet, tenths of a foot, and hundredths of a foot. As shown in *Figure 19*, each line and space marked on an engineer's rod is ¹⁄₁₀₀ foot wide. Metric rods are also available.

There are several accessories used with leveling rods. A movable red-and-white metal disk called a target (*Figure 20*) is used to help make more precise rod readings. The target's vernier scale is set parallel to and beside the primary scale of the leveling rod. Its use enables readings to the nearest sixty-fourth of a foot (architect's rod) or nearest thousandth of a foot (engineer's rod).

The target is moved up or down on the rod until the 0 on the vernier scale is lined up with the crosshairs of the leveling instrument. The target is then clamped in place. To read the vernier scale, count the number of vernier divisions up from the 0 (index mark) until one of the vernier divisions lines up exactly with a division on the rod scale itself. This number is added to the last division on the rod, just below the vernier's index mark.

A bull's-eye rod level is normally attached to a leveling rod for use in keeping the rod plumb for sighting and while the reading is being taken.

leveling rods are made to withstand the severity of everyday use, but they must be handled, stored, and used properly to avoid unnecessary damage. A damaged rod can contribute to errors. Guidelines for the proper care and handling of leveling rods are as follows:

- Clean the face, joints, and bottom of the rod frequently during use.
- Avoid touching the face of the rod. Over time, this can cause the numbers and markings to be worn off.
- Make sure all the rod hardware is securely fastened.

Leveling Rod

What is a quick way to tell if a leveling rod is a feet-and-inches rod (architect's rod) or a decimal-foot rod (engineer's rod)?

- When using a telescoping-type leveling rod, make sure that it is fully extended. Failure to extend a rod fully will result in major errors.
- Never throw a leveling rod into the back of a vehicle or leave it sticking out from a vehicle. When not in use, store it in its protective case.

5.4.0 Performing the Initial Setup and Adjustment of Leveling Instruments

The initial setup of a leveling instrument such as a builder's level, automatic level, or transit, is completed as follows:

Step 1 Select a location to set up the instrument, so that its horizontal line of sight will be at a correct height to intercept the level rod, as shown in *Figure 21*.

Step 2 Set up a tripod, making sure to spread its legs wide enough (at least 3 feet between the legs) to provide a firm foundation for the instrument. Push the legs firmly into the ground and fasten them securely. If setting up on sloping ground, make sure to place one leg of the tripod into the slope. Also, make sure the head of the tripod is horizontal. If the tripod head is too far out of level, there is little chance of correctly leveling the instrument on top of it.

Step 3 Carefully remove the leveling instrument from its case and loosen its horizontal clamp screw (*Figure 22*). If using a transit, also loosen the vertical clamp screw and place the telescope lock lever in the closed position. Attach the instrument securely to the tripod.

Step 4 If setting up over a point, use a plumb bob to center on the exact point. Otherwise, proceed to Step 5. To hang the plumb bob, attach the cord to the hook provided on the tripod. Move the tripod and attached instrument over the point, making sure that the tripod is set up firmly. Shift the instrument on the tripod head until the plumb bob is directly over the point.

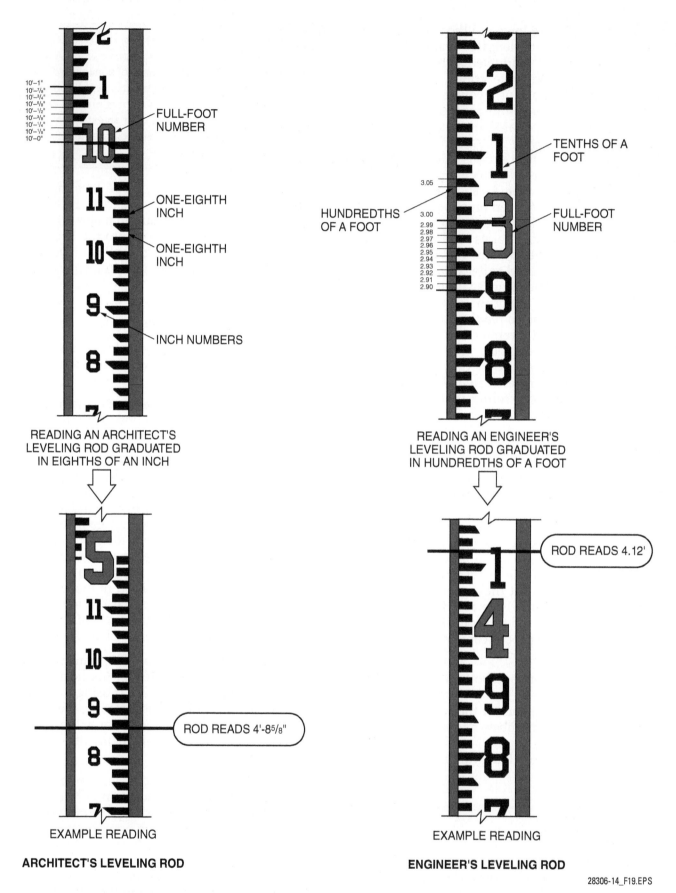

Figure 19 Reading a leveling rod.

28306-14_F19.EPS

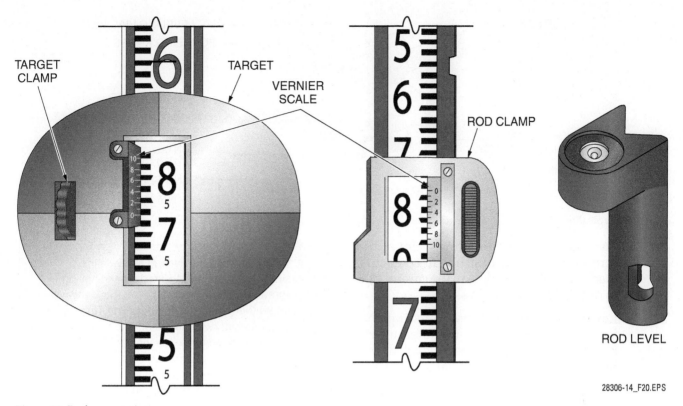

Figure 20 Rod accessories.

28306-14_F20.EPS

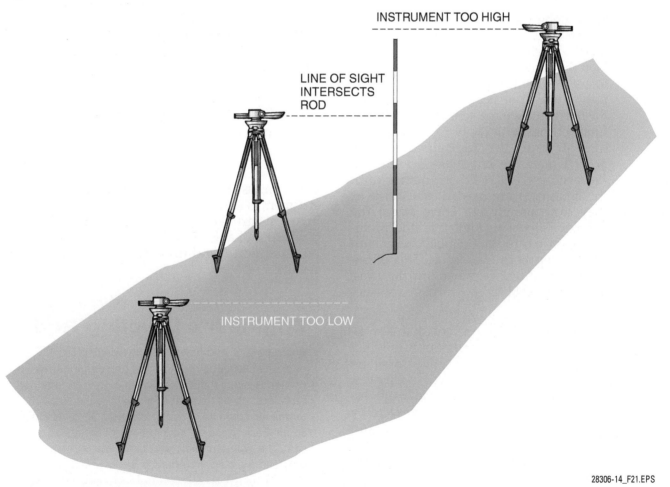

28306-14_F21.EPS

Figure 21 Set up the instrument so that the line of sight is at the correct level.

 28306-14 Site Layout—Distance Measurement and Leveling Module Six 23

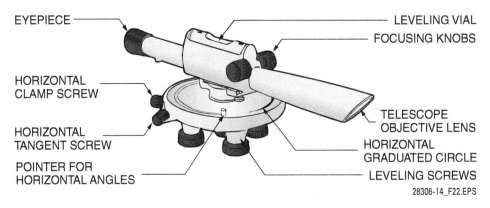

EYEPIECE

LEVELING VIAL

FOCUSING KNOBS

HORIZONTAL CLAMP SCREW

HORIZONTAL TANGENT SCREW

POINTER FOR HORIZONTAL ANGLES

TELESCOPE OBJECTIVE LENS

HORIZONTAL GRADUATED CIRCLE

LEVELING SCREWS

28306-14_F22.EPS

Figure 22 Typical builder's level operator controls.

Step 5 Turn down the instrument leveling screws by hand just until firm contact is made with the tripod head. Be careful not to overtighten the screws.

Step 6 Level the instrument by adjusting the leveling screws per the manufacturer's instructions. When properly leveled, the bubble in the leveling vial should remain exactly centered as the telescope is rotated in a complete circle around its base.

Step 7 Sight along the top of the telescope tube to aim the telescope in the direction of a distant leveling rod or other target, then look through the telescope and adjust the focus. When the crosshairs (*Figure 23*) are positioned on or near the target, tighten the horizontal clamp screw and make the final settings with the horizontal tangent screw to bring the crosshairs exactly on point. Focus the crosshairs by turning the eyepiece one way or another until the crosshairs are as dark and crisp as they can possibly be. Then, adjust the telescope's focusing knob until the graduations on the rod are legible, sharp, and crisp. Keep both eyes open. This eliminates squinting, does not tire the eyes, and gives the best view through the telescope.

Failure to focus the telescope crosshairs properly will cause parallax. Parallax occurs when there is an apparent movement in the crosshairs on the rod or object being viewed as the eye moves. If this occurs when reading a level rod, major errors can occur. Parallax can be easily checked by looking at the rod or object being viewed and moving slightly from side to side while looking at the crosshairs. If the crosshairs stay on the same spot, no parallax exists.

NOTE

When turning two leveling screws simultaneously (as required when leveling four-screw and three-screw instruments), always rotate them in opposite directions. Turn one in a counterclockwise direction and the other in a clockwise direction when viewed from above. Turn them at the same rate. When rotating the two leveling screws, the spirit-level bubble will always follow the direction of the left-hand thumb. That is, if the left thumb is turning the leveling screw in a counterclockwise direction, the bubble will move towards the left. Note that the left-thumb rule also applies if the left hand is used to adjust a single leveling screw, such as is necessary when leveling a three-screw system.

5.5.0 Testing the Calibration of Leveling Instruments

Field testing a leveling instrument for correct calibration and adjustment should be done when using an instrument for the first time or if the instrument is suspected of being out of adjustment. Two tests are recommended: a horizontal-crosshair test and a line-of-sight test. Each test should be repeated several times to make sure of your results.

Engineer's Leveling Rod

Some people who have difficulty reading an engineer's leveling rod find that thinking in terms of money helps them. For example, the red foot numbers can be thought of as dollars; the black tenths numbers as 10 cents, 20 cents, 30 cents and so on. Each space or black line width counts as one cent. The point on the longer black line midway between the black numbers is five cents.

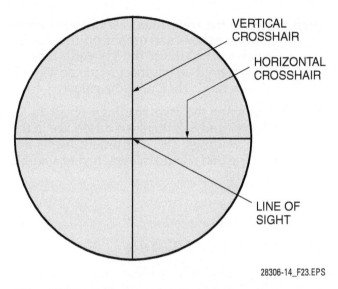

Figure 23 Crosshairs seen when looking through a telescope.

28306-14_F23.EPS

5.5.1 *Horizontal-Crosshair Test*

The object of the horizontal-crosshair test is to ensure that the instrument's horizontal crosshair is in a plane that is perpendicular to the vertical axis of the instrument. With a properly adjusted instrument, any part of the horizontal crosshair can be placed on the object or point being viewed with the telescope and the instrument will still provide an accurate reading.

The horizontal-crosshair test is simple to perform. First, level the instrument, then sight the horizontal-crosshair reticule on a distant nail head or other well-defined point (*Figure 24*). Once the crosshair is placed on the point, turn the instrument's horizontal tangent screw so that the instrument slowly rotates about its vertical axis. The crosshair should stay fixed on the point as the instrument is rotated. If any part of the crosshair moves above or below the reference point, the instrument needs adjustment and should be returned to a repair facility.

5.5.2 *Line-of-Sight Test*

The instrument's line of sight should be parallel to a horizontal line. The horizon line is represented by the axis of the leveling vial, or builder's level, or the axis through the compensator for an automatic level. There are several methods for testing an instrument's line of sight. One method is briefly outlined here.

Step 1 Locate an area to set up a tripod that is about 10 feet away from a wall or other permanent object and about 75–100 feet away (at a 90- to 180-degree angle) from another wall or permanent object.

Step 2 Mount an automatic or other high-accuracy instrument that is known to be in calibration and adjustment onto the tripod. Make sure the tripod is on a firm base, then level the instrument according to the manufacturer's instructions.

Step 3 Mark the locations of the tripod's leg points. These marks will be used later in this procedure.

Step 4 Sight the instrument on the near wall and place a section of rod ribbon on the wall so that the horizontal crosshair of the instrument intersects the middle of the ribbon. Mark a line-of-sight reference point on the ribbon.

Step 5 Rotate the instrument 90 to 180 degrees and sight on the far wall. Place a section of rod ribbon on the far wall so that the horizontal crosshair of the instrument intersects the middle of the ribbon. Mark the line-of-sight reference point on the ribbon.

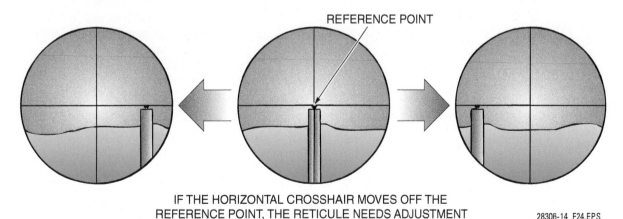

IF THE HORIZONTAL CROSSHAIR MOVES OFF THE
REFERENCE POINT, THE RETICULE NEEDS ADJUSTMENT

28306-14_F24.EPS

Figure 24 Horizontal-crosshair test.

Step 6 Carefully remove the calibrated instrument from the tripod, then mount and level the instrument to be checked onto the tripod. Make sure the positions of the tripod leg points have not moved from the locations marked in Step 3.

Step 7 Sight the instrument on the ribbon reference mark on the near wall. Have a person hold an engineer's rule on the reference mark, then read the rule.

Step 8 Rotate the instrument 90 to 180 degrees and sight the instrument on the ribbon reference mark on the far wall. Have a person hold an engineer's rule on the reference mark, then read the rule.

Step 9 Compare the two readings obtained in Steps 7 and 8. If there is a difference in the two readings, the instrument needs adjusting and should be returned to a repair facility.

Removing a Leveling Instrument from Its Case

When removing a leveling instrument from its case, pay close attention to how it sits in the case. The more moving parts an instrument has, the harder it will be to fit it back into the case if all the parts are not aligned properly.

Additional Resources

Basic Surveying Technology, Latest Edition. Stillwater, OK: The Mid-America Vocational Curriculum Consortium, Inc.

Code Check, Latest edition. Newtown, CT: Taunton Press.

Measuring, Marking, and Layout. Newtown, CT: Taunton Press.

Surveying Principles and Applications. 2008. Barry F. Kavanagh. Upper Saddle River, NJ: Prentice Hall.

Surveying with Construction Applications. 2009. Barry F. Kavanagh. Upper Saddle River, NJ: Prentice Hall.

5.0.0 Section Review

1. The most common magnification power used in builder's levels is _____.

    a. 12 power (12x)
    b. 18 power (18x)
    c. 20 power (20x)
    d. 32 power (32x)

2. Tripod legs are anchored to the ground using _____.

    a. metal leg points with foot pads
    b. plastic stakes
    c. rubber foot pads
    d. steel anchors with nylon loops

3. The width of each line and space on an architect's rod is _____.

    a. ¹⁄₃₂ inch
    b. ⅛ inch
    c. ¹⁄₁₀₀ inch
    d. ¹⁄₁₀ inch

4. Improperly focused telescope crosshairs will result in _____.

    a. deviation
    b. parallax
    c. triangulation
    d. stereopsis

5. A leveling instrument's line of sight should be _____.

    a. parallel to a horizontal line
    b. parallel to a vertical line
    c. in a plane that is perpendicular to the vertical axis of the instrument
    d. in a plane that is parallel to the vertical axis of the instrument

SECTION SIX

6.0.0 BASICS OF DIFFERENTIAL LEVELING

Objective

Explain the basics of differential leveling.
  a. Define differential-leveling terminology.
  b. Explain the differential-leveling procedure.
  c. Explain how field notes are recorded and used.

Trade Terms

Backsight (BS): A reading taken on a leveling rod held on a point of known elevation to determine the height of the leveling instrument.

Foresight (FS): A reading taken on a leveling rod held on a point in order to determine a new elevation.

Height of instrument (HI): The elevation of the line of sight of the telescope relative to a known elevation. It is determined by adding the backsight elevation to the known elevation.

Station: An instrument setting location in differential leveling.

Temporary benchmark: A point of known (reference) elevation determined from benchmarks through leveling, and permanent enough to last for the duration of a project.

Turning point (TP): A temporary point within an open or closed differential-leveling circuit whose elevation is determined by differential leveling. It is normally the leveling-rod location. Its elevation is determined by subtracting the foresight elevation from the height-of-the-instrument elevation.

The process of differential leveling is based on the measurement of vertical distances from a level line. Elevations are transferred from one point to another by using a leveling instrument to first read a rod held vertically on a point of known elevation, then to read a rod held on a point of unknown elevation (*Figure 25*).

Following this, the unknown elevation is calculated by adding or subtracting the readings. To determine elevations between two or more widely separated points or points on a sloping terrain, several repetitions of the same basic differential-leveling process are performed.

6.1.0 Understanding Differential-Leveling Terminology

Before describing the differential-leveling process, it is important to first review and/or introduce some related terms:

- *Elevation* – Elevation is the vertical distance above a datum point. For leveling purposes, a datum is normally based on the ocean's mean sea level (MSL). At numerous locations throughout the United States, the government has installed monuments marked with known elevations referenced to MSL. When readily available, such a monument would be used as the elevation reference. When no monument is readily available, which is the case at most construction sites, a point is established and an elevation arbitrarily assigned. Typically, an elevation of 100 feet, 500 feet, or 1,000 feet is used.
- *Benchmark (BM)* – A benchmark is a relatively permanent object with a known elevation located near or on a site. It can be iron stakes driven into the ground, a concrete monument with a brass disk in the middle, or a chiseled mark at the top of a concrete curb.
- *Backsight (BS)* – A reading taken on a leveling rod held on a point of known elevation to determine the height of the leveling instrument.
- *Foresight (FS)* – A reading taken on a leveling rod held on a point in order to determine the elevation.
- *Height of instrument (HI)* – The elevation of the line of sight of the telescope above the datum plane. It is determined by adding the backsight elevation to the known elevation.
- *Turning point (TP)* – A temporary point whose elevation is determined by differential leveling. The turning-point elevation is determined by subtracting the foresight elevation from the height-of-the-instrument elevation.
- *Closed loop* – Making a traverse consisting of a series of differential measurements that return to the point from which they began.

6.2.0 Using the Differential-Leveling Procedure

Before beginning the leveling process, select a benchmark that is closest to your work. If the exact location of the closest benchmark is not known, refer to the site plan. Determine the longest reasonable distance between the measurement points in order to shorten the amount of work that must be done by minimizing instrument setups.

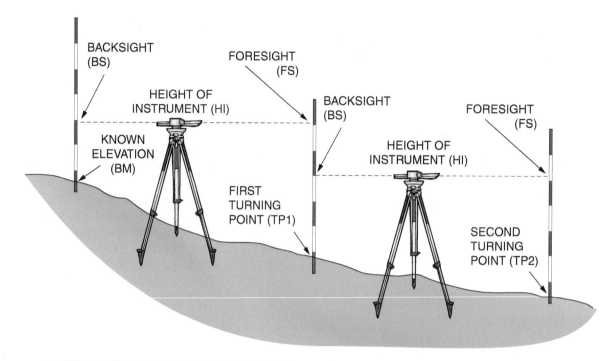

KNOWN ELEVATION (BM) + BACKSIGHT (BS) = HEIGHT OF INSTRUMENT (HI)
HEIGHT OF INSTRUMENT (HI) – FORESIGHT (FS) = TURNING POINT (TP) ELEVATION

28306-14_F25.EPS

Figure 25 Differential-leveling relationships.

Note that some sites are relatively flat, while others have steep slopes. Regardless of the slope involved, the procedure for leveling is done in the same way. However, when leveling at a site with a very steep slope, the procedure becomes more time consuming. This is because the line of sight of the instrument relative to intercepting a leveling rod is shorter, thus requiring more setups to cover the distance.

The differential-leveling procedure generally involves two people working together and communicating with each other. One person is designated as the rod person and the other the instrument person. Depending on the complexity of the task, recording of the collected measurement data in the field notes may be done by either person, both, or sometimes by a third person. An example of a typical differential-leveling procedure is described below, and its path (traverse) is shown in *Figure 26*.

Step 1 The procedure begins by recording the benchmark (BM) and its known elevation in the station and elevation columns of the field notes. For the example shown, the entries are BM (station) and 1,000 feet (elevation).

Step 2 The instrument person sets up the leveling instrument at Station 1 (STA 1) in preparation for the first measurement. It should be located so that a level rod placed on

the BM is in the line of sight of the level and the rod can be clearly read. Note that this same point should also allow the line of sight of the level to intercept a level rod held on the proposed location of the first turning point (TP1). Set this point equally distant between the two points and no farther away than 150–200 feet from either point of measurement.

Step 3 While the rod person holds the level rod plumb on the BM, a backsight rod reading is taken, then recorded in the field notes. In this example, the BS reading of 7.77 feet is recorded in the BS (+) column of the notes. Following this, the height of the instrument (HI) is calculated and recorded in the HI column of the field notes. In this example, the HI is recorded as 1,007.77 feet (HI = BM + BS = 1,000 feet + 7.77 feet).

Step 4 The rod person paces or otherwise measures the approximate distance between the BM and the leveling instrument, and then advances an equal distance beyond the level in the desired direction of the first turning point (TP1). This point must be located such that when the level rod is placed on it, the line of sight of the leveling instrument will intercept the rod. The rod person selects an appropriate solid

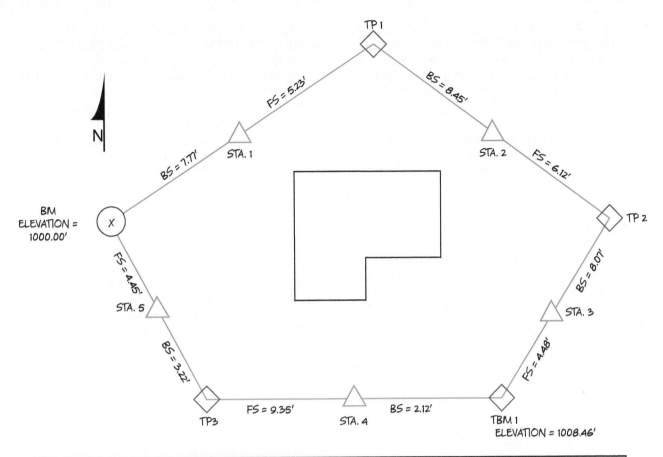

STATION (STA)	BS (+)	HI	FS (-)	ELEVATION
BENCH MARK (BM) TO TEMPORARY BENCH MARK 1 (TBM 1)				
BM	7.77'			1000.00'
STA. 1		1007.77'		
TP 1	8.45'		5.23'	1002.54'
STA. 2		1010.99'		
TP 2	8.07'		6.12'	1004.87'
STA. 3		1012.94'		
TBM 1	2.12'		4.48'	1008.46'
STA. 4		1010.58'		
TP 3	3.22'		9.35'	1001.23'
STA. 5		1004.45'		
BM			4.45'	1000.00'
Σ CHECK	29.63'		29.63'	

DIFFERENCE = 0.00'

Figure 26 Differential-leveling traverse and related field-notes data.

28306-14_F26.EPS

surface such as a sidewalk or large rock for the turning point. Note that an unmarked point on grass or soil should never be used as a turning point. If no natural solid object is available, a metal turning pin, railroad spike, or wooden stake driven in the ground can serve as a turning point. When a turning point on a solid surface such as a sidewalk or pavement is used, the point should be marked and identified by the turning-point number.

Step 5 While the rod person holds the level rod plumb on TP1, a foresight rod reading is taken, then recorded in the field notes. In this example, the FS reading of 5.23 feet is recorded in the FS (–) column. Following this, the elevation of TP1 is calculated and recorded in the elevation column of the field notes. In this example, the elevation is recorded as 1,002.54 feet (turning-point elevation = HI – FS = 1,007.77 feet – 5.23 feet).

Step 6 In preparation for the next set of backsight and foresight readings, the instrument person moves the leveling instrument to a point beyond TP1 and sets up the instrument at Station 2, which is approximately midway between TP1 and TP2.

Step 7 Once the leveling instrument is set up, backsight and foresight readings are taken between the points TP1 and TP2 in the same way as previously described in Steps 3 through 5, with the following exceptions. The known elevation of TP1 is used instead of the BM to calculate the instrument height (HI) at Station 2. Then, the new HI and the foresight reading on TP2 are used to calculate the elevation of TP2.

Step 8 Steps 3 through 7 are repeated as necessary to complete the differential-measurement loop from the TP2 to the temporary benchmark TBM1, then back via TP3 to the starting point at BM.

In the example shown, the leveling traverse is run back to the starting point at BM. This is called closing the loop or a closed loop. Any leveling survey should close back either on the starting benchmark or on some other point of known elevation in order to provide a check of the measurements taken.

Leveling notes should always be checked for arithmetic or calculator input errors. This is done by simply summing the backsight (BS) and foresight (FS) columns and comparing the difference

between them with the starting and ending elevations. As shown in *Figure 26*, the difference between the BS sum and the FS sum is 0 feet.

The difference between the starting elevation of 1,000 feet and the ending elevation of 1,000 feet is also 0 feet. Since the differences are equal, the arithmetic checks and the loop is properly closed. An error would exist if the differences were not equal or were not within the established accuracy standard or tolerances specified for the project. Using the same example, the calculations for the traverse between the BM and TBM1 can be checked in the same manner. This is shown in *Figure 27*.

When performing differential leveling, it is easy to make mistakes. However, mistakes can be eliminated by constantly checking and rechecking your work. Some common mistakes to avoid when performing differential leveling include the following:

• Backsight and foresight distances not equal
• Instrument not leveled
• Rod not plumb (if not using a level, the rod should be rocked forward and backward and from side to side, then the smallest reading recorded)
• Sections of an extended leveling rod not adjusted properly
• Dirt or ice accumulated on the base of the rod
• Misreading the rod
• Recording incorrect values in the field notes
• Moving the position of a turning point between backsight and foresight readings

6.3.0 Using Field Notes

In the differential-leveling procedure described in the previous section, constant reference has been made to recording measurement information in field notes. Writing a legible and accurate set of notes in a field book, whether in a print format (*Figure 28*) or an electronic one (*Figure 29*), is just as important as doing the leveling or layout work itself. This is because field notes provide a historical record of the work performed. They serve as a reference should there ever be a question about the correctness or integrity of your work, especially in a court of law. Field notes should leave no room for misinterpretation. Your notes should be written so that others can understand your work.

General guidelines for writing and keeping field notes include:

• All field notebooks should contain the name, address, and phone number of the owner.
• All pages should be numbered, and there should be a table-of-contents page.

STATION (STA)	BS (+)	HI	FS (−)	ELEVATION
BENCH MARK (BM) TO TEMPORARY BENCH MARK 1 (TBM 1)				
BM	7.77'			1000.00'
STA. 1		1007.77'		
TP 1	8.45'		5.23'	1002.54'
STA. 2		1010.99'		
TP 2	8.07'		6.12'	1004.87'
STA. 3		1012.94'		
TBM 1			4.48'	1008.46'
	24.29'		15.83'	

MATH CHECK: 1000.00
              +   24.29
              ───────────
                 1024.29
              −   15.83
              ───────────
                 1008.46

28306-14_F27.EPS

Figure 27 Example of a math check.

- Make neatly printed entries in the book using a suitable sharp pencil with hard lead (3H or 4H). Never use cursive script in a field book.
- Begin each new task on a new page. The left-hand pages are generally used for entering numerical data and the right-hand pages are for making sketches and notes.
- Always record the date, time, weather conditions, names of crew members and their assignments, and a list of the equipment used.
- Record each measurement in the field book immediately after it is taken. Do not trust it to memory.
- Record data exactly. Ideally, the data should be checked by two crew members at the time it is recorded.
- Make liberal use of sketches if needed for clarity. They should be neat and clearly labeled, including the approximate north direction. Do not crowd the sketches.

- Never erase. If a mistake is made, draw a single line through the incorrect entry and write the correct data above it.
- Draw a diagonal across the page and mark the word VOID on the tops of pages that, for one reason or another, are invalid. When marking the page, be careful not to make the voided information unreadable. The date and name of the person voiding the page should also be recorded.
- Mark the word COPY on the top of copied pages. Refer to the name and page number of the original document.
- Always keep the field book in a safe place on the job site. At night, lock it up in a fireproof safe. Original field books should never be destroyed, even if copied for one reason or another.

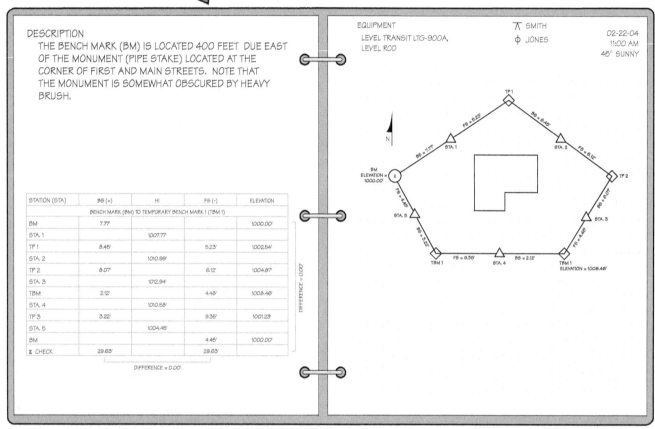

DESCRIPTION

THE BENCH MARK (BM) IS LOCATED 400 FEET DUE EAST
OF THE MONUMENT (PIPE STAKE) LOCATED AT THE
CORNER OF FIRST AND MAIN STREETS. NOTE THAT
THE MONUMENT IS SOMEWHAT OBSCURED BY HEAVY
BRUSH.

STATION (STA)	BS (+)	HI	FS (-)	ELEVATION
BENCH MARK (BM) TO TEMPORARY BENCH MARK 1 (TBM 1)				
BM	7.77'			1000.00'
STA. 1		1007.77'		
TP 1	8.45'		5.23'	1002.54'
STA. 2		1010.99'		
TP 2	8.07'		6.12'	1004.87'
STA. 3		1012.94'		
TBM	2.12'		4.48'	1008.46'
STA. 4		1010.58'		
TP 3	3.22'		9.35'	1001.23'
STA. 5		1004.45'		
BM			4.45'	1000.00'
Σ CHECK	29.63'		29.63'	

DIFFERENCE = 0.00'

DIFFERENCE = 0.00'

EQUIPMENT
LEVEL TRANSIT LTG-900A,
LEVEL ROD

SMITH
JONES

02-22-04
11:00 AM
45° SUNNY

28306-14_F28.EPS

Figure 28 Example of print field notes.

28306-14_F29.EPS

Figure 29 Electronic field book.

Side Shots

Some surveyors and masons take intermediate readings to points that are not part of the main differential-leveling loop. These readings are called side shots. It is important to make sure that your differential-leveling loop is properly closed before making any side shots. After closing the loop, side shots can be taken from established turning points.

Additional Resources

Construction Surveying and Layout: A Step-by-Step Engineering Methods Manual. 2002. Wesley G. Crawford. West Lafayette, IN: Creative Construction Publishing.

Measuring, Marking, and Layout. Newtown, CT: Taunton Press.

Surveying Principles and Applications. 2008. Barry F. Kavanagh. Upper Saddle River, NJ: Prentice Hall.

Surveying with Construction Applications. 2009. Barry F. Kavanagh. Upper Saddle River, NJ: Prentice Hall.

6.0.0 Section Review

1. A relatively permanent object with a known elevation located near or on a site is called a _____.

   a. turning point
   b. benchmark
   c. closed loop
   d. control point

2. Compared with the backsight distance, the foresight distance should be _____.

   a. equal
   b. proportional
   c. greater
   d. less

3. If a mistake is made in the field notes, correct it by _____.

   a. erasing the incorrect entry and writing the correct data over the erasure
   b. circling the incorrect entry and writing the correct data next to it
   c. drawing a single line through the incorrect entry and writing the correct data above it
   d. underlining the incorrect entry and adding an explanatory note, then writing the correct data on the next line

7.0.0 LEVELING APPLICATIONS

Objective

Identify leveling applications.
  a. Explain how to transfer elevations up a structure.
  b. Explain profile, cross-section, and grid leveling.

Performance Task

Use differential-leveling and distance-measurement procedures to transfer elevations up a structure.

Trade Term

Earthwork: All construction operations connected with excavating (cutting) or filling earth.

In addition to setting benchmarks or grade stakes, there are many applications involving leveling, including:

- Transferring elevations up a structure
- Profile leveling
- Cross-section leveling
- Grid leveling

7.1.0 Transferring Elevations up a Structure

When constructing multistory buildings and other tall structures, ground elevations frequently need to be transferred vertically up the structure as it is being built to maintain the design grades. One method for accomplishing this involves the use of both differential-leveling and taping skills.

The process begins by first establishing a temporary benchmark (see TBM1 in *Figure 30*) with a known elevation, at the base of the structure, by using differential-leveling methods. Following this, a tape is used to measure up from TBM1 the vertical distance needed to establish a second temporary benchmark (TBM2) on the floor or level of the structure on which the elevation(s) are needed. Once TBM2 has been established on the upper level, a leveling instrument can be set up and a height of instrument (HI) calculated in the normal way by backsighting on and reading a rod held on TBM2 (HI = BS + TBM2 elevation). Note that the elevation of TBM2 is equal to the eleva-

tion of TBM1 plus the tape distance. Following this, any subsequent leveling tasks are performed on the upper floor or level just as if the instrument were placed on the ground.

Sometimes, points that need elevations may be above the line of sight of the instrument, such as with elevations for the bottom of a beam and ceiling levels. Taking the elevation in these instances requires that the level rod be held upside down and placed against the beam or ceiling. This results in a positive foresight reading that must be added to the HI rather than subtracted as is normally done.

7.2.0 Performing Profile, Cross-Section, and Grid Leveling

Profile, cross-section, and grid leveling are all methods used to determine the profile of a terrain or surface. These methods are briefly described here. However, procedures for performing these leveling methods are beyond the scope of this module. Such procedures can be found in most surveying or field-engineering texts or reference books, some of which are listed in *Additional Resources and References* in this module.

7.2.1 Profile Leveling

Profile leveling is the process of determining the elevation of a series of points along the ground at approximately uniform intervals along a continuous center line, such as when determining the profile of the ground along the center line of a highway. The method and calculations used to perform profile leveling are the same as those used for the differential-leveling process.

Profile leveling consists of making a series of differential-leveling measurements in the usual manner while traversing the project center line. However, from the instrument's HI position at each station, a series of additional intermediate foresight readings are taken on several points, called profile points, along the center line to determine their elevations. These readings are taken at regular intervals or where the terrain changes abruptly, causing sudden changes in elevations to occur. After the fieldwork has been completed, this data can be used to plot the profile of the land along the center line.

7.2.2 Cross-Section Leveling

Cross-section leveling is basically the same as profile leveling. The difference is that rather than determining intermediate elevations of several profile points along a center line, cross-section leveling determines elevations for several profile points that are perpendicular to the center line. Note that for a specific project, there is only one center-line profile but there can be numerous cross sections. Cross-section profile plots derived from cross-section leveling data are used for estimating quantities of earthwork to be performed.

7.2.3 Grid Leveling

Grid leveling is one process that can be used to determine the existing topography of a building lot or other land area. It is also used when neces-sary to determine earthwork quantities related to an excavation (pit) or a mound. This is normally done when it is necessary to calculate the volume of material that has been excavated or placed.

Basically, this method requires that a rectangular profile grid be laid out on the building lot with grid intersections occurring at regular intervals spaced about 50 feet or 100 feet apart. Following this, differential leveling is done in a similar manner as for profile leveling, except that more intermediate elevation readings can be taken from one instrument position.

When performed in conjunction with earthwork, grid leveling is normally done both before and after the earthwork is accomplished. The difference between the original and final elevations is then used in a volume formula to calculate the volume of material excavated or filled.

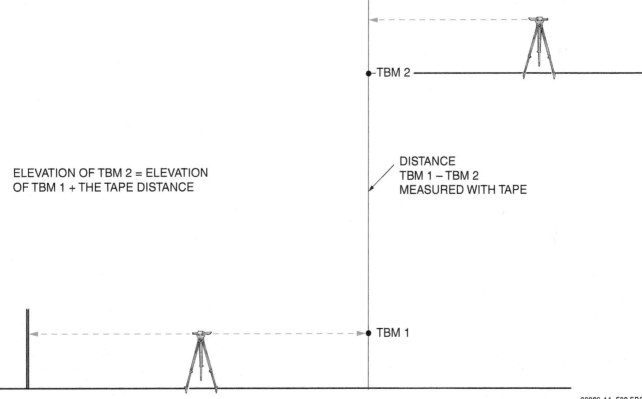

ELEVATION OF TBM 2 = ELEVATION OF TBM 1 + THE TAPE DISTANCE

TBM 2

DISTANCE TBM 1 – TBM 2 MEASURED WITH TAPE

TBM 1

28306-14_F30.EPS

Figure 30 Transferring elevations up a structure.

Additional Resources

Architectural Graphic Standards. 1998. The American Institute of Architects. New York: John Wiley & Sons, Inc.

Construction Surveying and Layout: A Step-by-Step Engineering Methods Manual. 2002. Wesley G. Crawford. West Lafayette, IN: Creative Construction Publishing.

Measuring, Marking, and Layout. Newtown, CT: Taunton Press.

Surveying Principles and Applications. 2008. Barry F. Kavanagh. Upper Saddle River, NJ: Prentice Hall.

Surveying with Construction Applications. 2009. Barry F. Kavanagh. Upper Saddle River, NJ: Prentice Hall.

7.0.0 Section Review

1. When transferring elevations up a structure, the first step is to _____.

   a. calculate a height of instrument
   b. measure the approximate distance between the benchmark and the leveling instrument
   c. take a foresight rod reading
   d. establish a temporary benchmark

2. The process of determining the elevation of a series of points along the ground at approximately uniform intervals along a continuous center line is called _____.

   a. cross-section leveling
   b. grid leveling
   c. differential leveling
   d. profile leveling

8.0.0 LAYING OUT BUILDING CORNERS

Objective

Describe how to lay out building corners.
  a. Explain how to construct batter boards.
  b. Describe how to use the 3-4-5 rule.

Performance Task

Check and/or establish 90-degree angles using the 3-4-5 rule.

Masons use their site-layout skills and tools to lay out right-angle building corners. Construction layout lines for corners can be prepared using batter boards or by using a mathematical formula commonly called the 3-4-5 rule.

8.1.0 Using Batter Boards

On some construction jobs, wooden frameworks called batter boards (*Figure 31*) are used to establish building and other construction layout lines. Used in pairs and with a string or wire attached and stretched between them, batter boards are used to create lines that mark the boundaries of a building or the center of column footings. They can also be used to set reference elevations such as elevations to the top of a footing or to the finish floor level of a building.

A batter board usually consists of a 2 × 4 or 2 × 6 horizontal board, called a ledger board, that is nailed or otherwise attached to stakes driven into the ground. Typically, the stakes are made from 2 × 4s.

The placement of batter boards is normally done after the exact locations of the building corners have been established. Placement involves driving the ledger-board support stakes firmly into the ground behind each building corner at a distance that allows enough working room between the batter boards and the immediate construction area. Depending on the type of job and the excavation equipment, this distance could be anywhere between 4 feet and 20 feet.

If it is necessary to drive the stakes into soft soil, or if the stakes extend 3 feet or more out of the ground, they should be braced to prevent any movement. Following this, a leveling instrument is used to sight and mark the stakes at the required elevation. Then, each ledger board is fastened to the outside of its support stakes so that its top edge is on the elevation mark. It is important to make sure that when all the related batter boards have been installed, the tops of the ledger boards are level with one another.

Once the batter boards are installed, the building corners can be transferred to the batter boards. One method for doing this involves the use of a plumb bob and nylon string. This is done by stretching a nylon string (line) between two opposite batter boards and directly over the building corner stakes. The plumb bob is used to locate the exact position of the line by suspending it directly over the center marker on each corner stake. When the line is accurately located on the two batter boards, a shallow saw cut (kerf) is made at this point on the outside top edge of each ledger board. This prevents the line from moving when being stretched and secured. The taut line is placed in the kerf and secured with a nail driven into the back of each ledger board. The procedure is repeated until all the building lines are in place.

After all the lines are installed between the batter boards, measurements should be made between the lines to make sure they are accurate. Also, the diagonals across the lines should be measured to make sure that they are equal. Equal-length diagonals indicate that the lines are square.

8.2.0 Using the 3-4-5 Rule

The 3-4-5 rule has been used in construction for centuries. It is a simple method that can be used for laying out or checking a 90-degree angle that does not require the use of a builder's level or transit. The numbers 3-4-5 represent dimensions in feet that describe the sides of a right triangle. The 3-4-5 rule is based on the Pythagorean theorem. It states that in any right triangle, the square of the longest side, called the hypotenuse (C), is equal to the sum of the squares of the two shorter sides (A and B). Stated mathematically:

$$C^2 = A^2 + B^2$$

Accordingly, for the 3-4-5 right triangle:

$$5^2 = 3^2 + 4^2$$
$$25 = 9 + 16$$
$$25 = 25$$

This theorem also applies if each number (3, 4, and 5) is multiplied by the same number. For example, if multiplied by the constant 3, it becomes a 9-12-15 triangle.

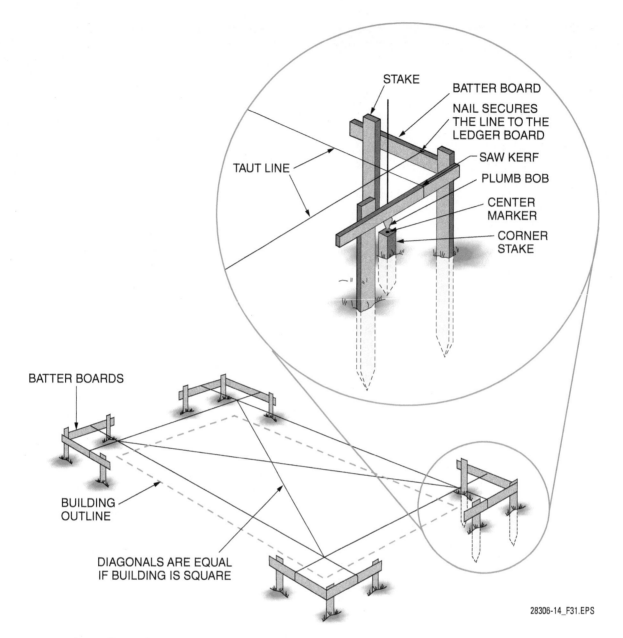

STAKE

BATTER BOARD

NAIL SECURES
THE LINE TO THE
LEDGER BOARD

SAW KERF

PLUMB BOB

CENTER
MARKER

CORNER
STAKE

TAUT LINE

BATTER BOARDS

BUILDING
OUTLINE

DIAGONALS ARE EQUAL
IF BUILDING IS SQUARE

28306-14_F31.EPS

Figure 31 Typical use of batter boards.

For most construction layout and checking, right triangles that are multiples of the 3-4-5 triangle are used (such as 9-12-15, 12-16-20, 15-20-25, and 30-40-50). The specific multiple is determined mainly by the relative distances involved in the job being laid out or checked. It is best to use the highest multiple that is practical. This is because when smaller multiples are used, any error made in measurement will result in a much greater angular error.

Figure 32 shows an example of the 3-4-5 rule involving the multiple 48-64-80. In order to square or check a corner as shown in the example, first measure 48 feet 0 inches down the line in one direction, then 64 feet 0 inches down the line in the other direction. The distance measured between the 48 foot

0 inch and 64 foot 0 inch points must be exactly 80 feet 0 inches if the angle is to be a perfect right angle. If the measurement is not exactly 80 feet 0 inches, the angle is not 90 degrees. This means that the direction of one of the lines or the corner point must be adjusted until a right angle exists.

Exact measurements are necessary to get the desired results when using the 3-4-5 method of laying out or checking a 90-degree angle. Any error in the measurements of the distances will result in not establishing a right angle as desired, or if an existing 90-degree angle is being checked, inaccurate measurements may cause an unnecessary adjustment to be made.

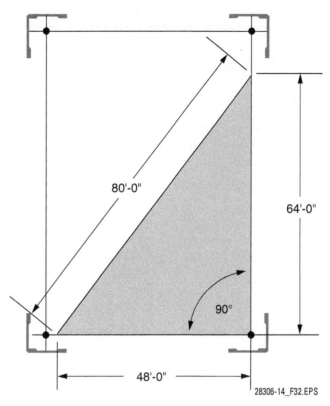

Figure 32 Example of checking the lines for square using the 48-64-80 multiple of the 3-4-5 rule.

Additional Resources

Basic Surveying Technology, Latest Edition. Stillwater, OK: The Mid-America Vocational Curriculum Consortium, Inc.

Construction Surveying and Layout: A Step-by-Step Engineering Methods Manual. 2002. Wesley G. Crawford. West Lafayette, IN: Creative Construction Publishing.

Measuring, Marking, and Layout. Newtown, CT: Taunton Press.

Plan Reading & Material Takeoff. Kingston, MA: R.S. Means Company.

Principles and Practices of Commercial Construction. 2008. Cameron K. Andres and Ronald C. Smith. Upper Saddle River, NJ: Prentice Hall.

Reading Architectural Plans: For Residential and Commercial Construction. 2002. Ernest R. Weidhaas. Upper Saddle River, NJ: Prentice Hall.

Surveying Principles and Applications. 2008. Barry F. Kavanagh. Upper Saddle River, NJ: Prentice Hall.

Surveying with Construction Applications. 2009. Barry F. Kavanagh. Upper Saddle River, NJ: Prentice Hall.

8.0.0 Section Review

1. In a batter board, the horizontal board that is attached to stakes driven into the ground is called the _____.

   a. ledger board
   b. leader board
   c. header board
   d. spanner board

2. The 3-4-5 rule is based on the _____.

   a. Euclidean theorem
   b. Pythagorean theorem
   c. Descartean theorem
   d. Newtonian theorem

Summary

The mason must have the knowledge to perform standard surveying measurements on the job site such as measuring distances, angles, and elevations. It is important to eliminate mistakes and reduce the size of errors in measurement. Do not use instruments that are damaged. Always handle measuring instruments carefully so that they are not damaged.

Masons need to be able to interpret construction site plans. This allows them to relate both natural and man-made features to the actual site's topography and layout. Masons should be able to use a wide range of distance measurement tools such as tapes, range poles, plumb bobs, gammon reels, and hand sight levels.

The ability to set up, adjust, and field-test leveling instruments, tripods, and leveling rods is an important skill for a mason. These tools, when used with proper leveling procedures, allow masons to determine site and building elevations and to record these data in field notes.

Finally, the mason should be able to use the 3-4-5 rule to lay out building corners. A skilled mason takes pride in using site layout skills in order to ensure that the masonry structure is built the best it can be.

1. The site plan for a project covering a large area would likely be drawn to a small scale such as _____.

   a. 1 inch = 10 feet
   b. 1 inch = 25 feet
   c. 1 inch = 50 feet
   d. 1 inch = 100 feet

2. Finish-grade references are keyed to a _____.

   a. benchmark
   b. level locator
   c. standard
   d. master point

3. The elevation of a primary control point may be arbitrarily chosen, or _____.

   a. the highest point on the site
   b. specified by code
   c. may be specified relative to sea level
   d. the lowest point on the site

4. The finish grade to be achieved during construction is shown on a topographic map by _____.

   a. dashed lines
   b. solid lines
   c. double lines
   d. dotted lines

5. On a topographic map, an elevation figure can be placed _____.

   a. below the contour line
   b. in a note
   c. above the contour line
   d. in an accompanying table

6. Contour lines that are straight and parallel to each other indicate _____.

   a. a man-made feature
   b. flat ground
   c. a streambed
   d. a cliff

7. If a building violates setback lines or ends up on the wrong property, legal liability rests with the _____.

   a. project engineer
   b. architect
   c. general contractor
   d. surveyor

8. On construction project drawings, dimensions are usually shown in feet and inches for _____.

   a. land measurements
   b. structures
   c. topographic features
   d. ground elevations

9. To convert a fraction of an inch to a decimal value, _____.

   a. divide the numerator by the denominator
   b. add the numerator to the denominator
   c. multiply the numerator by the denominator
   d. subtract the numerator from the denominator

10. A secondary control point, or hub stake, is typically a _____.

    a. length of ¾-inch rebar
    b. 3-inch-square concrete monument
    c. 2-inch cast-iron cylinder
    d. 1½-inch-square length of wood

11. Working control points are usually located with reference to _____.

    a. benchmarks
    b. contour lines
    c. secondary control points
    d. property corner markers

12. Information marked on a stake should be _____.

    a. written in all capital letters
    b. arranged from the bottom upward
    c. written in pencil
    d. abbreviated whenever possible

13. In addition to identifying control points or markers, color coding is used to _____.
    a. better organize the job site
    b. provide information about their purpose
    c. make replacement easier
    d. warn equipment operators of their presence

14. The measuring method referred to as taping is sometimes also called _____.
    a. ranging
    b. stepping
    c. chaining
    d. linking

28306-14_RQ01.EPS

Figure 1

15. The items shown in Review Question *Figure 1* are called _____.
    a. surveyor's weights
    b. taping pins
    c. control-point markers
    d. plumb bobs

16. Measuring tapes are graduated in feet and inches or in feet and _____.
    a. tenths of a foot
    b. eighths of a foot
    c. sixths of a foot
    d. thirds of a foot

17. The end of a steel tape has a heavy loop that allows the attachment of _____.
    a. a spring clamp
    b. tension handles
    c. a plumb bob
    d. weights

18. A device that is used to help maintain alignment for taping and to mark measurement points at the site so that they are more visible is called a _____.
    a. plumb bob
    b. leveling rod
    c. range pole
    d. transit level

19. A plumb bob typically weighs _____.
    a. 8 ounces
    b. 16 ounces
    c. 24 ounces
    d. 36 ounces

20. A hand sight level has a built-in _____.
    a. bubble level
    b. range finder
    c. motion compensator
    d. elevation indicator

21. If a mason paces off a distance of 180 feet and takes 72 paces, the average pace length would be _____.
    a. 2.05 feet
    b. 2.15 feet
    c. 2.50 feet
    d. 2.55 feet

22. An electro-optical EDMI measures distance by means of a reflected signal that has undergone a(n) _____.
    a. phase shift
    b. frequency compression
    c. amplitude increase
    d. polarity reversal

23. The advantage of a builder's level with a three-screw leveling system is _____.
    a. lower cost
    b. faster leveling
    c. simpler installation
    d. greater accuracy

24. A built-in compensator to maintain a true level line of sight is a feature of a(n) _____.
    a. laser level
    b. transit
    c. builder's level
    d. automatic level

25. The limit that a beam emitted by a laser level can be detected by a sensor is _____.
    a. 500 feet
    b. 750 feet
    c. 1,000 feet
    d. 2,500 feet

26. Tripod legs should be secured by connecting chains when setting up _____.
    a. on a smooth floor or pavement
    b. on a slope
    c. on soft or wet ground
    d. on rocky, uneven ground

27. A 10-foot leveling rod with alternating red-and-white stripes 0.10 foot wide is a _____.
    a. Chicago rod
    b. San Francisco rod
    c. Florida rod
    d. Philadelphia rod

28. The vernier scale on a target allows readings on an engineer's rod to the nearest _____.
    a. ten-thousandth of a foot
    b. thousandth of a foot
    c. hundredth of a foot
    d. tenth of a foot

29. Crosshairs of the instrument are focused (made as dark and crisp as possible) by adjusting the _____.
    a. horizontal tangent screw
    b. telescope's focusing knob
    c. eyepiece of the telescope
    d. instrument leveling screws

30. A leveling-rod reading taken from a point of known elevation to establish the height of the leveling instrument is called a _____.
    a. backsight
    b. baseline sight
    c. foresight
    d. calibration sight

31. Elevation of a turning point is determined by _____.
    a. subtracting the backsight elevation from the height-of-the-instrument elevation
    b. adding the foresight elevation to the height-of-the-instrument elevation
    c. adding the backsight elevation to the height-of-the-instrument elevation
    d. subtracting the foresight elevation from the height-of-the-instrument elevation

32. A leveling traverse that is run back to its starting point is referred to as a _____.
    a. wrap-up
    b. closed level loop
    c. connected survey
    d. completed traverse

33. A reference and historical record of work performed is represented by the surveyor's _____.
    a. project record
    b. journal
    c. final report
    d. field notes

34. Sketches in field notes should include the _____.
    a. time of day
    b. drawing scale
    c. approximate north direction
    d. weather conditions

35. Methods used to establish the profile of a terrain or surface include grid leveling, profile leveling, and _____.
    a. cross-section leveling
    b. contour leveling
    c. cut-and-fill leveling
    d. project leveling

36. In profile leveling, the additional intermediate foresight readings made along the center line are called _____.
    a. projection points
    b. profile points
    c. interpolated points
    d. center-line points

37. Grid leveling done in conjunction with earthwork is done _____.
    a. only before the earthwork is begun
    b. only after earthwork is completed
    c. at intervals throughout the earthwork
    d. both before and after the earthwork

38. Batter boards may be erected as far as 20 feet from the actual building corner to _____.
    a. conform to OSHA regulations
    b. avoid disturbed ground
    c. provide working room
    d. meet local code requirements

39. Batter-board stakes extending 3 or more feet out of the ground should be _____.
    a. marked with a red flag
    b. braced to prevent movement
    c. replaced with rebar
    d. driven the same distance deep

40. A triangle with sides in the proportion 3:4:5 is called a(n) _____.
    a. equilateral triangle
    b. classic triangle
    c. right triangle
    d. isosceles triangle

Trade Terms Quiz

Fill in the blank with the correct term that you learned from your study of this module.

1. The process of making horizontal and vertical distance measurements is called _____.

2. A(n) _____ is a set of lines, typically horizontal and vertical, placed in a telescope used for sighting purposes.

3. A reading taken on a leveling rod held on a point in order to determine a new elevation is called _____.

4. A(n) _____ is a vertical measuring device that consists of two or more movable sections with graduated markings.

5. A point of known (reference) elevation determined from benchmarks through leveling, and permanent enough to last for the duration of a project, is called a(n) _____.

6. The _____ is the elevation of the line of sight of the telescope relative to a known elevation.

7. A collapsible measuring instrument usually made of fiberglass, cloth, or stainless steel is called a(n) _____.

8. A(n) _____ is a horizontal or vertical point established in the field to serve as part of a known framework on the site.

9. The apparent movement of the crosshairs in a surveying instrument caused by movement of the eyes is called _____.

10. A(n) _____ is a temporary point within an open or closed differential-leveling circuit whose elevation is determined by differential leveling.

11. An instrument setting location in differential leveling is called a(n) _____.

12. _____ is a method of leveling used to determine the difference in elevation between two points.

13. All construction operations connected with excavating (cutting) or filling earth are called _____.

14. To _____ is to add soil or rock on site to achieve a required elevation.

15. A reference point established by the surveyor on or close to the property, usually at one corner of the lot, is called _____.

16. _____ are a permanent record of field measurement data and related information.

17. An optical instrument used in surveying is called a(n) _____.

18. To _____ is to remove soil or rock on site to achieve a required elevation.

19. A reading taken on a leveling rod held on a point of known elevation to determine the height of the leveling instrument is called a(n) _____.

20. To _____ a stake is to position it at a specified distance and direction from the control point to allow that area to be worked in without disturbing the stake.

Trade Terms

Backsight (BS)	Differential leveling	Height of instrument (HI)	Tape
Benchmark	Earthwork	Leveling rod	Taping
Control point	Field notes	Offset	Temporary benchmark
Crosshairs	Fill	Parallax	Transit level
Cut	Foresight (FS)	Station	Turning point (TP)

Trade Terms Introduced in This Module

Backsight (BS): A reading taken on a leveling rod held on a point of known elevation to determine the height of the leveling instrument.

Benchmark: A reference point established by the surveyor on or close to the property, usually at one corner of the lot.

Control point: A horizontal or vertical point established in the field to serve as part of a known framework for all points on the site.

Crosshairs: A set of lines, typically horizontal and vertical, placed in a telescope used for sighting purposes.

Cut: To remove soil or rock on site to achieve a required elevation.

Differential leveling: A method of leveling used to determine the difference in elevation between two points.

Earthwork: All construction operations connected with excavating (cutting) or filling earth.

Field notes: A permanent record of field measurement data and related information.

Fill: Adding soil or rock on site to achieve a required elevation.

Foresight (FS): A reading taken on a leveling rod held on a point in order to determine a new elevation.

Height of instrument (HI): The elevation of the line of sight of the telescope relative to a known elevation. It is determined by adding the backsight elevation to the known elevation.

Leveling rod: A vertical measuring device that consists of two or more movable sections with graduated markings.

Offset: To position a stake at a specified distance and direction from the control point to allow that area to be worked in without disturbing the stake. Offset stakes include the distance from, and direction to, the control point.

Parallax: The apparent movement of the crosshairs in a surveying instrument caused by movement of the eyes.

Station: An instrument setting location in differential leveling.

Tape: A measuring tape, usually made of fiberglass, cloth, or stainless steel.

Taping: The process of making horizontal and vertical distance measurements.

Temporary benchmark: A point of known (reference) elevation determined from benchmarks through leveling, and permanent enough to last for the duration of a project.

Transit level: An optical instrument used in surveying.

Turning point (TP): A temporary point within an open or closed differential-leveling circuit whose elevation is determined by differential leveling. It is normally the leveling-rod location. Its elevation is determined by subtracting the foresight elevation from the height-of-the-instrument elevation.

CONSTRUCTION ABBREVIATIONS

Above mean sea level	ABMSL
Abutment	abt.
Approximate	approx.
At	@
Avenue	Ave.
Average	avg.
Back of sidewalk	BSW
Back of walk	BW
Backsight	BS
Begin curb return	BCR
Benchmark	BM
Between	betw.
Bottom	bot.
Boulevard	Blvd.
Boundary	bndry.
Bridge	br.
Calculated	calc.
Cast-iron pipe	CIP
Catch basin	CB
Catch point	CP
Cement-treated base	CTB
Concrete block wall	CBW
Construction	const.
Control point	CP
County	Co.
Court	Ct.
Creek	cr.
Curb	cb.
Curb and gutter	C&G
Cut	C
Description	desc.
Destroyed	dest.
Detour	det.
Direct	D
Distance	dist.
Distance	D
Distance, horizontal	Dh
District	Dist.
Ditch	dit.
Drive	Dr.
Driveway	drwy.
Drop inlet	DI
Edge of gutter	EG
Edge of pavement	EP
Edge of shoulder	ES
Elevation	el.
End wall	EW

Equation	eqn.
Existing	exist.
Expressway	Exwy.
Fahrenheit	F
Fence	fe.
Fence post	FP
Feet	ft.
Field book	FB
Fill	f
Finish grade	FG
Fire hydrant	FH
Flow line	FL
Foot	ft.
Footing	ftg.
Foresight	FS
Found	fd.
Foundation	fdn.
Freeway	Fwy.
Galvanized	galv.
Galvanized steel pipe	GSP
Gas line	GL
Gas valve	GV
Geodetic	geod.
Grid	grd.
Ground	grnd.
Gutter	gtr.
Head wall	hdwl.
Height	ht.
Height of instrument	HI
Highway	Hwy.
Hub & tack	H&T
Inch	in.
Inside diameter	ID
Instrument	inst.
Intersection	int
Iron pipe	IP
Irrigation pipe	irr.P
Junction	jct.
Kilometer	km
Lane	ln.
Left	lt.
Manhole	MH
Marker	mkr.
Maximum	max.
Measured	meas.
Median	med.
Mile	mi.

Millimeter	mm	Right of way	R/W
Minimum	min.	River	Riv.
Minute	min.	Road	rd.
Monument	mon.	Roadway	rdwy.
Nail	N	Rock	rk.
North	N	Route	Rte.
Number	# or no.	Section	S
Offset	O/S	Sewer line (sanitary)	SS
Original ground	OG	Shoulder	shldr.
Outside diameter	OD	Sidewalk	SW
Overhead	OH	Slope stake	SS
Page	p.	South	S
Pages	pp.	Spike	spk.
Party chief	PC	Stake	stk.
Pavement	pvmt.	Standpipe	SP
Perforated metal pipe	PMP	Station	sta.
Pipe	P	Steel	stl.
Place	pl.	Storm drain	SDr.
Plastic	plas.	Street	St.
Point	pt.	Structure	str.
Point of intersection	PI	Subdivision	subd.
Portland cement concrete	PCC	Subgrade	SG
Power pole	PP	Tack	tk.
Pressure	press.	Telephone cable	tel.C.
Private	pvt.	Telephone pole	tel.P.
Project control survey	PCS	Temperature	temp.
Property line	PL	Temporary benchmark	TBM
Punch mark	PM	Top back of curb	TBC
Railroad	RR	Top of bank	TB
Railroad spike	RRspk.	Top of curb	TC
Read head nail	RH	Township	T
Record	rec.	Tract	tr.
Reference	ref.	Transmission tower	TT
Reference monument	RM	Turning point	TP
Reference point	RP	Water line	WL
Reinforced concrete pipe	RCP	Water valve	WV
Retaining wall	ret.W	Wing wall	WW
Right	rt.		

Additional Resources

This module presents thorough resources for task training. The following resource material is suggested for further study.

Architectural Drawing and Light Construction. 2009. Philip A. Grau III, Edward J. Muller, and James G. Fausett. Upper Saddle River, NJ: Prentice Hall.

Architectural Graphic Standards. 1998. The American Institute of Architects. New York: John Wiley & Sons, Inc.

Basic Surveying Technology, Latest Edition. Stillwater, OK: The Mid-America Vocational Curriculum Consortium, Inc.

Code Check, Latest Edition. Newtown, CT: Taunton Press.

Construction Surveying and Layout: A Step-by-Step Engineering Methods Manual. 2002. Wesley G. Crawford. West Lafayette, IN: Creative Construction Publishing.

Measuring, Marking, and Layout. Newtown, CT: Taunton Press.

Plan Reading & Material Takeoff. Kingston, MA: R.S. Means Company.

Principles and Practices of Commercial Construction. 2008. Cameron K. Andres and Ronald C. Smith. Upper Saddle River, NJ: Prentice Hall.

Reading Architectural Plans: For Residential and Commercial Construction. 2002. Ernest R. Weidhaas. Upper Saddle River, NJ: Prentice Hall.

Surveying Principles and Applications. 2008. Barry F. Kavanagh. Upper Saddle River, NJ: Prentice Hall.

Surveying with Construction Applications. 2009. Barry F. Kavanagh. Upper Saddle River, NJ: Prentice Hall.

Figure Credits

Section Review Answers

Answer	Section Reference	Objective
Section One		
1. d	1.1.0	1a
Section Two		
1. c	2.1.2	2a
2. d	2.2.0	2b
3. a	2.3.0	2c
4. c	2.4.0	2d
5. a	2.5.0	2e
Section Three		
1. a	3.1.0	3a
2. c	3.2.0	3b
3. b	3.3.0	3c
4. d	3.4.0	3d
Section Four		
1. a	4.1.0	4a
2. b	4.2.0	4b
Section Five		
1. c	5.1.1	5a
2. a	5.2.0	5b
3. b	5.3.0	5c
4. b	5.4.0	5d
5. a	5.5.2	5e
Section Six		
1. b	6.1.0	6a
2. a	6.2.0	6b
3. c	6.3.0	6c
Section Seven		
1. d	7.1.0	7a
2. d	7.2.1	7b
Section Eight		
1. a	8.1.0	8a
2. b	8.2.0	8b

NCCER CURRICULA — USER UPDATE

NCCER makes every effort to keep its textbooks up-to-date and free of technical errors. We appreciate your help in this process. If you find an error, a typographical mistake, or an inaccuracy in NCCER's curricula, please fill out this form (or a photocopy), or complete the online form at **www.nccer.org/olf**. Be sure to include the exact module ID number, page number, a detailed description, and your recommended correction. Your input will be brought to the attention of the Authoring Team. Thank you for your assistance.

Instructors – If you have an idea for improving this textbook, or have found that additional materials were necessary to teach this module effectively, please let us know so that we may present your suggestions to the Authoring Team.

NCCER Product Development and Revision
13614 Progress Blvd., Alachua, FL 32615

Email: curriculum@nccer.org
Online: www.nccer.org/olf

❏ Trainee Guide   ❏ Lesson Plans   ❏ Exam   ❏ PowerPoints   Other _____

Craft / Level: _____   Copyright Date: _____

Module ID Number / Title: _____

Section Number(s): _____

Description: _____

_____

_____

_____

_____

Recommended Correction: _____

_____

_____

_____

_____

Your Name: _____

Address: _____

_____

Email: _____   Phone: _____

28308-14

Stone Masonry

Natural and manufactured stone is widely used for a variety of structural and veneer applications. The tools and techniques that masons use to install stone are similar to those used for installing block and brick, but with some differences due to the nature of stone as a building material. This module introduces trainees to the types of stone used in masonry construction and how they are cut and finished. It also discusses the various hand and power tools, lifting devices, and anchors used to cut, shape, position, support, and install stone. The techniques for estimating stone volumes and stone veneers are discussed. Trainees are introduced to the techniques for installing stone using anchors and mortar, and for installing adhered stone veneer.

Module Seven

Trainees with successful module completions may be eligible for credentialing through the NCCER Registry. To learn more, go to **www.nccer.org** or contact us at **1.888.622.3720**. Our website has information on the latest product releases and training, as well as online versions of our *Cornerstone* magazine and Pearson's product catalog.

Your feedback is welcome. You may email your comments to **curriculum@nccer.org**, send general comments and inquiries to **info@nccer.org**, or fill in the User Update form at the back of this module.

This information is general in nature and intended for training purposes only. Actual performance of activities described in this manual requires compliance with all applicable operating, service, maintenance, and safety procedures under the direction of qualified personnel. References in this manual to patented or proprietary devices do not constitute a recommendation of their use.

28308-14
STONE MASONRY

Objectives

When you have completed this module, you will be able to do the following:

1. Describe types of stone and the quarrying, cutting, and finishing processes used on them.
   a. Identify the types of stone.
   b. Describe how stone is quarried.
   c. Explain how stone is cut and finished.
2. Identify the tools and devices used in stone masonry.
   a. Identify hand tools used in stone masonry.
   b. Identify power tools used in stone masonry.
   c. Identify lifting devices used in stone masonry.
   d. Identify fasteners and connectors used in stone masonry.
3. Describe how to estimate various types of stone.
   a. Describe how to estimate stone veneers.
   b. Describe how to perform stone volume estimates.
4. Identify stone installation techniques.
   a. Describe how to install stone using anchors.
   b. Describe how to install stone using mortar.
   c. Describe how to install adhered stone veneers.

Performance Task

Under the supervision of your instructor, you should be able to do the following:

1. Estimate quantities of stone and stone materials.

Trade Terms

Ashlar	Honed	Pumice	Shop ticket
Basalt	Igneous	Quarried	Slabbed
Cladding	Kerf	Quarry sap	Strap
Clamp	Lath	Random rectangular	Strapmaster
Dimension stone	Limestone	stone	Strata
Dress	Luster	Rubble	Substrate
Finishing	Manufactured stone	Sandstone	Vacuum cup
Granite	Metamorphic	Scratch coat	Vacuum lifter
Guillotine	Permeability	Season	Worm-drive saw
Hand clamp	Point loading	Sedimentary	

Industry-Recognized Credentials

If you're training through an NCCER-accredited sponsor, you may be eligible for credentials from NCCER's Registry. The ID number for this module is 28308-14. Note that this module may have been used in other NCCER curricula and may apply to other level completions. Contact NCCER's Registry at 888.622.3720 or go to **www.nccer.org** for more information.

Code Note

Codes vary among jurisdictions. Because of the variations in code, consult the applicable code whenever regulations are in question. Referring to an incorrect set of codes can cause as much trouble as failing to reference codes altogether. Obtain, review, and familiarize yourself with your local adopted code.

Contents ————————————————

Topics to be presented in this module include:

Figures and Tables

1.0.0 INTRODUCTION TO STONE

Objective

Describe types of stone and the quarrying, cutting, and finishing processes used on them.

a. Identify the types of stone.
b. Describe how stone is quarried.
c. Explain how stone is cut and finished.

Trade Terms

Ashlar: A square- or rectangular-cut stone masonry unit; or, a flat-faced surface having sawn or dressed bed and joint surfaces.

Basalt: A dark, durable form of igneous rock often used in walls and cobblestones.

Finishing: The process of honing, flame finishing, splitting, and polishing the exposed face or faces of stone.

Granite: A very hard and durable form of igneous rock widely used in masonry for exterior and interior installations.

Honed: To be lightly polished.

Igneous: A type of stone that is formed when molten rock or volcanic lava cools and solidifies.

Limestone: A type of sedimentary stone consisting primarily of calcite that is widely used in loadbearing and veneer masonry applications, and is also a key ingredient in concrete.

Luster: The level of reflective shine on the exposed surface of stone.

Metamorphic: Igneous or sedimentary stone that has been subjected to extreme heat or pressure over a long period of time, causing it to change its physical or chemical structure.

Permeability: The extent to which a substance allows liquids and gases to pass through it.

Pumice: A light-colored, powdery form of igneous stone rich in silica, often used to make concrete and cinder block, and also as an abrasive material.

Quarried: Mined or extracted.

Quarry sap: A brownish stain that forms on the surface of freshly quarried stone as water leaches from it.

Random rectangular stone: Stone of modular dimensions that has vertical and horizontal bed joints.

Rubble: Small, irregular stone debris left over from the quarrying process.

Sandstone: A type of sedimentary stone consisting primarily of layers of quartz and feldspar; it is used in ornamental and decorative stone masonry installations and to make grindstones.

Season: To allow freshly quarried stone to dry out.

Sedimentary: A type of stone that is created by the gradual settling and compression of minerals and organic particles into layers.

Shop ticket: A document assigned to a piece of slabbed stone, identifying the final dimensions and finish to be applied, and assigning the stone a number to aid in construction.

Slabbed: Sawn to a predetermined thickness from a larger quarried block.

Strata: Layers of sedimentary stone.

Stone masonry is one of humankind's oldest crafts, dating back to the creation of the first tools. Many of the oldest man-made structures on earth—ranging from the Egyptian pyramids to medieval cathedrals—were built by masons using stone. Today, stone is perhaps most commonly used as a decorative material, but it can also be used as structural facing (*Figure 1*). Stone is also used as a veneer over block wythes and structural elements. The brownstone buildings in New York City, the gray stone buildings of Paris, and many of the government buildings and monuments in Washington, D.C., are of stone construction. Stone is also used indoors; for example, on kitchen countertops and for decorative walls. Because stone is a naturally occurring material, it offers the advantages of durability and permanent color. Stone is also available in a wide variety of finishes.

28308-14_F01.EPS

Figure 1 Stone structural facing.

This section serves as a general introduction to stone masonry. In it, you will learn about the different types of stone, how stone is quarried, or mined, and how stone is cut and finished for use in architectural applications. Although masons are at the end of the long chain of people and processes required to prepare stone for use in construction, it is important for you to understand the context within which this takes place. It will help ensure that your work is a reflection of the quality not only of the material, but also of the craftsmanship of the many people who prepared the stone so that you can install it.

1.1.0 Types of Stone

Stone is a naturally occurring solid material that consists of minerals that have joined together through chemical bonds. Many types of stone contain silica, which forms crystals when it comes in contact with other minerals. Stone can vary according to the different types of minerals and chemicals it contains, the size and texture of the mineral particles that make up the stone, and the stone's permeability, or ability to allow water and gases to flow through it. Architects and designers select stone for use in masonry based on a combination of these factors, as well as on the visual aesthetic of the stone itself.

Geologists (scientists who study the history and structure of Earth) classify stone into three categories: igneous, sedimentary, and metamorphic. Because the minerals that make up stone can combine in an almost infinite variety of ways, these categories are meant to be very broad. Stone can exhibit the characteristics of more than one category. However, for the purposes of stone masonry, these categories are a useful way to think about the properties and uses of stone. The following sections briefly summarize each of the three categories of stone.

1.1.1 Igneous Stone

Igneous stone is formed when magma or lava, both molten rock, cools and becomes solid. Up to 95 percent of Earth's crust, to a depth of 10 miles, is made from igneous stone. Geologists have identified over 700 types of igneous stone.

Igneous stone that is formed belowground (magma) tends to cool slowly and has a coarser mineral grain, while stone that is formed on the surface (lava) cools quickly and has a much finer mineral grain. Granite is a type of igneous rock that is formed belowground. Granite is a very hard and durable stone, and is widely used in masonry for exterior and interior walls, floor-

ing, steps, and countertops. Pumice and basalt are two common types of igneous stone that are formed on Earth's surface. Pumice is a light-colored, powdery stone, rich in silica, that is often used in concrete and cinder block. It is also used as an abrasive material for polishing other types of stone. Basalt is darker and more durable. It is often used as a structural element in walls and to make cobblestones in high-traffic areas.

Igneous rock can weather quickly. Its porous surface traps water, minerals, and chemicals, which can discolor the stone. It is also rich in iron, which oxidizes when exposed to water. These weathering effects add character to an architectural installation, giving a wall or building the appearance of being much older than it really is.

> **NOTE**
> The word *igneous* comes from the Latin word *ignis*, which means fire. It is also the source of the English words *ignite* and *ignition*.

1.1.2 Sedimentary Stone

Sedimentary stone is created by the gradual accumulation of minerals and organic particles as they settle on the ground and in bodies of water. These particles are then slowly compressed into a solid by the weight of additional layers of sediment that form above them. These layers are called strata (*Figure 2*). Sedimentary stone accounts for less than 10 percent of Earth's crust. Sedimentary stone is softer than igneous stone, and is easier to cut and shape.

Limestone and sandstone are types of sedimentary stone that are widely used in stone masonry construction. Limestone is made primarily from the mineral calcite. In powder form, it is a key ingredient in concrete. In its solid form, it is used in both loadbearing and veneer applications

28308-14_F02.EPS

Figure 2 The layers of Arizona's Grand Canyon are examples of sedimentary stone strata.

(*Figure 3*). Limestone is highly regarded by architects and designers for its warm colors as well as its strength, which allows it to be used in any masonry application. Limestone is also durable and requires little maintenance beyond occasional repointing of joints.

Sandstone, which forms in layers, is made from the minerals quartz and feldspar. It is widely used in ornamental and decorative features such as mantelpieces and fountains. In addition to its many masonry uses, sandstone is also widely used to make grindstones.

1.1.3 Metamorphic Stone

Metamorphic stone is made of igneous or sedimentary stone that has been subjected to extreme heat or pressure, which causes it to change either physically or chemically over a long period of time. The resulting stone can incorporate veins and layers of many different colors, resulting in a pleasing visual aesthetic as well as great strength. Types of metamorphic rock commonly used in masonry include marble, slate, and travertine, which all began as different types of sedimentary stone.

Marble is widely used as a structural and decorative material. It is vulnerable to mild acids, including those used in common household cleaners. It is also susceptible to damage by abrasion. Serpentinite, which is another type of metamorphic stone, has a similar appearance to marble but is more resistant to acids and abrasion and is widely used in countertops.

28308-14_F03.EPS

Figure 3 Limestone is used extensively on the exterior of the National Museum of the American Indian in Washington, D.C.

1.2.0 Quarrying Stone

Most natural stone used in masonry applications is mined from large, naturally occurring veins of stone that run deep into the earth. The mining process is called quarrying, and the pits from which the stone is extracted are called quarries. Quarried stone is typically cut into large block units and allowed to sit for anywhere from six weeks to eight months in order to allow it to season, or dry out. Freshly quarried stone has high water content and is not suitable for installation right away. The seasoning process allows this water to leach out of the stone and form a brownish stain called quarry sap. The stain caused by quarry sap is not permanent, and it will gradually disappear as the stone seasons.

Seasoned block is then slabbed, or sawn to a predetermined thickness. Common widths for slabbed marble and granite are ½ inch, ⅞ inch, 1¼ inches, and 1½ inches. Limestone and sandstone is typically slabbed to thicknesses between 2 and 6 inches (*Figure 4*). The slabbed stone is then loaded onto flatbed trucks and shipped to finishing shops where it is honed (lightly polished), flame finished, split, and polished according to the project specifications.

The stone debris left over when shaped block has been removed is called rubble. Rubble is irregular with sharp edges. It is often used as filler and in structures where a rustic or natural look is desired. Rubble can be roughly squared with a brick hammer to make it fit more easily.

1.3.0 Cutting and Finishing Stone

When slabbed stone arrives at the finishing shop, the contractor's shop drawings are used to create shop tickets for each piece of stone, specifying how the stone should be cut and finished for use in the final project. Each shop ticket assigns a number to the individual stone, and provides the dimensions and angles for the final result. A

GOING GREEN

Stone Is Recyclable

Because stone is a natural product that lasts thousands of years, it can be reused over and over again. When a stone building is torn down, for example, the stone can be used to construct a new building or repurposed to make walls or pavers. Old stone can also be used to restore historic buildings, cutting down on potential environmental disruption caused by new construction.

6" THICK × 5'-0" × 18'-0"
5" THICK × 5'-0" × 14'-0"
4" THICK × 5'-0" × 11'-0"
3" THICK × 4'-0" × 9'-0"
2" THICK × 3'-0" × 5'-0"

28308-14_F04.EPS

Figure 4 Typical dimensions for slabbed limestone.

stonecutter uses a specialized power saw to cut the stone according to the shop tickets (*Figure 5*). Stone-cutting saws are fitted with a water spray to cool the saw blade.

Once the stone has been cut to the desired shape, it is then prepared for finishing. Finishing involves honing, flame finishing, splitting, and polishing the face or faces of the stone that will be exposed to view according to the project specifications. There are five commonly used types of finishes used on stone:

- *Split face* – A concave or convex surface created by cracking the stone along its grain.
- *Rock face* – The natural surface of the stone.
- *Honed finish* – A smooth texture with little or no luster, or reflective shine.
- *Flame finish* – A rough, irregular lusterless finish used on granite.
- *Polished finish* – A smooth, glass-like finish applied to marble and granite that gives it a high luster.

Rectangular and square stone that has been cut to the same dimensions and finished with smooth and flat bed and joint surfaces is called ashlar. Because its finished shape and surfaces allow it to be stacked closely together, ashlar stone is used for a wide range of applications and can be stacked dry or installed with anchors or mortar. Stone of modular dimensions that has vertical and horizontal bed joints is called random rectangular stone. The use of random rectangular stone results in walls and structures that have irregular geometric patterns. Random rectangular stone can be installed with adjustable anchors or mortar, but is less frequently installed dry.

28308-14_F05.EPS

Figure 5 Stone-cutting saw.

NCCER – *Masonry Level Three* 28308-14

While most natural stone is obtained from quarries, stone masonry construction also makes use of stone that has been broken up and weathered naturally. This type of stone is called fieldstone. Depending on the type of stone and the environmental conditions to which it has been exposed, fieldstone can be smooth or rough, and can be any shape, size, and color. Depending on the application, a mason may choose to finish the stone to varying degrees or leave it in its original state. Masons can achieve aesthetically pleasing effects by combining various sizes, shapes, colors, and textures of fieldstone in an irregular pattern.

Once the stone has been cut and finished at the finishing shop, it is packaged to prevent chipping and scratching, and loaded onto flatbed trucks for delivery to the project site. There, it is then installed by the mason using a variety of hand and power tools, lifting devices, fasteners, and connectors.

The Biltmore Estate

Completed in 1895 for millionaire George Vanderbilt II, the 250-room Biltmore House in Asheville, North Carolina, is the largest privately owned house in the United States. The house was constructed from nearly 5,000 tons of Indiana limestone and it took six years to build. It includes 65 fireplaces, 43 bathrooms, and three separate kitchens. Each of the limestone slabs used in the 102-step spiral grand staircase extends through the building's wall, allowing the weight of the wall to hold the steps in place—a feat of mechanical engineering that eliminated the need for heavy (and visible) supports.

Additional Resources

Indiana Limestone Handbook, Latest Edition. Bedford, IN: Indiana Limestone Institute of America, Inc.

1.0.0 Section Review

1. Up to 95 percent of Earth's crust to a depth of 10 miles consists of _____.

    a. sedimentary stone
    b. igneous stone
    c. marble and serpentinite
    d. sandstone and limestone

2. Seasoning of freshly quarried stone can take up to _____.

    a. two weeks
    b. four weeks
    c. six months
    d. eight months

3. The stone finish consisting of a concave or convex surface created by cracking the stone along its grain is called _____.

    a. split face
    b. flame finish
    c. rock face
    d. honed finish

SECTION TWO

2.0.0 STONE MASONRY TOOLS AND DEVICES

Objective

Identify the tools and devices used in stone masonry.

    a. Identify hand tools used in stone masonry.

    b. Identify power tools used in stone masonry.

    c. Identify lifting devices used in stone masonry.

    d. Identify fasteners and connectors used in stone masonry.

Trade Terms

Clamp: A mechanical lifting device that grips the sides of stone using friction.

Guillotine: A hand tool consisting of two sharp spring-loaded segmental blades that split stone simultaneously from above and below.

Hand clamp: A hand tool that is used to grip stone and to serve as a handle.

Kerf: A slot cut into stone and designed to receive the end of a strap anchor.

Strap: A mechanical lifting device that uses slings to support stone from underneath.

Strapmaster: A hand tool that uses a lever-operated ratchet to bend, cut, punch, twist, and shape metal sheet and rod.

Vacuum cup: A hand tool used to grip rough or irregularly shaped stone through the use of mechanically induced suction.

Vacuum lifter: A mechanical lifting device used to grip rough or irregularly shaped stone through the use of several mechanically or hydraulically operated suction surfaces.

Worm-drive saw: A power saw that uses a cylindrical gear to spin the blade slowly and with more torque than other types of power saws.

Masons use a variety of hand and power tools, lifting devices, fasteners, and connectors when working with stone. Some of these tools, such as trowels, hammers, and chisels, are already familiar to you. The fasteners and connectors used in stone masonry are similar to those used in block and brick masonry, but have been specially designed for stone application. Other specialized tools and equipment that you will learn about are designed to be used only

with stone. As with other tools that masons use, the quality of the tool directly affects the quality of the work.

This section introduces the tools and equipment that you will need to lay stone masonry units. Some tools are used more than others. You may not use every tool on every job. By the end of this section, you will be able to identify each item and explain how it is used in stone masonry.

2.1.0 Using Hand Tools in Stone Masonry

Masons use hand tools to cut, carry, clean, align, and level stone. The following sections describe the typical masonry hand tools that you will use to perform stone masonry construction.

You are responsible for keeping your hand tools in good repair. Repair or replace defective tools immediately. Clean all tools that touch mortar or grout immediately after use. Use a bucket of water to soak tools that you will use again in a few minutes. Keep wooden handles out of the water. Remember, if mortar dries on a tool, it will harden and make the tool unusable.

Clean tools thoroughly with a rag and wire brush. Be sure that you remove all mortar or grout completely. After cleaning, check that tool handles are secure and free from cracks or splinters. Sharpen blades and cutting edges when they become dull or nicked.

2.1.1 Trowels, Hammers, and Chisels

As with block and brick masonry, trowels (*Figure 6*) are used in stone masonry to place, move, and shape mortar; to mix, scrape, and shape mortar; and to clean mortar from masonry units and tools. The trowel's handle is often used to tap units into place.

Trowels come in different shapes and sizes for different purposes. Trowels can range in width from about 4 to 7 inches and can be up to 13 inches long. Some trowels have specialized uses and will not be used as often as a standard brick trowel. Some trowels are particular to certain parts of the country.

Hammers used in stone masonry (*Figure 7*) are typically made from carbide steel and have concave faces. Some stone masonry hammers are

Check Tools Daily

Cleaning and checking your tools at the end of each day is a good habit. This will keep them in good working order and prevent loss.

6

NCCER – *Masonry Level Three* 28308-14

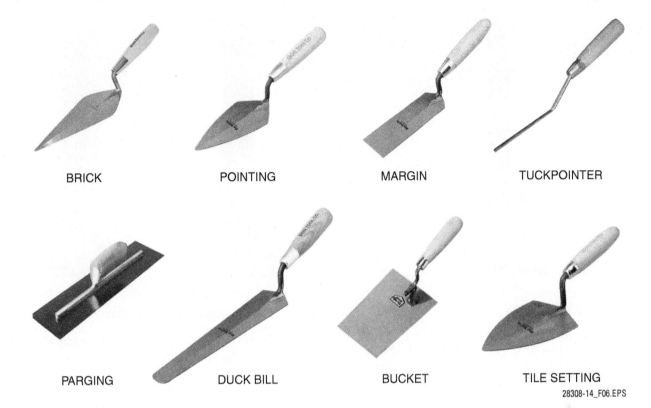

BRICK   POINTING   MARGIN   TUCKPOINTER

PARGING   DUCK BILL   BUCKET   TILE SETTING

28308-14_F06.EPS

Figure 6 Different types of trowels.

also fitted with a chisel blade. When the faces and blades become worn from use, they should be shaped and sharpened using a power grinder fitted with a silicon carbide wheel. Leather hammers, also called rawhide hammers or rawhide mallets, use rolled leather for the striking surfaces. Leather hammers are used to tap and set stone into position without marking, chipping, or cracking the stone. The leather striking surface can be replaced when worn.

Various types of chisels are used with stone hammers to cut and split stone. The mason's steel chisel (*Figure 8*) is typically used to cut veined stone. Rubber-grip mason's chisels (*Figure 9*) are used to cut stone of all types. The tooth chisel (*Figure 10*) has a toothed edge designed to cut soft stone and shape it to fit. It should not be used for hard stone. The pitching tool (also shown in *Figure 10*) is used for sizing, trimming, and facing hard stone.

Remember that chisel heads flatten after long use. The striking head mushrooms out, and metal burrs form at the edges of the head. These deformed edges can fly off and may cause injury.

(A) COMBINATION STONE HAMMER   (B) LEATHER STONE HAMMER

28308-14_F07.EPS

Figure 7 Stone masonry hammers.

Inspect chisels every day for dullness or deformation. To prevent injury, grind off the deformed part of the chisel head on a grinding wheel. While

Hold the Trowel Properly

When holding a trowel, keep your thumb along the top of the handle, not the shank. If your thumb is on the shank, it can get coated with mortar as you work, causing skin irritation.

28308-14_F08.EPS

Figure 8 Mason's steel chisel.

28308-14_F10.EPS

Figure 10 Tooth chisel and pitching tool.

grinding, cool the chisel with water to keep it from overheating. If it gets overheated, the steel loses its temper and may unexpectedly shatter. Do not attempt to repair or sharpen a chisel blade yourself; instead, have a blacksmith perform the repair or sharpening.

2.1.2 Hand Lifting Devices

Masons use wheeled dollies and vacuum cups to move and place small stone by hand. Dollies may be simple flatbed or angle-bed carts, or they may have restraints to hold the stone in place while being moved. Dollies are typically fitted with lock-

able wheels to prevent the dolly from moving during loading and unloading.

Vacuum cups (*Figure 11*) work similarly to the small plastic suction cups you attach to bathroom tile and other smooth surfaces in your home. The cup is placed on the surface of stone and the air is then mechanically pulled or squeezed out. The cup's rubber or silicon lip maintains a seal to prevent air from leaking back in. The suction creates a strong grip on rough or irregularly shaped stone, allowing the mason to lift it much more easily and with less strain than would be possible by hand alone. Suction cups that are used to lift marble should be returned to the manufacturer at least yearly for rebuilding. Hand clamps (*Figure 12*) are used to grip stone and serve as a handle that the mason can use to carry the stone from place to place.

28308-14_F09.EPS

Figure 9 Rubber-grip mason's chisel.

28308-14_F11.EPS

Figure 11 Vacuum cup used to lift stone by hand.

Figure 12 Hand clamp used to lift stone.

Masons use mechanical lifting devices to move stone that is heavier or larger than what can be moved by one or two people. These devices are covered in the section titled, *Using Mechanical Lifting Devices in Stone Masonry.*

> **WARNING!**
>
> Never allow yourself or another worker to stand or walk underneath stone that is being lifted using any mechanical means. If the stone falls, it will cause serious, and potentially fatal, injury.

2.1.3 Stone Guillotine

Guillotines use two sharp spring-loaded segmental blades to split stone simultaneously from above and below (*Figure 13*). Guillotines are available in models that can be operated by hand or by foot. They are designed to be portable so that they can be moved around the job site easily. Most guillotines are designed to allow the mason to cut stone at an angle as well as vertically. The height of the cutting table on a typical guillotine is adjustable for comfort. Long handles or foot pedals provide greater leverage to make the cutting process easier.

Always wear appropriate personal protective equipment and follow the manufacturer's recommended safety instructions when operating a guillotine, as it can cause severe and even fatal injury. Check the tightness of the blades regularly, as they can shift and come loose with repeated use. Never operate a guillotine with loose blades,

Figure 13 Guillotine used to split stone.

as it can cause the bolts holding the blades to weaken and break under pressure.

2.1.4 Strapmasters

In stone masonry construction, as with block and brick masonry, fasteners and connectors are often used to anchor stone to a backing wythe. In many cases, the mason will need to bend, cut, punch, twist, and shape these fasteners and connectors to enable them to be placed and aligned properly. A special hand tool called a Strapmaster is often used to bend, cut, punch, twist, and shape the fastener or connector (*Figure 14*). The Strapmaster is designed to cut metal bar, tube, and strap stock of various widths and thicknesses. It is available in two sizes; the smaller size can be used on mild steel up to $3/16$ inch thick, while the larger size can be used on mild and stainless steel up to $1/4$ inch thick.

The Strapmaster's punch and die can be quickly removed and replaced when worn or damaged. They can also be swapped for larger or smaller punches and dies, depending on the thickness of the metal being worked. However, the punches

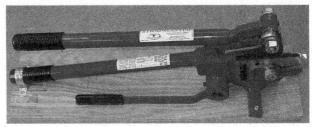

Figure 14 Strapmaster.

and dies for the various models of Strapmaster are not interchangeable. Before using, ensure that the cutting blades are properly aligned and that the teeth on the ratchet drive are engaging the lever. Always be sure to follow the manufacturer's instructions when servicing the various components of the tool. Connectors and fasteners used to anchor stone are discussed in the section titled, *Using Fasteners and Connectors in Stone Masonry.*

2.2.0 Using Power Tools in Stone Masonry

Just as with block and brick masonry, stone masonry requires the use of power tools and equipment to save time and labor. Because stone is often large and heavy, power tools are essential. When using power tools and equipment, always follow power-tool safety rules. Inspect items before using them to make sure they are clean and functional. Disconnect power cords, and turn off engines before inspecting or repairing power equipment. There are certain safety guidelines that you should follow when using gasoline powered tools:

- Be sure there is proper ventilation before operating gasoline-powered equipment indoors.
- Use caution to prevent contact with hot manifolds and hoses.
- Be sure the equipment is out of gear before starting it.
- Always keep the appropriate fire extinguishers near when filling, starting, and operating gasoline-powered equipment. OSHA (Occupational Safety and Health Administration) requires that gasoline-powered equipment be turned off prior to filling.
- Do not pour gasoline into the carburetor or cylinder head when starting the engine.
- Never pour gasoline into the fuel tank when the engine is hot or when the engine is running.
- Do not operate equipment that is leaking gasoline.

2.2.1 Stone Saws

Masons use handheld circular saws to cut stone (*Figure 15*). The diameter of a blade on a handheld saw can be up to 14 inches. Handheld masonry saws with 12-inch blades are sometimes called

28308-14_F15.EPS

Figure 15 Circular saw.

cutoff saws. These smaller saws can be powered by gasoline, electricity, or hydraulics.

Stone masonry often requires the use of saws with a slower rotational speed or higher torque than what conventional saws can provide. A worm-drive saw (*Figure 16*) uses a special type of cylindrical gear that resembles a large threaded bolt. This allows the saw blade to spin more slowly and offer greater torque. Worm-drive saws are durable and are easy to service on the job. The blades used on worm-drive saws are generally smaller than those used on standard circular saws. Many worm-drive saws are equipped with bevels to allow the blade to cut at an angle.

Use smooth-edged diamond blades on stone saws. This will reduce the chance of kickback, which is the reaction caused by a pinched, misaligned, or snagged blade that causes it to stop momentarily, propelling the saw away from the cutting surface and toward the operator. Diamond blades can be irrigated to prevent them from overheating and burning up. The irrigating water

28308-14_F16.EPS

Figure 16 Worm-drive saw.

wets the masonry unit, cools the blade, and controls dust.

When using handheld saws of any size to cut stone, review the saw's control operations before starting to cut. The guidelines for the safe use of saws include the following:

- Wear a hard hat and eye protection to guard against flying chips.
- Wear rubber boots and gloves to reduce the chance of electric shock.
- Wear earplugs and/or other ear protection.
- Wear a respirator when using a dry-cut saw.
- Check the guards to ensure they move and close freely. Never operate a saw with damaged or missing guards.
- Check all adjustment levers before cutting to make sure they are set at the correct bevel and depth.
- Always hold the saw firmly when operating, and support the work being cut to prevent loss of control.
- Between cuts, always release the trigger and hold the saw until the blade comes to a complete stop.
- Check that the blade is properly mounted and tightened before starting to cut.
- Do not cut stone with excessive pressure per pass. Make repeated passes using a light, forward pressure to achieve the desired depth.
- Clean stone dust from the saw's air vents frequently. Always disconnect the saw's plug from the power source and let the blade come to a complete stop before cleaning the air vents.

> **WARNING!**
> Make sure electric saws are grounded, especially if you are using water to cool the blade. When plugging in an electric saw, unless the blade is not in contact with the surface to be cut, hold the saw away from your body and have a co-worker plug in the saw. This will prevent the blade from skipping and possibly cutting your fingers.

2.2.2 Stone Grinders

Stonemasons use small 4½-inch grinders to cut notches and anchor slots in stone (*Figure 17*). When using a grinder, use both hands to grip the handles and position your body and arms so that you can absorb kickback from the grinder. Kickback can also cause damage to the surface of the stone, and in some cases can cause abrasive wheels to shatter with explosive force.

Before restarting a grinder, ensure that the wheel is centered in the kerf, or groove, but not

28308-14_F17.EPS

Figure 17 Grinder for use in stone masonry.

bound in the material. Also, ensure that the wheel depth lever is secured, to prevent the grinder from digging too deep. Before adjusting the grinder, always ensure that the power cord has been disconnected from the power source and that the wheel has come to a complete stop.

To change a wheel once it has come to a complete stop and the power has been disconnected, follow the manufacturer's instructions to remove the guard cover and push the spindle lock that keeps the wheel from rotating. Then loosen and remove the wheel locknut and the wheel. Clean the inside of the guard housing before installing the new wheel. Place the wheel on the spindle according to the manufacturer's directions, install the locknut, and replace the guard cover. Always allow the grinder to reach full speed before touching the wheel to the surface being ground. Always use appropriate personal protective equipment when operating a grinder, and follow the manufacturer's instructions when servicing it.

> **WARNING!**
> Wear a hard hat, gloves, and eye protection when using a grinder.

Observe the following rules for the safe use of a grinder:

- Ensure the grinding head is in good condition and has been properly secured.
- Wear gloves and other personal protective equipment when appropriate.
- Follow the manufacturer's instructions for installing shrouds and other dust-collection devices to help capture the dust created by the grinding process.
- Make sure the power switch is off before plugging in the grinder. Otherwise, there is a risk that stone dust on the electrical components can cause an electrical short, causing the grinder to start up.
- Remove all safety keys before starting the grinder.

- Use the proper grinding head for the material being ground.
- Ensure that the wheel guard is able to move freely and close during operation.
- Grip the grinder tightly on the main and auxiliary handles when using the grinder. Hold the grinder only on the insulated gripping surfaces.
- Ensure that the grinder is clean before using it, and clean it periodically while using it. Always allow the grinder wheel to come to a complete stop and unplug the grinder before cleaning it.
- Never force the grinder along as you work or apply excessive pressure. Use smooth, straight strokes. Maintain a firm grip on the handles when operating.
- Be sure the grinder head has stopped turning before putting the grinder down.

2.2.3 Hammer Drills

Mason's use ⅜-inch and ½-inch hammer drills to drill holes in masonry materials to allow the placement of fasteners and connectors (*Figure 18*). Hammer drills are more effective than other drills at this function because they provide a combination of power, speed, and hammering action. Carbide-tipped bits for hammer drills are available in several standard sizes.

> **WARNING!**
>
> Most hammer drills have enough torque to break your wrist. Make sure that you have a firm grip on the side handle when using a hammer drill. You should never hold on to just the main handle. Use both hands to equalize the rotation of the drill.

2.3.0 Using Mechanical Lifting Devices in Stone Masonry

For stone that is too large or heavy to lift, move, and set by hand, masons use specialized lifting devices attached to boom forklifts and cranes. These lifting devices can be grouped into three broad categories: clamps, thin cables, and straps. Clamps

28308-14_F18.EPS

Figure 18 Hammer drill.

grip the sides of stone using friction, while cables and straps use slings to support stone from underneath. All three devices have their advantages for different types, sizes, and shapes of stone. Always refer to the manufacturer's instructions before using a lifting device to ensure that it is suitable for the type of application.

Clamps (*Figure 19*), also called lifters, use mechanical pressure to grip the sides of slabbed stone. Cushioned pads on the gripping surfaces prevent them from scratching the stone. A hook is used to attach the clamp to the boom or crane. Clamps are available in a variety of shapes and sizes for different thicknesses of stone, and can be used in combination with other clamps when attached to a spreader bar.

Thin cables are made from metal that does not react with or discolor the stone. Straps are durable slings, made from nylon or padded chain, attached to spreader bars to hoist stone from underneath (*Figure 20*). When used with spreader bars, straps provide multiple points of support for the weight of the stone. This allows them to be used to lift stone that is long, or stone that has a complex shape.

Power-Saw Safety

When using power saws of any size, always follow these rules:

- Wear a hard hat and eye protection to guard against flying chips.
- Wear rubber boots and gloves to reduce the chance of electric shock.
- Wear earplugs and/or other ear protection.
- Wear a respirator when using a dry-cut saw.
- Check that the blade is properly mounted and tightened before starting to cut.

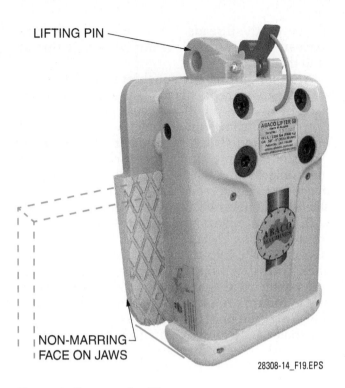

LIFTING PIN

NON-MARRING FACE ON JAWS

28308-14_F19.EPS

Figure 19 Clamp used to lift stone.

Another type of stone-lifting device is the **vacuum lifter** (*Figure 21*). Vacuum lifters work similarly to the hand-operated vacuum cups discussed in *Hand Lifting Devices*, but on a much larger scale. The vacuum lifter consists of several large suction plates fitted with rubber or silicon seals. Each plate is placed tightly against the stone and the air inside is mechanically or hydraulically removed. This creates a powerful attachment that can be used to support the weight of the stone. Vacuum lifters can be used on rough or irregularly shaped stone with sides that have been at least honed, and on stone that requires the lifting stresses to be more evenly distributed across its surface.

Always follow the manufacturer's safety instructions and wear appropriate personal protective equipment when using mechanical lifting devices. Many manufacturers require operators of lifting equipment to undergo training in their equipment prior to use. Be sure to select the proper-size clamp or strap for the weight of the stone slab. Never attempt to lift stone that exceeds the working load limit of the clamp or strap. To ensure proper grip on a clamp, inspect the clamp's rubber pads prior to use to ensure that they are free from dirt and grease. Never attempt to lift a wet slab with a clamp, as the clamp will not be able to grip the stone securely.

Regardless of the type of lifting device used, never attempt to lift more than one stone in a single load. Ensure that the load is balanced before attempting to lift it; an unbalanced load can slip and fall, or can put strain on the stone, causing it to crack or break apart. Never jerk a clamp or strap with the boom or crane. Lift the boom or crane arm slowly so that it takes up the slack gradually.

Prior to use, always inspect the shackles used to connect a lifting device to a boom or crane. Replace shackles that appear worn or damaged. Never use a worn or damaged clamp or strap to lift stone. Inspect clamps and straps regularly according to the manufacturer's instructions, and have a qualified person repair or replace worn or damaged equipment. When lifting and moving a load, ensure that all workers are safely clear of the load. Never leave a lifted load unsupervised, and never allow workers to move underneath a lifted load.

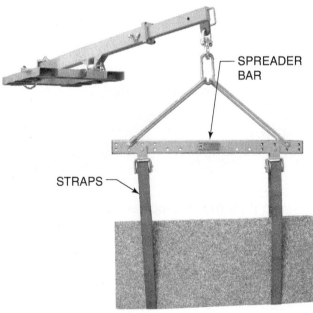

SPREADER BAR

STRAPS

28308-14_F20.EPS

Figure 20 Straps and spreader bar being used to lift stone.

28308-14_F21.EPS

Figure 21 Vacuum lifter.

Always store lifting equipment according to the manufacturer's instructions to protect against exposure to weather, dirt, and harmful or corrosive chemicals. Job-site procedures may require written records of clamp and shackle inspections; check with the site supervisor.

2.4.0 Using Fasteners and Connectors in Stone Masonry

The project engineer designs the anchoring system to be used on a particular project. In the module *Masonry Openings and Metal Work*, you learned how to use joint reinforcement to tie a single masonry wythe together, how to use anchors to tie two back-to-back masonry wythes together, and how to use ties and bolts to tie a masonry wythe to a structural element. The principle for stone masonry construction is essentially the same, though the types of fasteners and connectors used to tie stone to masonry are designed especially for use in stone construction. Installing these systems is discussed in the section titled *Installing Stone with Anchors*.

The types of fasteners and connectors used in a stone masonry installation will depend on the type of application (for example, interior or exterior), the size and type of stone being installed, and the backing to which the stone will be anchored. *Figure 22* illustrates common strap anchors that can be installed in slots. Strap anchors are typically ⅛ inch to ½ inch thick and between ¾ inch and 3 inches wide. They are typically finished in either uncoated carbon steel, hot-dip galvanized steel, or stainless steel. Strap anchors can be cut, bent, and shaped as required, using a Strapmaster or other cutting and bending tool. Rod and plate anchors (*Figure 23*) are also widely used to join stone to backing wythes and structural elements.

The type of fastener and connector used will depend in part on the stresses to which the stone structure will be exposed. Strap anchors and rods resist compression and shear stresses.

When the locations of the anchors have been set, use mortar to form a base under each anchor; use masonry scraps if the anchor protrudes into a core. Fill mortar around the bolt tightly, then use your trowel to pack the mortar. Make sure the anchor is plumb or level, depending on the location. When installing rod anchors, try to keep the

mortar off the threads. When you finish, be sure to clean the threads so the nut will easily thread onto the bolt.

Another important consideration when installing stone anchors is to ensure that the metal used in an anchor will not cause galvanic corrosion when it comes into contact with other metal anchoring and structural elements. Two or more pieces of the same kind of metal can safely come into contact with each other, but dissimilar metals may eventually corrode. Aluminum and stainless steel should never be allowed to come into contact with anchors or structural elements that are made from dissimilar metals. Cast iron and galvanized steel can come into contact without a risk of galvanic corrosion. Copper and bronze can all come into contact with each other safely. Many codes and standards require the use of metal anchors that are coated or painted with a layer of nonreactive material. Refer to the manufacturer's instructions and project specifications when selecting the type of anchor for the application.

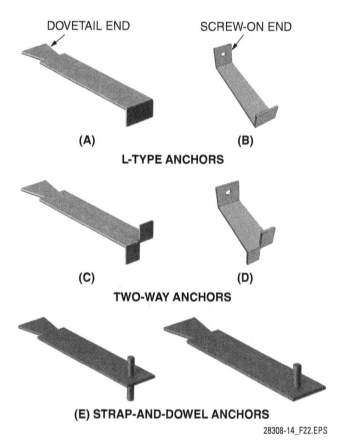

(A)   (B)
L-TYPE ANCHORS

(C)   (D)
TWO-WAY ANCHORS

(E) **STRAP-AND-DOWEL ANCHORS**

28308-14_F22.EPS

Figure 22 Stone strap anchors.

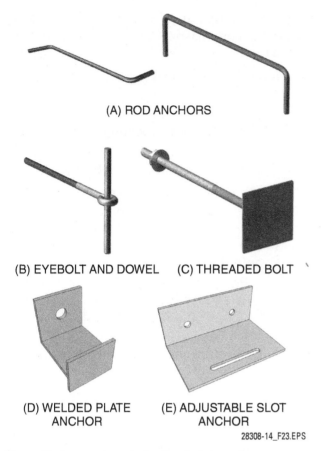

(A) ROD ANCHORS

(B) EYEBOLT AND DOWEL   (C) THREADED BOLT

(D) WELDED PLATE
ANCHOR

(E) ADJUSTABLE SLOT
ANCHOR

28308-14_F23.EPS

Figure 23 An assortment of rod and plate anchors.

Lewis Pins and the Box Lewis

Before the advent of clamp and vacuum lifters, large stone was typically lifted into place using a pair of devices called lewis pins. Lewis pins are metal shafts inserted into holes (called lewis holes) drilled into the top of the stone slab at a 30-degree angle, at an equal distance apart and facing each other. A hook on the end of the pin was then attached to the boom or crane, or to a bar spreader if more than one pin was required. While lewis pins are still manufactured and sometimes used to lift extremely heavy loads, masons do not use them as often because clamps and vacuum lifters are faster and easier to use, and they do not require the stone to be drilled.

A similar lifting device, the box lewis, is no longer permitted in masonry construction. The box lewis was a dovetail-shaped device inserted into a similarly shaped slot cut into the top of the stone. When renovating or demolishing old stone construction, you may see slots for a box lewis in the tops of old stone.

Additional Resources

ASTM A153, *Standard Specification for Zinc Coating (Hot-Dip) on Iron and Steel Hardware*, Latest Edition. West Conshohocken, PA: ASTM International.

ASTM C1242, *Standard Guide for Selection, Design, and Installation of Dimension Stone Attachment Systems*, Latest Edition. West Conshohocken, PA: ASTM International.

Marble and Stone Slab Veneer, Second Edition. 1989. James E. Amrhein and Michael W. Merrigan. Los Angeles, CA: Masonry Institute of America.

2.0.0 Section Review

1. The hand tool that masons often use to bend, cut, punch, twist, and shape fasteners and connectors is called the _____.

   a. Diemaker
   b. Big Hawg
   c. Strapmaster
   d. Grout Grunt™

2. Handheld masonry saws with 14-inch blades are also known as _____.

   a. rapid-cut saws
   b. worm-drive saws
   c. cutoff saws
   d. target saws

3. Straps provide multiple points of support for the weight of the stone when used with _____.

   a. splitter bars
   b. spreader bars
   c. vacuum lifters
   d. lewis pins

4. Strap anchors are typically finished in any of the following coatings *except* _____.

   a. uncoated carbon steel
   b. stainless steel
   c. hot-dip galvanized steel
   d. high-carbon powdered steel

3.0.0 ESTIMATING STONE

Objective

Describe how to estimate various types of stone.
a. Describe how to estimate stone veneers.
b. Describe how to perform stone volume estimates.

Performance Task

Estimate quantities of stone and stone materials.

Estimating stone masonry is difficult because each type of stone has its own particular characteristics. As a rule, stone masonry is estimated using the same units in which it is purchased, but these units vary from square feet to tons, depending on the type of stone. Use the following guidelines when dealing with stone estimating:

- Veneer and facing stone up to 8-inches thick are estimated in square feet of the required thickness.
- Stone panels over 8-inches thick are estimated in cubic feet.
- Trim, sills, jambs, and other stone less than 12-inches wide in the smallest face dimension are estimated in linear feet.
- Trim and feature panels greater than 12-inches wide in the smallest face dimension are estimated in square feet of the required thickness.
- Steps, buttresses, rubble walls, and entrance walls are estimated in cubic feet.
- Flagstone and other paving materials are estimated in square feet of the required thickness.
- Special features such as cornerstones and columns are estimated by the unit, fully described.

3.1.0 Estimating for Adhered Stone Veneer

Adhered stone veneer is always estimated in terms of square feet. The thickness of the veneer must be clearly stated. Include any trim pieces and anchors required. An accurate estimate is extremely important in stone veneers because they are usually very expensive.

The estimating process is simplified somewhat because very detailed shop drawings are normally required for fabrication and erection. These drawings are sent along with the estimate to the quarry or stonemason's yard. To illustrate, consider a marble veneer wall as detailed in *Figure 24*.

Step 1 Determine the square feet of the wall area.

$$A = \text{wall length} \times \text{height}$$
$$A = 10.0 \times 6.0 = 60 \text{ square feet}$$

Step 2 Determine the area of each panel.

$$Pa = \text{panel length} \times \text{panel height}$$
$$Pa = 5.0 \times 1.5 = 7.5 \text{ square feet}$$

Step 3 Determine the number of panels required.

$$Pn = \text{square feet of wall} \div \text{square feet per panel}$$
$$Pn = 60 \div 7.5 = 8 \text{ panels}$$

Step 4 Determine the requirements for special trim, if any is specified. Note that the panels have an inset border trim along the top edge, as shown in the Panel Elevation and Section A of the drawing. This design detail does not require any additional pieces of marble. No separate trim is required for this project.

Step 5 Determine anchorage requirements. Typically, assume four anchors per panel.

$$An = 4 \times Pn$$
$$An = (4 \times 8) = 32 \text{ anchors}$$

Step 6 Check all drawings for possible omissions, and check all calculations for accuracy.

The final listing should reflect that 8 polished marble panels are required. Each panel is 5 feet × 1 foot 6 inches, and 1¼-inches thick with a 1½-inch × ½-inch relief band along the top edge of each panel. The type of anchors required should also be described in detail.

3.2.0 Performing Stone Volume Estimates

Many types of stone are estimated by volume rather than square foot or unit measure. Bulk stone is estimated in either cubic yards or tons. Because unfinished stone is irregular in shape, it would be impractical to estimate or measure except by a cubic measure or weight (how much of it will fill a truck).

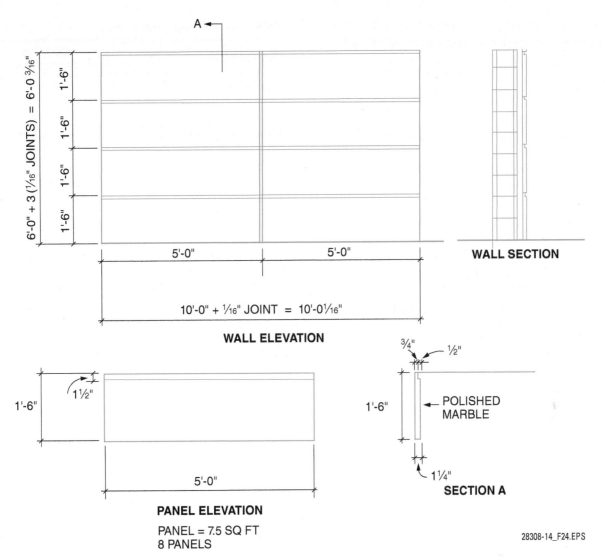

Figure 24 Stone-veneer wall elevation and details.

3.2.1 The Cubic-Yard Method

The cubic-yard system is a simple and effective way of estimating stone. The dimensions of the wall are used to determine the total volume in cubic feet. This is then converted to a cubic-yard figure.

For example, consider a garden wall that is to be constructed of rubble or fieldstone. The wall dimensions are 64-feet long, 8-feet high, and 18-inches thick.

Step 1 Determine the total cubic feet in the structure:

$$V = \text{length} \times \text{height} \times \text{width}$$

$$V = 64 \times 8 \times 1.5 = 768 \text{ cubic feet}$$

Step 2 Convert the cubic feet to cubic yards. Remember, 1 cubic yard equals 27 cubic feet.

$$\text{Cubic yards} = \text{cubic feet} \div 27$$

$$\text{Cubic yards} = 768 \div 27 = 28.44 \text{ cubic yards}$$

Step 3 Add a percentage for waste. This will usually range from 5 to 15 percent.

$$\text{Total cubic yards} = \text{cubic yards} \times (1 + \text{waste percentage})$$

$$\text{Total cubic yards} = 28.44 \times 1.10$$

$$\text{Total cubic yards} = 31.29 \text{ cubic yards}$$

Step 4 Recheck the accuracy of all figures.

If the stone wall has openings, subtract the cubic-foot volume of these openings from the total before converting to cubic yards.

3.2.2 The Ton Method

Many types of stone are estimated by the ton. Specific kinds of stone have specific weights, as shown in *Table 1*.

To estimate stone by weight, determine the volume in cubic feet, then convert this to pounds.

Use the appropriate conversion factor from *Table 1*. Consider the following example, using the same stone-wall measurements used in the previous section:

Step 1   Determine the total volume of the wall:
V = length × height × width
V = 64 × 8 × 1.5 = 768 cubic feet

Step 2   Determine the weight per cubic foot of the stone material. In this case, the wall is to be built of rubble.
Rubble = 130 pounds per cubic foot

Step 3   Determine the total weight of the stone:
TW = total cubic feet ×
weight of stone per cubic foot
TW = 768 × 130 = 99,840 pounds

Step 4   Convert to tons:
T = pounds ÷ 2,000
T = 99,840 ÷ 2,000 = 49.92 tons

Step 5   Add a percentage for waste, usually 10 percent:
TW = tons × (1 + waste percentage)
TW = 49.92 × 1.10 = 54.91 tons

Step 6   Check all calculations for accuracy.

Table 1 Stone Weights by Type

Type of Stone	Weight In Pounds per Cubic Foot
Ashlar	130 to 160
Common Brick	119 to 128
Common Stone	158
Fireclay Brick	150
Granite	168 to 172
Limestone	150 to 180
Marble	160 to 174
Porcelain Stone	159
Portland Stone	150 to 157
Quartz	165
Rip-rap	100
Sandstone	140 to 145
Slate	168

In addition to estimating the quantity of stone, you must also determine the quantity of mortar required. Check the documents for dowels, anchors, and any reinforcement that may be listed on a specification sheet or simply indicated on the drawings. A complete estimate must include all items.

Over Forty-Three Tons of Stone

The Pepsi Center is an ultramodern sports arena and the home of the Denver Nuggets and the Colorado Avalanche. Miles of granite and ceramic tile decorate the floors and walls of this arena.

Over $2 million worth of tile and stone were installed. The stonework includes approximately 60,000 sq ft of granite, 18,000 sq ft of travertine, and 9,000 sq ft of slate. This is 43½ tons of stone. Each of the pieces was prepared and installed to exact specifications.

Additional Resources

Estimating Building Costs for the Residential and Light Commercial Construction Professional. 2012. Wayne J. DelPico. New York: John Wiley & Sons.

Estimating in Building Construction. 1999. Frank R. Dagostino and Leslie Feigenbaum. New York: Pearson Education.

3.0.0 Section Review

1. Stone that is less than 12-inches wide in the smallest face dimension, such as trim, sills, and jambs, is estimated in _____.

   a. square feet
   b. pounds
   c. linear feet
   d. square inches

2. When estimating stone using the cubic-yard method, remember to subtract from the total the volume of _____.

   a. mortar
   b. openings
   c. anchors, rebar, and horizontal joint reinforcement
   d. floor space

Section Four

4.0.0 Stone Installation

Objective

Identify stone installation techniques.
  a. Describe how to install stone using anchors.
  b. Describe how to install stone using mortar.
  c. Describe how to install adhered stone veneers.

Trade Terms

Cladding: A general term for a layer of nonbearing stone installed over a masonry wythe or structural backing.

Dimension stone: A common term for stone that has been cut, shaped, and finished for use in masonry construction.

Dress: To square up the edges of stone to make them rectangular and to straighten any jagged edges.

Lath: Corrosion-resistant metal mesh attached to the substrate that serves as a base for mortar or plaster.

Manufactured stone: Premade stone made from cast cementitious material with pigments and other added materials that give the appearance of natural stone.

Point loading: An uneven distribution of structural load in a masonry structure that can cause cracking or buckling.

Scratch coat: A layer of mortar that has been scored or scratched with a raking tool in order to allow stone units to be adhered.

Substrate: The backing material to which natural and manufactured stone veneer is attached.

Stone masonry units—often referred to as dimension stone—can be installed using anchors, with mortar, or as a veneer. The techniques that masons use for each of these methods will already be familiar to you, so the following sections will review only the broad general requirements for each method. Engineered-stone masonry systems are designed by the project engineer. The installation of engineered systems is not covered in this module.

Local codes typically establish only the minimum requirements for installing stone masonry.

The installation method will depend on a variety of factors, including:

- Durability
- Ease of maintenance
- Design flexibility
- Appearance
- Weather and moisture resistance
- Sustainability
- Cost

When installing stone coping, pitch the stone toward the roof to avoid discoloring the exposed face and walls. To prevent condensation from entering a wall, vent attics over porticos. Washes and drips should be provided on stone that projects from the structural surface. Ensure the back faces of stone walls and columns do not come into physical contact with each other, as expansion can force the stone out from its original position.

When receiving stone at the job site, always inspect the stone to ensure that it has not been damaged during transit. Do not use fiber transportation pads to store stone. When storing stone, follow the manufacturer's recommended procedures for preventing damage and discoloration. *Figure 25* illustrates proper storage techniques for limestone slabs, for example. Do not store stone directly on yellow or red subsoil, as this can cause permanent staining. Protect projecting courses, sills, and openings against traffic and mortar droppings during construction. Protect stone against runoff or washes from work being performed overhead. At the end of the workday and during inclement weather, cover walls and openings to protect them.

4.1.0 Installing Stone with Anchors

Stone cladding, which is a layer of nonbearing stone installed over a masonry wythe or structural backing, can be held in place by strap or rod anchors. The techniques used to anchor stone using these connectors are similar to those used to anchor block and brick, but there are some differences. Always refer to the local applicable code and standards in your area when using anchors in stone masonry construction.

4.1.1 Installing Strap Anchors

Anchors are designed to hold stone in position while resisting lateral and horizontal loads. The bent ends of strap anchors are designed to be inserted into matching slots cut into the stone using a stone hammer and chisel. These slots are called kerfs (*Figure 26*). The kerf is filled with mortar in

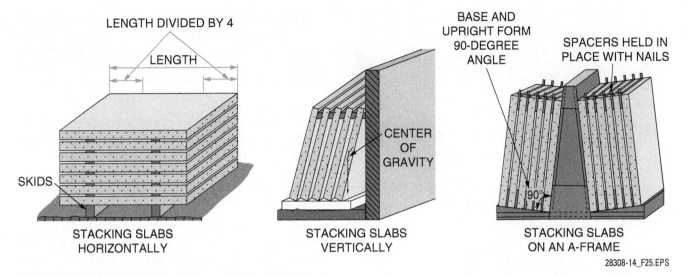

LENGTH DIVIDED BY 4

LENGTH

SKIDS

**STACKING SLABS
HORIZONTALLY**

BASE AND
UPRIGHT FORM
90-DEGREE
ANGLE

CENTER
OF
GRAVITY

**STACKING SLABS
VERTICALLY**

SPACERS HELD IN
PLACE WITH NAILS

90°

**STACKING SLABS
ON AN A-FRAME**

28308-14_F25.EPS

Figure 25 Proper storage of limestone slabs.

order to hold the strap in place and to prevent it from coming into contact with the sides or floor of the kerf. The mortar also prevents moisture from seeping in. Moisture can cause discoloration, and can also split stone when the moisture freezes. Always refer to the project specifications for the type of mortar to be used to fill the kerf. Some projects may call for a filler material other than mortar.

Because strap anchors will transfer structural load to the stone, it is important to ensure that kerfs are cut deep enough to ensure that the anchor remains engaged in the event of movement or flexing brought on by expansion or contraction. The straps must also be strong enough to maintain their shape under the structural load. If a strap anchor distorts under load, it can be forced into direct contact with the walls or floor of the kerf, resulting in **point loading**. Point loading is a condition in which structural loads are transmitted unevenly through the structure, which can cause cracking or buckling. To prevent deflection or rotation, at least two anchorage points must be provided for each stone unit.

4.1.2 *Installing Rod Anchors*

Rod anchors are attached similarly to strap anchors, but require round holes rather than rectangular kerfs to anchor them (*Figure 27*). These can be drilled rather than chiseled, which can result in cost as well as time savings. Another advantage of rod anchors is that the round holes used to anchor them are better at distributing mechanical stresses evenly than rectangular kerfs.

The number and placement of rod anchors will vary depending on the structural load. As a rule of thumb, at least four rod anchors will be required for a given stone panel, with at least one on each side of the panel. Typically, when the design calls for two anchors per edge, the anchor holes will be inset from the edge by ⅕ of the edge's total length. When placing more than two anchors, locate the holes in the center third of the stone's thickness, while maintaining a consistent distance from the face of the stone. The diameter of the anchor should not exceed ¼ the thickness of the stone. The depth of the hole should be at least four times the rod's diameter, and the hole's diameter should exceed the rod's diameter by ¹⁄₁₆ inch to ⅛ inch.

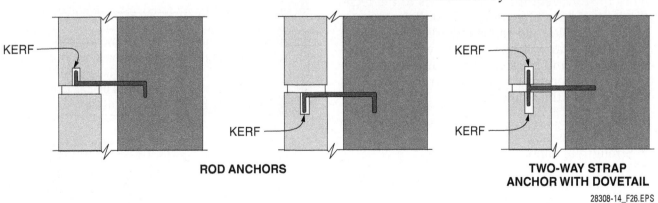

KERF

ROD ANCHORS

KERF

KERF

KERF

**TWO-WAY STRAP
ANCHOR WITH DOVETAIL**

28308-14_F26.EPS

Figure 26 Strap anchors embedded in kerfs.

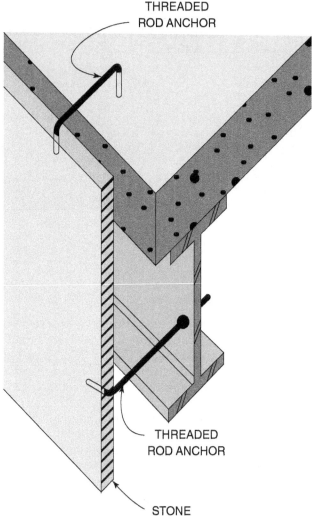

THREADED
ROD ANCHOR

THREADED
ROD ANCHOR

STONE

28308-14_F27.EPS

Figure 27 Rod anchors embedded in holes.

4.2.0 Installing Stone with Mortar

You are already familiar with the basic materials, tools, and techniques that are used to lay loadbearing and nonbearing block and brick with mortar. The techniques required to lay stone with mortar are broadly similar, though the unique characteristics of stone require some changes to the techniques you have already practiced. Refer to your local applicable code and standards for guidance on using mortar with stone masonry. As with mortar for block and brick construction, the desirable qualities in good hardened mortar for stone masonry include the following:

• Durability
• Compressive strength
• Strength of bond
• Volume change
• Appearance

The project specifications will identify the type of mortar mix to be used. Typically, unless otherwise specified, the mortar used to lay and point stone masonry units will likely be a mixture of 3 parts mortar sand with 1 part of any of the following:

• Masonry cement
• A fifty-fifty blend of portland cement and masonry cement (for use with rubble or natural cleft paving)
• A fifty-fifty blend of portland cement and hydrated lime (for use with sandstone, limestone, and other soft stone)

Set up the mixing area as you would for a block or brick project. Place the mixer and material stockpile so that they can be used efficiently during the project. If mortar will be moved in wheelbarrows, do not set up the mixing area downhill from the work area, if possible. Store mortar materials in a shed, or if they are to be kept outside, ensure they are covered with plastic sheets or canvas tarps, and ensure they are elevated off the ground.

Use a mechanical mortar mixer (*Figure 28*) to mix the mortar. When using a power mixer, place the materials near the mixing area. The mixer should be at the center of these materials in such a way that all materials can be easily reached. Leave a clear path for the wheelbarrow or other equipment used to move the finished mortar. The mixer should be securely supported and the wheels blocked to avoid tire wear. Review the operation manual for the machine used, and review safety procedures for working with power equipment. Have the mix formula written down, including the number of cubic feet of cement and lime as well as the cubic feet or shovelfuls of sand needed for the size of the batch to be mixed.

> **WARNING!**
> Wear eye protection and other appropriate personal protective equipment when using a power mixer. Never place any part of your body in the mixer.

> **WARNING!**
> Keep your hands out of the mouth of the mixer. If a torn bag falls into the mixer, do not try to remove it while the mixer is turning. Turn the machine off and allow its blades to stop completely before reaching inside.

Do not use stone that has cracks, seams, or other defects that could affect its ability to resist weather. Ensure that the stone is clean and free of debris or dust. The surfaces should be suitable for

Figure 28 Mortar mixer.

good strength of bond. Quarried, ashlar, and random rectangular stone is especially well suited for use with mortar, but fieldstone and rubble can also be used if they are suitably flat or if they can be trimmed appropriately. The project engineer may specify the minimum and maximum dimensions for the height, width, and length of stone that can be used. Stone used as headers should be larger than stone used for the wall itself.

Dress the stone appropriately before laying. Dressing involves squaring up the edges to make them rectangular and straightening any jagged edges. Use a hammer and chisel to finish corners and angles. Use large stone for the bottom row or foundation. Make full mortar joints as you would with block or brick, ensuring the stone is well bedded. Joints and beds should be no more than 1-inch thick. Should a joint be broken while laying, remove the stone unit and clean it thoroughly before resetting it in fresh mortar. When placing heavy stone with mortar, insert setting pads in the mortar joints to bear the weight of the stone until the mortar has set. This will prevent cracks due to settling.

Point face joints before the mortar sets, or if this is not possible, rake the joints to a depth of at least 2 inches before the mortar sets. Then, once the mortar has set, wet the joint thoroughly with potable water and fill with mortar. Ensure that the mortar has been driven into the joints, and then use a pointing tool to finish the joint. Once the mortar has set, clean the wall and the surrounding area using the techniques you have already learned.

4.3.0 Installing Adhered Stone Veneers

Adhered stone veneers are attached to backing material, called a substrate, with mortar. They are used for interior applications such as fireplace fronts and wall accents, as well as for exterior building walls and freestanding structures (*Figure 29*). While natural stone is sometimes used in veneer applications, manufactured stone, also called cast stone or concrete stone, is more commonly used. Manufactured stone is made from cast cementitious material with pigments and other added materials that give the veneer the appearance of natural stone. Veneer made from manufactured stone is also called adhered concrete masonry veneer (ACMV).

Stone veneer can be installed using mortar, usually Type N or Type S, on wood frame walls with rigid sheathing, as well as on other types of walls including masonry walls, poured-in-place concrete walls, concrete tilt-up panels, and even existing masonry surfaces, provided that the walls have been prepared in accordance with the manufacturer's instructions. Because it is a nonbearing veneer, the individual units in an adhered veneer system are thinner than block or brick. Typical interior veneer units are ¾-inch to 2½-inches thick, while veneer units designed for outdoor use may be 3- to 4-inches thick. Manufactured stone veneer should be installed at least 4 inches above grade.

Figure 29 Manufactured stone veneer installed on a house.

The substrate of an adhered veneer installation consists of a water-resistant barrier (WRB) and lath (corrosion-resistant metal mesh attached to the substrate that serves as a base for mortar or plaster) that are attached to the wythe or structural element using corrosion-resistant fasteners, covered by a scratch coat of mortar. This is a layer of mortar that has been roughed or scratched with a raking tool when firm, to allow the adhesion of stone units. *Table 2* lists the various permitted substrate materials for various types of wall systems. Always refer to the local applicable code and project specifications for the appropriate and approved substrate materials for the project. These techniques can also be used to install adhered stone veneer over continuous insulation up to ½-inch thick. In most cases, an engineered fastening system will be required to install adhered stone veneer over insulation thicker than ½ inch.

Flashing should be made from corrosion-resistant materials and integrated into the water-resistant barrier. Never install stone veneer over movement joints. Once the substrate has been installed, apply a nominal ½-inch-thick layer of mortar over the substrate so that the lath is completely covered. When the mortar is thumbprint-hard, scratch or score it horizontally. Dampen the mortar again before applying stone units.

On a tarp or other protected surface, lay out a minimum 25-square-foot section of stone on the ground, ensuring an aesthetic mix of colors, shapes, and sizes. Install veneer stone starting with the corners and then proceeding from the top down, to minimize cleanup requirements. Wet the back of each stone unit with water, butter it to a nominal thickness of ½ inch, and then work the unit firmly into the scratch coat using gentle sliding and twisting motions to ensure the stone has been set in place. Stop moving the stone as soon as you feel the mortar begin to "grab" the stone unit after a few seconds. This action should result in a sufficient amount of mortar to squeeze out from the back of the unit to form a full setting bed that completely covers the scratch coat. Alternately, mortar may be applied directly to the scratch coat, or a combination of these techniques may be used.

Repeat these steps until the section is complete. Prepare additional sections and install them using the same techniques. If grouting is required, wait for the mortar to set completely before applying the grout. *Figure 30* shows a typical wall section of adhered stone veneer installed in an external application.

> **CAUTION**
>
> If an adhered stone veneer unit is accidentally moved after setting, do not attempt to force the unit back into place or to pack it with additional mortar. Remove the stone and completely remove the mortar from the unit and the scratch coat. Apply fresh mortar to the stone or scratch coat, and reposition the stone using gentle sliding and twisting motions.

Do not install manufactured stone veneer where it will come into frequent water contact from sources such as lawn sprinklers, downspouts, and drainage pipes. Deicing materials, salt, and harsh proprietary cleaners should not be used on manufactured stone veneer. Prolonged exposure to them can discolor and damage the surface.

Another Name for Manufactured Stone Veneer

On the job, you may hear masons refer to "lick and stick" or "lick 'em and stick 'em." This is a common way to refer to manufactured stone veneer (also called adhered concrete masonry veneer, or ACMV).

Table 2A Substrate Materials for Adhered Stone Veneer on Various Wall Systems

Wall System	Water-Resistant Barrier	Lath	Fastening	Scratch Coat
Wall Type: Wood or steel stud, no more than 16 inches on center **Rigid Sheathing:** Gypsum wall board Plywood Oriented strand board Concrete board Fiber board Note: Continuous insulation over rigid sheathing is limited to maximum ½-inch thick	Minimum two separate layers	2.5-lb or 3.4-lb self-furred corrosion-resistant lath (ASTM C847) Or 18-gauge woven wire mesh (ASTM C1032) Or Alternate lath acceptable with a product evaluation acceptance report showing compliance to ICC-ES AC 275.	Corrosion-resistant fasteners (ASTM C1063), minimum ¾-inch nail penetration into wood framing member, or minimum ¾-inch staple penetration into wood framing member, or minimum ¾-inch penetration of metal framing member.	Mortar, nominal ½-inch thick, Type N or Type S meeting ASTM C270. "Scratch" surface when somewhat firm.
Wall Type: "Open stud" construction Wood or steel, no more than 16 inch on-center No sheathing or insulation board only (open studs) Note: Nonrigid insulation board over rigid sheathing is limited to maximum ½-inch thick	Minimum two separate layers	3.4-lb self-furring ⅜-inch ribbed corrosion-resistant lath (ASTM C847) Or 18-gauge woven wire mesh (ASTM C1032) Or Alternate lath acceptable with a product evaluation acceptance report showing compliance to ICC-ES AC 275.	Corrosion-resistant fasteners (ASTM C1063), minimum ¾-inch nail penetration into wood framing member, or minimum ¾-inch staple penetration into wood framing member, or minimum ⅜-inch penetration of metal framing member.	Mortar, nominal ½-inch thick, Type N or Type S meeting ASTM C270. "Scratch" surface when somewhat firm.
Wall Type: Clean concrete, masonry/CMU, stucco scratch coat, (1st layer of cement plaster), or stucco brown coat (2nd layer of cement plaster) Note: Walls/surfaces must be clean and free from release agents, paints, stains, sealers, or other bond-break materials, that may reduce strength of mortar adhesion	Note: A WRB may be needed to prevent moisture from penetrating the wall.	Install lath if question or concern regarding ability of veneer to adhere to wall: 2.5-lb or 3.4-lb self-furring ⅜-inch ribbed corrosion resistant lath (ASTM C847) Or 18-gauge woven wire mesh (ASTM C1032) Alternate lath acceptable with a product evaluation acceptance report showing compliance to ICC-ES AC 275.	If lath is applied, use corrosion-resistant fasteners (ASTM 1063).	If a scratch coat is required use a nominal ½-inch thick, Type N or Type S mortar, meeting ASTM 270. "Scratch" surface when somewhat firm.

Table 2B Substrate Materials for Adhered Stone Veneer on Various Wall Systems

Wall System	Water-Resistant Barrier	Lath	Fastening	Scratch Coat
Wall Type: Existing concrete, masonry/CMU, stucco, or brick (structurally sound) (e.g., painted or not clean) If the wall system is effectively cleaned and with adequate surface roughness, see the table above	Note: A WRB may be needed to prevent moisture from penetrating the wall.	2.5-lb or 3.4-lb self-furring ³⁄₈-inch ribbed corrosion-resistant lath (ASTM C847) Or 18-gauge woven wire mesh (ASTM C1032) Alternate lath acceptable with a product evaluation acceptance report showing compliance to ICC-ES AC 275.	Use corrosion-resistant fasteners (ASTM C1063).	Nominal ½-inch thick, Type N or Type S mortar, meeting ASTM C270. "Scratch" surface when somewhat firm.
Wall Type: Metal buildings or other surfaces/wall construction not listed above. See manufacturer for recommendations regarding sheathing	See manufacturer for recommendations.			

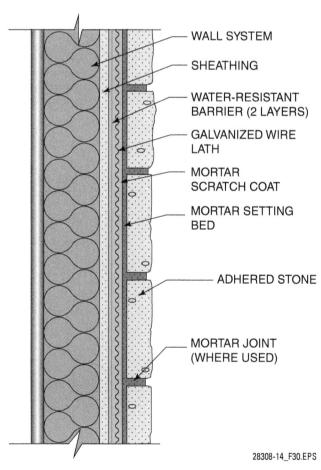

WALL SYSTEM

SHEATHING

WATER-RESISTANT BARRIER (2 LAYERS)

GALVANIZED WIRE LATH

MORTAR SCRATCH COAT

MORTAR SETTING BED

ADHERED STONE

MORTAR JOINT (WHERE USED)

28308-14_F30.EPS

Figure 30 Section view of a typical external installation of adhered stone veneer.

Additional Resources

ASTM C1670, *Standard Specification for Adhered Manufactured Stone Masonry Veneer (AMSMV) Units*, Latest Edition. West Conshohocken, PA: ASTM International.

ASTM C1780, *Standard Practice for Installation Methods for Adhered Manufactured Stone Masonry Veneer*, Latest Edition. West Conshohocken, PA: ASTM International.

Masonry Cement Mortar: Its Proper Use in Construction. 1993. Pat Howley. Murray Hill, NJ: ESSROC Materials.

Technical Note TN1, *Cold and Hot Weather Construction*. 2006. Reston, VA: The Brick Industry Association. **www.gobrick.com**

TEK 5-1B, *Concrete Masonry Veneer Details*. 2003. Herndon, VA: National Concrete Masonry Association. **www.ncma.org**

4.0.0 Section Review

1. Kerfs are holes designed to receive the bent ends of _____.

   a. rod anchors
   b. strap anchors
   c. dovetail anchors
   d. continuous joint reinforcement

2. When using mortar to install stone, one typical formula for mortar mix is 3 parts mortar sand with 1 part _____.

   a. water
   b. hydrated lime
   c. portland cement
   d. masonry cement

3. In most cases, an engineered fastening system will be required when installing adhered stone veneer over insulation thicker than _____.

   a. 2 inches
   b. 1½ inches
   c. 1 inch
   d. ½ inch

Summary

This module provided an overview of the various aspects of stone masonry that masons need to know in order to be familiar with the terms, tools, and techniques of stone masonry. Masons can pursue specific training in the various aspects of stone masonry as part of their continued career development.

Stone is classified into three broad categories: igneous, sedimentary, and metamorphic. Granite, pumice, and basalt are forms of igneous rock that are commonly used in stone masonry construction. Sedimentary stone is created by the gradual buildup of minerals and organic particles. Sandstone and limestone are examples of this type of stone. Marble and serpentinite are examples of metamorphic stone.

Most natural stone is quarried from large pits and cut and finished for various applications. Ashlar and random rectangular stone are common shapes for quarried stone. Fieldstone, which is stone that has been broken up and weathered naturally, is also used in stone masonry.

Common masonry hand tools include trowels, hammers, and chisels to shape stone; dollies, carts, and suction cups to move and lift stone; guillotines to split stone; and Strapmasters to cut, punch, punch, twist, and bend anchors. Commonly used power tools include saws, grinders, and hammer drills. To lift and position large stone units, masons use mechanical lifting devices such as clamps, thin cables, and straps. Strap and rod anchors are used to fasten stone to masonry wythes and structural members.

Masons are responsible for estimating the quantities of stone to be used in stone masonry construction. Stone for adhered stone veneer installations is always estimated in square feet. Bulk stone is estimated using either the cubic-yard or ton method.

Stone is installed using anchors, or with mortar, or as part of an adhered veneer system. Strap and rod anchors are used to attach stone to masonry wythes or structural backing. Mortar can be used to build stone structures using techniques that are similar to those used to build block and brick masonry structures. Adhered stone veneer requires that mortar be applied to vertical surfaces to attach thin sections of natural or manufactured stone to a substrate. Each of these methods, when performed properly, can result in aesthetically pleasing and durable stone masonry installations that will last for generations.

1. Types of igneous stone identified by geologists total more than _____.

   a. 550
   b. 700
   c. 1,000
   d. 1,750

2. Limestone is a sedimentary stone formed mostly from the mineral _____.

   a. calcium
   b. lithium
   c. lignite
   d. calcite

3. A metamorphic stone often used for countertops because it resembles marble but is more durable is called _____.

   a. serpentinite
   b. basalt
   c. selenite
   d. granite

4. To help cool the cutting blade, stone saws are often equipped with a(n) _____.

   a. vacuum system
   b. intercooler
   c. water spray
   d. pumped coolant system

5. The finishing method used to give stone a smooth texture with little or no luster is known as a _____.

   a. flame finish
   b. honed finish
   c. semigloss finish
   d. polished finish

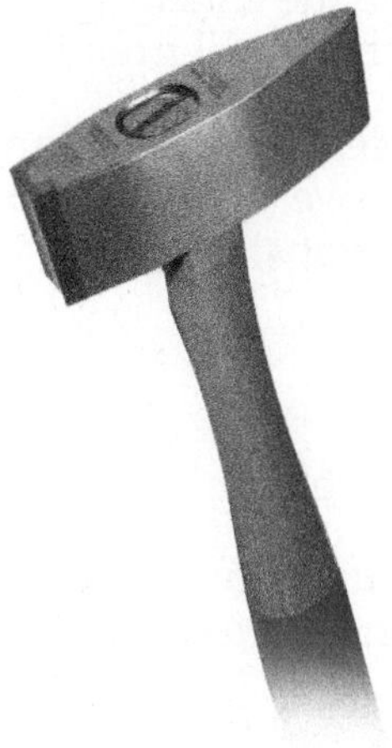

28308-14_RQ01.EPS

Figure 1

6. The tool shown in Review Question *Figure 1* is a _____.

   a. stone-trimming hammer
   b. pitching hammer
   c. chisel-face hammer
   d. combination stone hammer

7. A stone guillotine should never be operated with a loose blade because it _____.

   a. can break the blade
   b. will produce ragged cuts
   c. can cause the attaching bolts to weaken and break under pressure
   d. dulls the cutting surfaces

8. A cutting tool that provides more torque and slower blade rotation than conventional saws is the _____.

   a. worm-drive saw
   b. power bevel saw
   c. high-torque saw
   d. friction-drive saw

9. To cut notches and anchor slots in stone, masons typically use a _____.

   a. combination stone hammer
   b. handheld grinder
   c. cutoff saw
   d. hammer drill

10. When changing a wheel on a grinder, press the spindle lock to _____.

    a. free the locknut
    b. prevent kickback
    c. disengage the clutch
    d. prevent the wheel from rotating

11. Lifting devices used with cranes or boom forklifts to move stone slabs are classified as thin cables, straps, or _____.

    a. clamps
    b. spreaders
    c. jack pins
    d. lifting dowels

12. The devices used to connect lifting devices to a boom or crane are called _____.

    a. turnbuckles
    b. shackles
    c. lewis pins
    d. skyhooks

13. Strap anchors and rods resist shear stresses and _____.

    a. drilling
    b. twisting
    c. compression
    d. tension

14. Strap anchors typically have a width of ¾ inch to _____.

    a. 1 inch
    b. 1½ inches
    c. 2½ inches
    d. 3 inches

15. A measure commonly used to estimate and purchase stone masonry units is the _____.

    a. cubic liter
    b. ounce
    c. ton
    d. hectare

16. Estimates for flagstone and other paving materials are expressed in _____.

    a. square feet of the required thickness
    b. cubic feet of the required thickness
    c. linear feet of the required thickness
    d. square meters, regardless of thickness

17. When estimating stone by weight, begin by determining _____.

    a. the length in linear feet
    b. the volume in cubic feet
    c. the weight in pounds
    d. the number of stone required

18. To prevent discoloration of walls and the exposed face of stone coping, the coping should be _____.

    a. provided with a drip edge
    b. installed with a pitch away from the roof
    c. carefully leveled
    d. installed with a pitch toward the roof

19. As a rule of thumb, the number of rod anchors required per panel is at least _____.

    a. two
    b. four
    c. six
    d. eight

20. Adhered stone veneers are usually installed over a mortar layer known as a(n) _____.

    a. scratch coat
    b. underlay
    c. leveling base
    d. bedding layer

Trade Terms Quiz

Fill in the blank with the correct term that you learned from your study of this module.

1. _____ is a type of sedimentary stone consisting primarily of calcite that is widely used in loadbearing and veneer masonry applications, and is also a key ingredient in concrete.

2. A hand tool that uses a lever-operated ratchet to bend, cut, punch, twist, and shape metal sheet and rod is called a(n) _____.

3. A light-colored, powdery form of igneous stone rich in silica, often used to make concrete and cinder block and as an abrasive material is called _____.

4. _____ is a very hard and durable form of igneous rock widely used in masonry for exterior and interior installations.

5. The brownish stain that forms on the surface of freshly quarried stone as water leaches from the stone is called _____.

6. To be _____ is to be sawn to a predetermined thickness from a larger quarried block.

7. The extent to which a substance allows liquids and gases to pass through it is called _____.

8. _____ is a square- or rectangular-cut stone masonry unit or a flat-faced surface having sawn or dressed bed and joint surfaces.

9. An uneven distribution of structural load in a masonry structure that can cause cracking or buckling is called _____.

10. _____ is a type of sedimentary stone consisting primarily of layers of quartz and feldspar; it is used in ornamental and decorative stone masonry installations and to make grindstones.

11. A hand tool that is used to grip stone and to serve as a handle is called a(n) _____.

12. _____ is the backing material to which natural and manufactured stone veneer is attached.

13. Igneous or sedimentary stone that has been subjected to extreme heat or pressure over a long period of time, causing it to change its physical or chemical structure, is called _____.

14. A(n) _____ is a slot cut into stone and designed to receive the end of a strap anchor.

15. Stone of modular dimensions that has vertical and horizontal bed joints is called _____.

16. _____ is the process of honing, flame finishing, splitting, and polishing the exposed face or faces of stone.

17. The level of reflective shine on the exposed surface of stone is called _____.

18. A(n) _____ is a power saw that uses a cylindrical gear to spin the blade slowly and with more torque than other types of power saws.

19. A mechanical lifting device that grips the sides of stone using friction is called a(n) _____.

20. A(n) _____ is a mechanical lifting device used to grip rough or irregularly shaped stone through the use of several mechanically or hydraulically operated suction surfaces.

21. The general term for a layer of nonbearing stone installed over a masonry wythe or structural backing is _____.

22. A(n) _____ is a mechanical lifting device that uses slings to support stone from underneath.

23. _____ is another word for mined or extracted.

24. The type of stone that is formed when molten rock or volcanic lava cools and solidifies is _____.

25. A(n) _____ is a layer of mortar that has been scored or scratched with a raking tool in order to allow stone units to be adhered.

26. A common term for stone that has been cut, shaped, and finished for use in masonry construction is _____.

27. _____ is small, irregular stone debris left over from the quarrying process.

28. To allow freshly quarried stone to dry out is to _____.

29. A(n) _____ is a document assigned to a piece of slabbed stone identifying the final dimensions and finish to be applied, and assigning the stone a number to aid in construction.

30. Layers of sedimentary stone are called _____.

31. _____ is premade stone made from cast cementitious material with pigments and other added materials that give the appearance of natural stone.

32. Stone that is created by the gradual settling and compression of minerals and organic particles into layers is called _____.

33. A(n) _____ is a hand tool consisting of two sharp spring-loaded segmental blades that split stone simultaneously from above and below.

34. A dark, durable form of igneous rock often used in walls and cobblestones is _____.

35. A(n) _____ is a hand tool used to grip rough or irregularly shaped stone through the use of mechanically induced suction.

36. Corrosion-resistant metal mesh attached to the substrate that serves as a base for mortar or plaster is called _____.

37. To be _____ is to be lightly polished.

38. To square up the edges of stone to make them rectangular and to straighten any jagged edges is called _____.

Trade Terms

Ashlar	Guillotine	Manufactured stone	Random rectangular	Slabbed
Basalt	Hand clamp	Metamorphic	stone	Strap
Cladding	Honed	Permeability	Rubble	Strapmaster
Clamp	Igneous	Point loading	Sandstone	Strata
Dimension stone	Kerf	Pumice	Scratch coat	Substrate
Dress	Lath	Quarried	Season	Vacuum cup
Finishing	Limestone	Quarry sap	Sedimentary	Vacuum lifter
Granite	Luster		Shop ticket	Worm-drive saw

Trade Terms Introduced in This Module

Ashlar: A square- or rectangular-cut stone masonry unit; or, a flat-faced surface having sawn or dressed bed and joint surfaces.

Basalt: A dark, durable form of igneous rock often used in walls and cobblestones.

Cladding: A general term for a layer of nonbearing stone installed over a masonry wythe or structural backing.

Clamp: A mechanical lifting device that grips the sides of stone using friction.

Dimension stone: A common term for stone that has been cut, shaped, and finished for use in masonry construction.

Dress: To square up the edges of stone to make them rectangular and to straighten any jagged edges.

Finishing: The process of honing, flame finishing, splitting, and polishing the exposed face or faces of stone.

Granite: A very hard and durable form of igneous rock widely used in masonry for exterior and interior installations.

Guillotine: A hand tool consisting of two sharp spring-loaded segmental blades that split stone simultaneously from above and below.

Hand clamp: A hand tool that is used to grip stone and to serve as a handle.

Honed: To be lightly polished.

Igneous: A type of stone that is formed when molten rock or volcanic lava cools and solidifies.

Kerf: A slot cut into stone and designed to receive the end of a strap anchor.

Lath: Corrosion-resistant metal mesh attached to the substrate that serves as a base for mortar or plaster.

Limestone: A type of sedimentary stone consisting primarily of calcite that is widely used in loadbearing and veneer masonry applications, and is also a key ingredient in concrete.

Luster: The level of reflective shine on the exposed surface of stone.

Manufactured stone: Premade stone made from cast cementitious material with pigments and other added materials that give the appearance of natural stone.

Metamorphic: Igneous or sedimentary stone that has been subjected to extreme heat or pressure over a long period of time, causing it to change its physical or chemical structure.

Permeability: The extent to which a substance allows liquids and gases to pass through it.

Point loading: An uneven distribution of structural load in a masonry structure that can cause cracking or buckling.

Pumice: A light-colored, powdery form of igneous stone rich in silica, often used to make concrete and cinder block and as an abrasive material.

Quarried: Mined or extracted.

Quarry sap: A brownish stain that forms on the surface of freshly quarried stone as water leaches from the stone.

Random rectangular stone: Stone of modular dimensions that has vertical and horizontal bed joints.

Rubble: Small, irregular stone debris left over from the quarrying process.

Sandstone: A type of sedimentary stone consisting primarily of layers of quartz and feldspar; it is used in ornamental and decorative stone masonry installations and to make grindstones.

Scratch coat: A layer of mortar that has been scored or scratched with a raking tool in order to allow stone units to be adhered.

Season: To allow freshly quarried stone to dry out.

Sedimentary: A type of stone that is created by the gradual settling and compression of minerals and organic particles into layers.

Shop ticket: A document assigned to a piece of slabbed stone identifying the final dimensions and finish to be applied, and assigning the stone a number to aid in construction.

Slabbed: Sawn to a predetermined thickness from a larger quarried block.

Strap: A mechanical lifting device that uses slings to support stone from underneath.

Strapmaster: A hand tool that uses a lever-operated ratchet to bend, cut, punch, twist, and shape metal sheet and rod.

Strata: Layers of sedimentary stone.

Substrate: The backing material to which natural and manufactured stone veneer is attached.

Vacuum cup: A hand tool used to grip rough or irregularly shaped stone through the use of mechanically induced suction.

Vacuum lifter: A mechanical lifting device used to grip rough or irregularly shaped stone through the use of several mechanically or hydraulically operated suction surfaces.

Worm-drive saw: A power saw that uses a cylindrical gear to spin the blade slowly and with more torque than other types of power saws.

Additional Resources

This module presents thorough resources for task training. The following resource material is suggested for further study.

ASTM A153, *Standard Specification for Zinc Coating (Hot-Dip) on Iron and Steel Hardware*, Latest Edition. West Conshohocken, PA: ASTM International.

ASTM C1242, *Standard Guide for Selection, Design, and Installation of Dimension Stone Attachment Systems*, Latest Edition. West Conshohocken, PA: ASTM International.

ASTM C1670, *Standard Specification for Adhered Manufactured Stone Masonry Veneer (AMSMV) Units*, Latest Edition. West Conshohocken, PA: ASTM International.

ASTM C1780, *Standard Practice for Installation Methods for Adhered Manufactured Stone Masonry Veneer*, Latest Edition. West Conshohocken, PA: ASTM International.

Estimating Building Costs for the Residential and Light Commercial Construction Professional. 2012. Wayne J. DelPico. New York: John Wiley & Sons.

Estimating in Building Construction. 1999. Frank R. Dagostino and Leslie Feigenbaum. New York: Pearson Education.

Indiana Limestone Handbook, Latest Edition. Bedford, IN: Indiana Limestone Institute of America, Inc.

Marble and Stone Slab Veneer, Second Edition. 1989. James E. Amrhein and Michael W. Merrigan. Los Angeles, CA: Masonry Institute of America.

Masonry Cement Mortar: Its Proper Use in Construction. 1993. Pat Howley. Murray Hill, NJ: ESSROC Materials.

Technical Note TN1, *Cold and Hot Weather Construction*. 2006. Reston, VA: The Brick Industry Association. **www.gobrick.com**

TEK 5-1B, *Concrete Masonry Veneer Details*. 2003. Herndon, VA: National Concrete Masonry Association. **www.ncma.org**

Figure Credits

Courtesy of Park Industries, CO01, Figure 5

Courtesy of Dennis Neal, FMA&EF, Figure 1

Courtesy of S4Carlisle Publishing Services, Figure 2

M.J. Lazun, Figure 3

Courtesy of Bon Tool Co., Figure 6, Figure 7a, Figures 8–10, Figure 12, Figure 15, RQ01, E01

Steven Fechino, Figure 7b

Photo courtesy of Wood's Powr-Grip Co., Inc., **WPG.com**, Figure 11

Courtesy of PAVE TECH, INC., Figure 13

Courtesy of **www.strapmaster.biz**, Figure 14

Courtesy of RIGID®, Figures 16–17

Courtesy of DEWALT Industrial Tool Co., Figure 18

Courtesy of Abaco Machines, USA, Inc., Figures 19–21

Courtesy of Hohmann & Barnard, Inc., Figures 22a, c, e and f, Figures 23a–c

Courtesy of Heckmann Building Products, Inc., Figures 22b and d, Figures 23d–e

Courtesy of MULTIQUIP INC., Figure 28

National Concrete Masonry Association, Figure 29, Table 2, Figure 30

Answer	Section Reference	Objective
Section One		
1. b	1.1.1	1a
2. d	1.2.0	1b
3. a	1.3.0	1c
Section Two		
1. c	2.1.4	2a
2. c	2.2.1	2b
3. b	2.3.0	2c
4. d	2.4.0	2d
Section Three		
1. c	3.0.0	3
2. b	3.2.1	3b
Section Four		
1. b	4.1.1	4a
2. d	4.2.0	4b
3. d	4.3.0	4c

NCCER CURRICULA — USER UPDATE

NCCER makes every effort to keep its textbooks up-to-date and free of technical errors. We appreciate your help in this process. If you find an error, a typographical mistake, or an inaccuracy in NCCER's curricula, please fill out this form (or a photocopy), or complete the online form at **www.nccer.org/olf**. Be sure to include the exact module ID number, page number, a detailed description, and your recommended correction. Your input will be brought to the attention of the Authoring Team. Thank you for your assistance.

Instructors – If you have an idea for improving this textbook, or have found that additional materials were necessary to teach this module effectively, please let us know so that we may present your suggestions to the Authoring Team.

NCCER Product Development and Revision

13614 Progress Blvd., Alachua, FL 32615

Email: curriculum@nccer.org
Online: www.nccer.org/olf

❑ Trainee Guide   ❑ Lesson Plans   ❑ Exam   ❑ PowerPoints   Other _____

Craft / Level: _____ Copyright Date: _____

Module ID Number / Title: _____

Section Number(s): _____

Description: _____

_____

_____

_____

_____

Recommended Correction: _____

_____

_____

_____

_____

Your Name: _____

Address: _____

_____

Email: _____ Phone: _____

Fundamentals of Crew Leadership

46101-11

NCCER

President: Don Whyte
Director of Product Development: Daniele Dixon
Fundamentals of Crew Leadership Project Manager: Patty Bird
Senior Manager of Production: Tim Davis
Quality Assurance Coordinator: Debie Hicks
Editor: Chris Wilson
Desktop Publishing Coordinator: James McKay
Production Specialist: Megan Casey

Editorial and production services provided by Topaz Publications, Liverpool, NY
Lead Writer/Project Manager: Tom Burke
Desktop Publisher: Joanne Hart
Art Director: Megan Paye
Permissions Editors: Andrea LaBarge, Alison Richmond

FOREWORD

Work gets done most efficiently if workers are divided into crews with a common purpose. When a crew is formed to tackle a particular job, one person is appointed the leader. This person is usually an experienced craftworker who has demonstrated leadership qualities. To become an effective leader, it helps if you have natural leadership qualities, but there are specific job skills that you must learn in order to do the job well.

This module will teach you the skills you need to be an effective leader, including the ability to communicate effectively; provide direction to your crew; and effectively plan and schedule the work of your crew.

As a crew member, you weren't required to think much about project cost. However, as a crew leader, you need to understand how to manage materials, equipment, and labor in order to work in a cost-effective manner. You will also begin to view safety from a different perspective. The crew leader takes on the responsibility for the safety of the crew, making sure that workers follow company safety polices and have the latest information on job safety issues.

As a crew leader, you become part of the chain of command in your company, the link between your crew and those who supervise and manage projects. As such, you need to know how the company is organized and how you fit into the organization. You will also focus more on company policies than a crew member, because it is up to you to enforce them within your crew. You will represent your team at daily project briefings and then communicate relevant information to your crew. This means learning how to be an effective listener and an effective communicator.

Whether you are currently a crew leader wanting to learn more about the requirements, or a crew member preparing to move up the ladder, this module will help you reach your goal.

This program consists of a Participant's Manual and Lesson Plans for the Instructor. The Participant's Manual contains the material that the participant will study, along with self-check exercises and activities, to help in evaluating whether the participant has mastered the knowledge needed to become an effective crew leader. The Lesson Plans include instructional outlines, suggested classroom activities, and homework based on the material in the Participant's Manual.

For the participant to gain the most from this program, it is recommended that the material be presented in a formal classroom setting, using a trained and experienced instructor. If the student is so motivated, he or she can study the material on a self-learning basis by using the material in both the Participant's Manual and the Lesson Plans. Recognition through the National Registry is available for the participants provided the program is delivered through an Accredited Sponsor by a Master Trainer or ICTP instructor. More details on this program can be received by contacting NCCER at **www.nccer.org**.

Participants in this program should note that some examples provided to reinforce the material may not apply to the participant's exact work, although the process will. Every company has its own mode of operation. Therefore, some topics may not apply to every participant's company. Such topics have been included because they are important considerations for prospective crew leaders throughout the industries supported by NCCER.

Industry-Recognized Credentials

If you are training through an NCCER-accredited sponsor, you may be eligible for credentials from NCCER's Registry. The ID number for this module is 46101-11. Note that this module may have been used in other NCCER curricula and may apply to other level completions. Contact NCCER's Registry at 888.622.3720 or go to **www.nccer.org** for more information.

Contents

Topics to be presented in this module include:

Contents (continued) ——————

Contents (continued)

Contents (continued) ———————————————

Figures and Tables ———————————————

Acknowledgments

This curriculum was revised as a result of the farsightedness
and leadership of the following sponsors:

ABC South Texas Chapter
HB Training & Consulting
Turner Industries Group, LLC

University of Georgia
Vision Quest Academy
Willmar Electric Service

This curriculum would not exist were it not for the dedication and unselfish energy of
those volunteers who served on the Authoring Team. A sincere thanks is extended to the following:

John Ambrosia
Harold (Hal) Heintz
Mark Hornbuckle
Jonathan Liston

Jay Tornquist
Wayne Tyson
Antonio "Tony" Vazquez

NCCER Partners

American Fire Sprinkler Association
Associated Builders and Contractors, Inc.
Associated General Contractors of America
Association for Career and Technical Education
Association for Skilled and Technical Sciences
Carolinas AGC, Inc.
Carolinas Electrical Contractors Association
Center for the Improvement of Construction
Management and Processes
Construction Industry Institute
Construction Users Roundtable
Construction Workforce Development Center
Design Build Institute of America
GSSC – Gulf States Shipbuilders Consortium
Manufacturing Institute
Mason Contractors Association of America
Merit Contractors Association of Canada
NACE International
National Association of Minority Contractors
National Association of Women in Construction
National Insulation Association
National Ready Mixed Concrete Association
National Technical Honor Society
National Utility Contractors Association
NAWIC Education Foundation
North American Technician Excellence

Painting & Decorating Contractors of America
Portland Cement Association
SkillsUSA®
Steel Erectors Association of America
U.S. Army Corps of Engineers
University of Florida, M. E. Rinker School of
Building Construction
Women Construction Owners & Executives, USA

Objectives

Upon completion of this section, you should be able to:

1. Describe the opportunities in the construction and power industries.
2. Describe how workers' values change over time.
3. Explain the importance of training and safety for the leaders in the construction and power industries.
4. Describe how new technologies are beneficial to the construction and power industries.
5. Identify the gender and minority issues associated with a changing workforce.
6. Describe what employers can do to prevent workplace discrimination.
7. Differentiate between formal and informal organizations.
8. Describe the difference between authority, responsibility, and accountability.
9. Explain the purpose of job descriptions and what they should include.
10. Distinguish between company policies and procedures.

1.0.0 INDUSTRY TODAY

Today's managers, supervisors, and crew leaders face challenges different from those of previous generations of leaders. To be a leader in industry today, it is essential to be well prepared. Today's crew leaders must understand how to use various types of new technology. In addition, they must have the knowledge and skills needed to manage, train, and communicate with a culturally diverse workforce whose attitudes toward work may differ from those of earlier generations and cultures. These needs are driven by changes in the workforce itself and in the work environment, and include the following:

- A shrinking workforce
- The growth of technology
- Changes in employee attitudes and values
- The emphasis on bringing women and minorities into the workforce
- The growing number of foreign-born workers
- Increased emphasis on workplace health and safety
- Greater focus on education and training

1.1.0 The Need for Training

Effective craft training programs are necessary if the industry is to meet the forecasted worker demands. Many of the skilled, knowledgeable craftworkers, crew leaders, and managers—the so-called baby boomers—have reached retirement age. In 2010, these workers who were born between 1946 and 1964, represented 38 percent of the workforce. Their departure leaves a huge vacuum across the industry spectrum. The Department of Labor (DOL) concludes that the best way for industry to reduce shortages of skilled workers is to create more education and training opportunities. The DOL suggests that companies and community groups form partnerships and create apprenticeship programs. Such programs could provide younger workers, including women and minorities, with the opportunity to develop job skills by giving them hands-on experience.

When training workers, it is important to understand that people learn in different ways. Some people learn by doing, some people learn by watching or reading, and others need step-by-step instructions as they are shown the process. Most people learn best by a combination of styles. It is important to understand what kind of a learner you are teaching, because if you learn one way, you tend to teach the way you learn. Have you ever tried to teach somebody and failed, and then another person successfully teaches the same thing in a different way? A person who acts as a mentor or trainer needs to be able to determine what kind of learner they are addressing and teach according to those needs.

The need for training is not limited to craftworkers. There must be supervisory training to ensure there are qualified leaders in the industry to supervise the craftworkers.

1.1.1 Motivation

As a supervisor or crew leader, it is important to understand what motivates your crew. Money is often thought to be a good motivator. Although that may be true to some extent, it has been proven to be a temporary solution. Once a person has reached a level of financial security, other factors come into play. Studies show that many people tend to be motivated by environment and conditions. For those people, a great workplace may provide more satisfaction than pay. If you give someone a raise, they tend to work harder for a period of time. Then the satisfaction dissipates and they may want another raise. People are often motivated by feeling a sense of accomplishment. That is why setting and working toward recognizable goals tends to make employees more productive. A person with a feeling of involvement or a sense of achievement is likely to be better motivated and help to motivate others.

1.1.2 Understanding Workers

Many older workers grew up in an environment in which they were taught to work hard and stay with the job until retirement. They expected to stay with a company for a long time, and companies were structured to create a family-type environment.

Times have changed. Younger workers have grown up in a highly mobile society and are used to rapid rewards. This generation of workers can sometimes be perceived as lazy and unmotivated, but in reality, they simply have a different perspective. For such workers, it may be better to give them small projects or break up large projects into smaller pieces so that they feel repetitively rewarded, thus enhancing their perception of success.

- *Goal setting* – Set short-term and long-term goals, including tasks to be done and expected time frames. Help the trainees understand that things can happen to offset the short-term goals. This is one reason to set long-term goals as well. Don't set them up for failure, as this leads to frustration, and frustration can lead to reduced productivity.
- *Feedback* – Timely feedback is important. For example, telling someone they did a good job last year, or criticizing them for a job they did a month ago, is meaningless. Simple recognition isn't always enough. Some type of reward should accompany positive feedback, even if it is simply recognizing the employee in a public way. Constructive feedback should be given in private and be accompanied by some positive action, such as providing one-on-one training to correct a problem.

1.1.3 Craft Training

Craft training is often informal, taking place on the job site, outside of a traditional training classroom. According to the American Society for Training and Development (ASTD), craft training is generally handled through on-the-job instruction by a qualified co-worker or conducted by a supervisor.

The Society of Human Resources Management (SHRM) offers the following tips to supervisors in charge of training their employees:

- *Help crew members establish career goals.* Once the goals are established, the training required to meet the goals can be readily identified.
- *Determine what kind of training to give.* Training can be on the job under the supervision of a co-worker. It can be one-on-one with the supervisor. It can involve cross-training to teach a new trade or skill, or it can involve delegating new or additional responsibilities.
- *Determine the trainee's preferred method of learning.* Some people learn best by watching, others from verbal instructions, and others by doing. When training more than one person at a time, try to use all three methods.

Communication is a critical component of training employees. The SHRM advises that supervisors do the following when training their employees:

- *Explain the task, why it needs to be done, and how it should be done.* Confirm that the trainees understand these three areas by asking questions. Allow them to ask questions as well.
- *Demonstrate the task.* Break the task down into manageable parts and cover one part at a time.
- *Ask trainees to do the task while you observe them.* Try not to interrupt them while they are doing the task unless they are doing something that is unsafe and potentially harmful.
- *Give the trainees feedback.* Be specific about what they did and mention any areas where they need to improve.

1.1.4 Supervisory Training

Given the need for skilled craftworkers and qualified supervisory personnel, it seems logical that companies would offer training to their employees through in-house classes, or by subsidizing outside training programs. While some contractors have their own in-house training programs or participate in training offered by associations and other organizations, many contractors do not offer training at all.

There are a number of reasons that companies do not develop or provide training programs, including the following:

- Lack of money to train
- Lack of time to train
- Lack of knowledge about the benefits of training programs
- High rate of employee turnover
- Workforce too small

- Past training involvement was ineffective
- The company hires only trained workers
- Lack of interest from workers
- Lack of company interest in training

For craftworkers to move up into supervisory and managerial positions, it will be necessary for them to continue their education and training. Those who are willing to acquire and develop new skills have the best chance of finding stable employment. It is therefore critical that employees take advantage of training opportunities, and that companies employ training as part of their business culture.

Your company has recognized the need for training. Your participation in a leadership training program such as this will begin to fill the gap between craft and supervisory training.

1.2.0 Impact of Technology

Many industries, including the construction industry, have made the move to technology as a means of remaining competitive. Benefits include increased productivity and speed, improved quality of documents, greater access to common data, and better financial controls and communication. As technology becomes a greater part of supervision, crew leaders need to be able to use it properly. One important concern with electronic communication is to keep it brief, factual, and legal. Because the receiver has no visual or auditory clues as to the sender's intent, the sender can be easily misunderstood. In other words, it is more difficult to tell if someone is just joking via e-mail because you can't see their face or hear the tone of their voice.

Cellular telephones, voicemail, and handheld communication devices have made it easy to keep in touch. They are particularly useful communication sources for contractors or crew leaders who are on a job site, away from their offices, or constantly on the go.

Cellular telephones allow the users to receive incoming calls as well as make outgoing calls. Unless the owner is out of the cellular provider's service area, the cell phone may be used any time to answer calls, make calls, and send and receive voicemail or email. Always check the company's policy with regard to the use of cell phones on the job.

Handheld communication devices known as smart phones allow supervisors to plan their calendars, schedule meetings, manage projects, and access their email from remote locations. These computers are small enough to fit in the palm of the hand, yet powerful enough to hold years of information from various projects. Information can be transmitted electronically to others on the project team or transferred to a computer.

2.0.0 GENDER AND CULTURAL ISSUES

During the past several years, the construction industry in the United States has experienced a shift in worker expectations and diversity. These two issues are converging at a rapid pace. At no time has there been such a generational merge in the workforce, ranging from The Silent Generation (1925–1945), Baby Boomers (1946–1964), Gen X (1965–1979), and the Millennials, also known as Generation Y (1980–2000).

This trend, combined with industry diversity initiatives, has created a climate in which companies recognize the need to embrace a diverse workforce that crosses generational, gender, and ethnic boundaries. To do this effectively, they are using their own resources, as well as relying on associations with the government and trade organizations. All current research indicates that industry will be more dependent on the critical skills of a diverse workforce—a workforce that is both culturally and ethnically fused. Across the United States, construction and other industries are aggressively seeking to bring new workers into their ranks, including women and racial and ethnic minorities. Diversity is no longer solely driven by social and political issues, but by consumers who need hospitals, malls, bridges, power plants, refineries, and many other commercial and residential structures.

Some issues relating to a diverse workforce will need to be addressed on the job site. These issues include different communication styles of men and women, language barriers associated with cultural differences, sexual harassment, and gender or racial discrimination.

2.1.0 Communication Styles of Men and Women

As more and more women move into construction, it becomes increasingly important that communication barriers between men and women are broken down and that differences in behaviors are understood so that men and women can work together more effectively. The Jamestown, New York Area Labor Management Committee (JALMC) offers the following explanations and tips:

- *Women tend to ask more questions than men do.* Men are more likely to proceed with a job and figure it out as they go along, while women are more likely to ask questions first.
- *Men tend to offer solutions before empathy; women tend to do the opposite.* Both men and women should say what they want up front, whether it's the solution to a problem, or simply a sympathetic ear. That way, both genders will feel understood and supported.
- *Women are more likely to ask for help when they need it.* Women are generally more pragmatic when it comes to completing a task. If they need help, they will ask for it. Men are more likely to attempt to complete a task by themselves, even when assistance is needed.
- *Men tend to communicate more competitively, and women tend to communicate more cooperatively.* Both parties need to hear one another out without interruption.

This does not mean that one method is more effective than the other. It simply means that men and women use different approaches to achieve the same result.

2.2.0 Language Barriers

Language barriers are a real workplace challenge for crew leaders. Millions of workers speak languages other than English. Spanish is commonly spoken in the United States. As the makeup of the immigrant population continues to change, the number of non-English speakers will rise dramatically, and the languages being spoken will also change. Bilingual job sites are increasingly common.

Companies have the following options to overcome this challenge:

- Offer English classes either at the work site or through school districts and community colleges.
- Offer incentives for workers to learn English.

As the workforce becomes more diverse, communicating with people for whom English is a second language will be even more critical. The following tips will help when communicating across language barriers:

- Be patient. Give workers time to process the information in a way that they can comprehend.
- Avoid humor. Humor is easily misunderstood and may be misinterpreted as a joke at the worker's expense.
- Don't assume that people are unintelligent simply because they don't understand what you are saying.
- Speak slowly and clearly, and avoid the tendency to raise your voice.
- Use face-to-face communication whenever possible. Over-the-phone communication is often more difficult when a language barrier is involved.
- Use pictures or drawings to get your point across.
- If a worker speaks English poorly but understands reasonably well, ask the worker to demonstrate his or her understanding through other means.

2.3.0 Cultural Differences

As workers from a multitude of backgrounds and cultures are brought together, there are bound to be differences and conflicts in the workplace.

To overcome cultural conflicts, the SHRM suggests the following approach to resolving cultural conflicts between individuals:

- *Define the problem from both points of view.* How does each person involved view the conflict? What does each person think is wrong? This involves moving beyond traditional thought processes to consider alternate ways of thinking.
- *Uncover cultural interpretations.* What assumptions are being made based on cultural programming? By doing this, the supervisor may realize what motivated an employee to act in a particular manner.

- *Create cultural synergy.* Devise a solution that works for both parties involved. The purpose is to recognize and respect other's cultural values, and work out mutually acceptable alternatives.

2.4.0 Sexual Harassment

In today's business world, men and women are working side-by-side in careers of all kinds. As women make the transition into traditionally male industries, such as construction, the likelihood of sexual harassment increases. Sexual harassment is defined as unwelcome behavior of a sexual nature that makes someone feel uncomfortable in the workplace by focusing attention on their gender instead of on their professional qualifications. Sexual harassment can range from telling an offensive joke or hanging a poster of a swimsuit-clad man or woman, to making sexual comments or physical advances.

Historically, sexual harassment was thought to be an act performed by men of power within an organization against women in subordinate positions. However, the number of sexual harassment cases over the years, have shown that this is no longer the case.

Sexual harassment can occur in a variety of circumstances, including but not limited to the following:

- The victim as well as the harasser may be a woman or a man. The victim does not have to be of the opposite sex.
- The harasser can be the victim's supervisor, an agent of the employer, a supervisor in another area, a co-worker, or a non-employee.
- The victim does not have to be the person harassed, but could be anyone affected by the offensive conduct.
- Unlawful sexual harassment may occur without economic injury to or discharge of the victim.
- The harasser's conduct must be unwelcome.

The Equal Employment Opportunity Commission (EEOC) enforces sexual harassment laws within industries. When investigating allegations of sexual harassment, the EEOC looks at the whole record, including the circumstances and the context in which the alleged incidents occurred. A decision on the allegations is made from the facts on a case-by-case basis. A crew leader who is aware of sexual harassment and does nothing to stop it can be held responsible. The crew leader therefore should not only take action to stop sexual harassment, but should serve as a good example for the rest of the crew.

Prevention is the best tool to eliminate sexual harassment in the workplace. The EEOC encourages employers to take steps to prevent sexual harassment from occurring. Employers should clearly communicate to employees that sexual harassment will not be tolerated. They do so by developing a policy on sexual harassment, establishing an effective complaint or grievance process, and taking immediate and appropriate action when an employee complains.

Both swearing and off-color remarks and jokes are not only offensive to co-workers, but also tarnish a worker's image. Crew leaders need to emphasize that abrasive or crude behavior may affect opportunities for advancement. If disciplinary action becomes necessary, it should be covered by company policy. A typical approach is a three-step process in which the perpetrator is first given a verbal reprimand. In the event of further violations, a written reprimand and warning are given. Dismissal typically accompanies subsequent violations.

2.5.0 Gender and Minority Discrimination

More attention is being placed on fair recruitment, equal pay for equal work, and promotions for women and minorities in the workplace. Consequently, many business practices, including the way employees are treated, the organization's hiring and promotional practices, and the way people are compensated, are being analyzed for equity.

Once a male-dominated industry, construction companies are moving away from this image and are actively recruiting and training women, younger workers, people from other cultures, and workers with disabilities. This means that organizations hire the best person for the job, without regard for race, sex, religion, age, etc.

To prevent discrimination cases, employers must have valid job-related criteria for hiring, compensation, and promotion. These measures must be used consistently for every applicant interview, employee performance appraisal, and hiring or promotion decision. Therefore, all workers responsible for recruitment, selection, and supervision of employees, and evaluating job performance, must be trained on how to use the job-related criteria legally and effectively.

3.0.0 BUSINESS ORGANIZATIONS

An organization is the relationship among the people within the company or project. The crew leader needs to be aware of two types of organizations. These are formal organizations and informal organizations.

A formal organization exists when the activities of the people within the work group are directed toward achieving a common goal. An example of a formal organization is a work crew consisting of four carpenters and two laborers led by a crew leader, all working together toward a common goal.

A formal organization is typically documented on an organizational chart, which outlines all the positions that make up an organization and shows how those positions are related. Some organizational charts even depict the people within each position and the person to whom they report, as well as the people that the person supervises. *Figures 1* and *2* show examples of organization charts for fictitious companies. Note that each of these positions represents an opportunity for advancement in the construction industry that a crew leader can eventually achieve.

An informal organization allows for communication among its members so they can perform as a group. It also establishes patterns of behavior that help them to work as a group, such as agreeing to use a specific training program.

An example of an informal organization is a trade association such as Associated Builders and Contractors (ABC), Associated General Contractors (AGC), and the National Association of Women in Construction (NAWIC). Those, along with the thousands of other trade associations in the U.S., provide a forum in which members with common concerns can share information, work on issues, and develop standards for their industry.

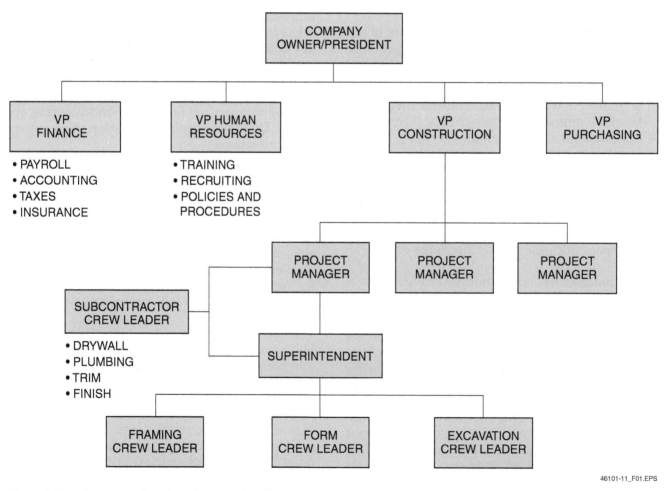

46101-11_F01.EPS

Figure 1 Sample organization chart for a construction company.

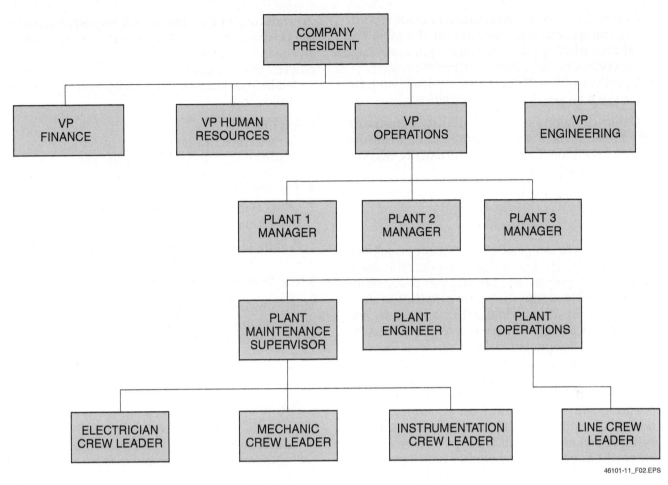

Figure 2 Sample organization chart for an industrial company.

Both types of organizations establish the foundation for how communication flows. The formal structure is the means used to delegate authority and responsibility and to exchange information. The informal structure is used to exchange information.

Members in an organization perform best when each member:

- Knows the job and how it will be done
- Communicates effectively with others in the group
- Understands his or her role in the group
- Recognizes who has the authority and responsibility

3.1.0 Division of Responsibility

The conduct of a business involves certain functions. In a small organization, responsibilities may be divided between one or two people. However, in a larger organization with many different and complex activities, responsibilities may be grouped into similar activity groups, and the responsibility for each group assigned to department managers. In either case, the following major departments exist in most companies:

- *Executive* – This office represents top management. It is responsible for the success of the company through short-range and long-range planning.
- *Human Resources* – This office is responsible for recruiting and screening prospective employees; managing employee benefits programs; advising management on pay and benefits; and developing and enforcing procedures related to hiring practices.
- *Accounting* – This office is responsible for all record keeping and financial transactions, including payroll, taxes, insurance, and audits.
- *Contract Administration* – This office prepares and executes contractual documents with owners, subcontractors, and suppliers.
- *Purchasing* – This office obtains material prices and then issues purchase orders. The purchasing office also obtains rental and leasing rates on equipment and tools.

- *Estimating*: This office is responsible for recording the quantity of material on the jobs, the takeoff, pricing labor and material, analyzing subcontractor bids, and bidding on projects.
- *Operations*: This office plans, controls, and supervises all project-related activities.

Other divisions of responsibility a company may create involve architectural and engineering design functions. These divisions usually become separate departments.

3.2.0 Authority, Responsibility, and Accountability

As an organization grows, the manager must ask others to perform many duties so that the manager can concentrate on management tasks. Managers typically assign (delegate) activities to their subordinates. When delegating activities, the crew leader assigns others the responsibility to perform the designated tasks.

Responsibility means obligation, so once the responsibility is delegated, the person to whom it is assigned is obligated to perform the duties.

Along with responsibility comes authority. *Authority* is the power to act or make decisions in carrying out an assignment. The type and amount of authority a supervisor or worker has depends on the company for which he or she works. Authority and responsibility must be balanced so employees can carry out their tasks. In addition, delegation of sufficient authority is needed to make an employee accountable to the crew leader for the results.

Accountability is the act of holding an employee responsible for completing the assigned activities. Even though authority and responsibility may be delegated to crew members, the final responsibility always rests with the crew leader.

3.3.0 Job Descriptions

Many companies furnish each employee with a written job description that explains the job in detail. Job descriptions set a standard for the employee. They make judging performance easier, clarify the tasks each person should handle, and simplify the training of new employees.

Each new employee should understand all the duties and responsibilities of the job after reviewing the job description. Thus, the time it takes for the employee to make the transition from being a new and uninformed employee to a more experienced member of a crew is shortened.

A job description need not be long, but it should be detailed enough to ensure there is no misun-

derstanding of the duties and responsibilities of the position. The job description should contain all the information necessary to evaluate the employee's performance.

A job description should contain, at minimum, the following:

- Job title
- General description of the position
- Minimum qualifications for the job
- Specific duties and responsibilities
- The supervisor to whom the position reports
- Other requirements, such as qualifications, certifications, and licenses

A sample job description is shown in *Figure 3*.

3.4.0 Policies and Procedures

Most companies have formal policies and procedures established to help the crew leader carry out his or her duties. A *policy* is a general state-

---

Position:
Crew Leader

General Summary:
First line of supervision on a construction crew installing concrete formwork.

Reports To:
Job Superintendent

Physical and Mental Responsibilities:
- Ability to stand for long periods
- Ability to solve basic math and geometry problems

Duties and Responsibilities:
- Oversee crew
- Provide instruction and training in construction tasks as needed
- Make sure proper materials and tools are on the site to accomplish tasks
- Keep project on schedule
- Enforce safety policies and procedures

Knowledge, Skills, and Experience Required:
- Extensive travel throughout the Eastern United States, home base in Atlanta
- Ability to operate a backhoe and trencher
- Valid commercial driver's license with no DUI violations
- Ability to work under deadlines with the knowledge and ability to foresee problem areas and develop a plan of action to solve the situation

46101-11_F03.EPS

Figure 3 Example of a job description.

---

ment establishing guidelines for a specific activity. Examples include policies on vacations, breaks, workplace safety, and checking out tools.

Procedures are the ways that policies are carried out. For example, a procedure written to implement a policy on workplace safety would include guidelines for reporting accidents and general safety procedures that all employees are expected to follow.

A crew leader must be familiar with the company policies and procedures, especially with regard to safety practices. When OSHA inspectors visit a job site, they often question employees and crew leaders about the company policies related to safety. If they are investigating an accident, they will want to verify that the responsible crew leader knew the applicable company policy and followed it.

Review Questions

1. The construction industry should provide training for craftworkers and supervisors _____.
   a. to ensure that there are enough future workers
   b. to avoid discrimination lawsuits
   c. in order to update the skills of older workers who are retiring at a later age than they previously did
   d. even though younger workers are now less likely to seek jobs in other areas than they were 10 years ago

2. Companies traditionally offer craftworker training _____.
   a. that a supervisor leads in a classroom setting
   b. that a craftworker leads in a classroom setting
   c. in a hands-on setting, where craftworkers learn from a co-worker or supervisor
   d. on a self-study basis to allow craftworkers to proceed at their own pace

3. One way to provide effective training is to _____.
   a. avoid giving negative feedback until trainees are more experienced in doing the task
   b. tailor the training to the career goals and needs of trainees
   c. choose one training method and use it for all trainees
   d. encourage trainees to listen, saving their questions for the end of the session

4. One way to prevent sexual harassment in the workplace is to _____.
   a. require employee training in which the potentially offensive subject of stereotypes is carefully avoided
   b. develop a consistent policy with appropriate consequences for engaging in sexual harassment
   c. communicate to workers that the victim of sexual harassment is the one who is being directly harassed, not those affected in a more indirect way
   d. educate workers to recognize sexual harassment for what it is—unwelcome conduct by the opposite sex

5. Employers can minimize all types of workplace discrimination by hiring based on a consistent list of job-related requirements.

   a. True
   b. False

6. Members tend to function best within an organization when they _____.

   a. are allowed to select their own style of clothing for each project
   b. understand their role
   c. do not disagree with the statements of other workers or supervisors
   d. are able to work without supervision

7. A formal organization is defined as a group of individuals who work independently, but share the same goal

   a. True
   b. False

8. A formal organization uses an organizational chart to _____.

   a. depict all companies with which it conducts business
   b. show all customers with which it conducts business
   c. track projects between departments
   d. show the relationships among the existing positions in the company

9. Which of the following is a function typically performed by the operations department of a company?

   a. Purchase materials
   b. Plan projects
   c. Prepare payrolls
   d. Recruiting and screening new hires

10. The company department that manages employee benefits and personnel recruiting is _____.

    a. Engineering
    b. Human Resources
    c. Purchasing
    d. Contract Administration

11. The power to make decisions and act on them in carrying out an assignment is _____.

    a. delegating
    b. responsibility
    c. decisiveness
    d. authority

12. Accountability is defined as _____.

    a. the power to act or make decisions in carrying out assignments
    b. giving an employee a particular task to perform
    c. the act of an employee responsible for the completion and results of a particular duty
    d. having the power to promote someone

13. A good job description should include _____.

    a. a complete organization chart
    b. any information needed to judge job performance
    c. the company dress code
    d. the company's sexual harassment policy

14. The purpose of a policy is to _____.

    a. establish company guidelines regarding a particular activity
    b. specify what tools and equipment are required for a job
    c. list all information necessary to judge an employee's performance
    d. inform employees about the future plans of the company

15. One example of a procedure would be the rules for taking time off.

    a. True
    b. False

LEADERSHIP SKILLS

Objectives

Upon completion of this section, you should be able to:

1. Describe the role of a crew leader.
2. List the characteristics of effective leaders.
3. Be able to discuss the importance of ethics in a supervisor's role.
4. Identify the three styles of leadership.
5. Describe the forms of communication.
6. Describe the four parts of verbal communication.
7. Describe the importance of active listening.
8. Explain how to overcome the barriers to communication.
9. List ways that leaders can motivate their employees.
10. Explain the importance of delegating and implementing policies and procedures.
11. Distinguish between problem solving and decision making.

1.0.0 INTRODUCTION TO LEADERSHIP

For the purpose of this program, it is important to define some of the positions that will be discussed. The term *craftworker* refers to a person who performs the work of his or her trade(s). The crew leader is a person who supervises one or more craftworkers on a crew. A superintendent is essentially an on-site supervisor who is responsible for one or more crew leaders or front-line supervisors. Finally, a project manager or general superintendent may be responsible for managing one or more projects. This training will concentrate primarily on the role of the crew leader.

Craftworkers and crew leaders differ in that the crew leader manages the activities that the skilled craftworkers on the crews actually perform. In order to manage a crew of craftworkers, a crew leader must have first-hand knowledge and experience in the activities being performed. In addition, he or she must be able to act directly in organizing and directing the activities of the various crew members.

This section explains the importance of developing effective leadership skills as a new crew leader. Effective ways to communicate with all levels of employees and co-workers, build teams, motivate crew members, make decisions, and resolve problems are covered in depth.

2.0.0 THE SHIFT IN WORK ACTIVITIES

The crew leader is generally selected and promoted from a work crew. The selection will often be based on that person's ability to accomplish tasks, to get along with others, to meet schedules, and to stay within the budget. The crew leader must lead the team to work safely and provide a quality product.

Making the transition from a craftworker to a crew leader can be difficult, especially when the new crew leader is in charge of supervising a group of peers. Crew leaders are no longer responsible for their work alone; rather, they are accountable for the work of an entire crew of people with varying skill levels and abilities, a multitude of personalities and work styles, and different cultural and educational backgrounds. Crew leaders must learn to put their personal relationships aside and work for the common goals of the team.

New crew leaders are often placed in charge of workers who were formerly their friends and peers on a crew. This situation can create some conflicts. For example, some of the crew may try to take advantage of the friendship by seeking special favors. They may also want to be privy to information that should be held closely. These problems can be overcome by working with the crew to set mutual performance goals and by freely communicating with them within permitted limits. Use their knowledge and strengths along with your own so that they feel like they are key players on the team.

As an employee moves from a craftworker position to the role of a crew leader, he or she will find that more hours will be spent supervising the work of others than actually performing the technical skill for which he or she has been trained. *Figure 4* represents the percentage of time craftworkers, crew leaders, superintendents, and project managers spend on technical and supervisory work as their management responsibilities increase.

The success of the new crew leader is directly related to the ability to make the transition from crew member into a leadership role.

3.0.0 BECOMING A LEADER

A crew leader must have leadership skills to be successful. Therefore, one of the primary goals of a person who wants to become a crew leader should be to develop strong leadership skills and learn to use them effectively.

There are many ways to define a leader. One straightforward definition is a person who influences other people in the achievement of a goal.

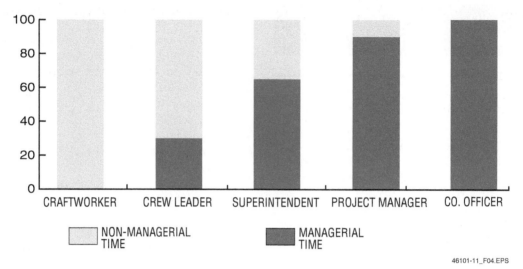

Figure 4 Percentage of time spent on technical and supervisory work.

Some people may have inherited leadership qualities or may have developed traits that motivate others to follow and perform. Research shows that people who possess such talents are likely to succeed as leaders.

3.1.0 Characteristics of Leaders

Leadership traits are similar to the skills that a crew leader needs in order to be effective. Although the characteristics of leadership are many, there are some definite commonalities among effective leaders.

First and foremost, effective leaders lead by example. In other words, they work and live by the standards that they establish for their crew members or followers, making sure they set a positive example.

Effective leaders also tend to have a high level of drive and determination, as well as a stick-to-it attitude. When faced with obstacles, effective leaders don't get discouraged; instead, they identify the potential problems, make plans to overcome them, and work toward achieving the intended goal. In the event of failure, effective leaders learn from their mistakes and apply that knowledge to future situations. They also learn from their successes.

Effective leaders are typically effective communicators who clearly express the goals of a project to their crew members. Accomplishing this may require that the leader overcome issues such as language barriers, gender bias, or differences in personalities to ensure that each member of the crew understands the established goals of the project.

Effective leaders have the ability to motivate their crew members to work to their full potential and become effective members of the team. Crew leaders try to develop crew member skills and encourage them to improve and learn as a means to contribute more to the team effort. Effective leaders strive for excellence from themselves and their team, so they work hard to provide the skills and leadership necessary to do so.

In addition, effective leaders must possess organizational skills. They know what needs to be accomplished, and they use their resources to make it happen. Because they cannot do it alone, leaders enlist the help of their team members to share in the workload. Effective leaders delegate work to their crew members, and they implement company policies and procedures to ensure that the work is completed safely, effectively, and efficiently.

Finally, effective leaders have the authority and self-confidence that allows them to make decisions and solve problems. In order to accomplish their goals, leaders must be able to calculate risks, absorb and interpret information, assess courses of action, make decisions, and assume the responsibility for those decisions.

3.1.1 Leadership Traits

There are many other traits of effective leaders. Some other major characteristics of leaders include the following:

- Ability to plan and organize
- Loyalty to their company and crew
- Ability to motivate
- Fairness

- Enthusiasm
- Willingness to learn from others
- Ability to teach others
- Initiative
- Ability to advocate an idea
- Good communication skills

3.1.2 Expected Leadership Behavior

Followers have expectations of their leaders. They look to their leaders to:

- Lead by example
- Suggest and direct
- Plan and organize the work
- Communicate effectively
- Make decisions and assume responsibility
- Have the necessary technical knowledge
- Be a loyal member of the group
- Abide by company policies and procedures

3.2.0 Functions of a Leader

The functions of a leader will vary with the environment, the group being led, and the tasks to be performed. However, there are certain functions common to all situations that the leader will be called upon to perform. Some of the major functions are:

- Organize, plan, staff, direct, and control work
- Empower group members to make decisions and take responsibility for their work
- Maintain a cohesive group by resolving tensions and differences among its members and between the group and those outside the group
- Ensure that all group members understand and abide by company policies and procedures
- Accept responsibility for the successes and failures of the group's performance
- Represent the group
- Be sensitive to the differences of a diverse workforce

3.3.0 Leadership Styles

There are three main styles of leadership. At one extreme is the autocratic or commander style of leadership, where the crew leader makes all of the decisions independently, without seeking the opinions of crew members. At the other extreme is the hands-off or facilitator style, where the crew leader empowers the employees to make decisions. In the middle is the democratic or collaborative style, where the crew leader seeks crew member opinions and makes the appropriate decisions based on their input.

The following are some characteristics of each of the three leadership styles:

Commander types:

- Expect crew members to work without questioning procedures
- Seldom seek advice from crew members
- Insist on solving problems alone
- Seldom permit crew members to assist each other
- Praise and criticize on a personal basis
- Have no sincere interest in creatively improving methods of operation or production

Partner types:

- Discuss problems with their crew members
- Listen to suggestions from crew members
- Explain and instruct
- Give crew members a feeling of accomplishment by commending them when they do a job well
- Are friendly and available to discuss personal and job-related problems

Facilitator types:

- Believe no supervision is best
- Rarely give orders
- Worry about whether they are liked by their crew members

Effective leadership takes many forms. The correct style for a particular situation or operation depends on the nature of the crew as well as the work it has to accomplish. For example, if the crew does not have enough experience for the job ahead, then a commander style may be appropriate. The autocratic style of leadership is also effective when jobs involve repetitive operations that require little decision-making.

However, if a worker's attitude is an issue, a partner style may be appropriate. In this case, providing the missing motivational factors may increase performance and result in the improvement of the worker's attitude. The democratic style of leadership is also used when the work is of a creative nature, because brainstorming and exchanging ideas with such crew members can be beneficial.

The facilitator style is effective with an experienced crew on a well-defined project.

The company must give a crew leader sufficient authority to do the job. This authority must be commensurate with responsibility, and it must be made known to crew members when they are hired so that they understand who is in charge.

A crew leader must have an expert knowledge of the activities to be supervised in order to be effective. This is important because the crew members need to know that they have someone to turn to when they have a question or a problem, when they need some guidance, or when modifications or changes are warranted by the job.

Respect is probably the most useful element of authority. Respect usually derives from being fair to employees, by listening to their complaints and suggestions, and by using incentives and rewards appropriately to motivate crew members. In addition, crew leaders who have a positive attitude and a favorable personality tend to gain the respect of their crew members as well as their peers. Along with respect comes a positive attitude from the crew members.

3.4.0 Ethics in Leadership

The crew leader should practice the highest standards of ethical conduct. Every day the crew leader has to make decisions that may have ethical implications. When an unethical decision is made, it not only hurts the crew leader, but also other workers, peers, and the company for which he or she works.

There are three basic types of ethics:

1. Business or legal
2. Professional or balanced
3. Situational

Business, or legal, ethics concerns adhering to all laws and regulations related to the issue.

Professional, or balanced, ethics relates to carrying out all activities in such a manner as to be honest and fair to everyone concerned.

Situational ethics pertains to specific activities or events that may initially appear to be a gray area. For example, you may ask yourself, "How will I feel about myself if my actions are published in the newspaper or if I have to justify my actions to my family, friends, and colleagues?"

The crew leader will often be put into a situation where he or she will need to assess the ethical consequences of an impending decision. For instance, should a crew leader continue to keep one of his or her crew working who has broken into a cold sweat due to overheated working conditions just because the superintendent says the activity is behind schedule? Or should a crew leader, who is the only one aware that the reinforcing steel placed by his or her crew was done incorrectly, correct the situation before the concrete is placed in the form? If a crew leader is ever asked to carry through on an unethical decision, it is up to him or her to inform the superintendent of the unethical nature of the issue, and if still requested to follow through, refuse to act.

4.0.0 COMMUNICATION

Successful crew leaders learn to communicate effectively with people at all levels of the organization. In doing so, they develop an understanding of human behavior and acquire communication skills that enable them to understand and influence others.

There are many definitions of communication. Communication is the act of accurately and effectively conveying or transmitting facts, feelings, and opinions to another person. Simply stated, communication is the method of exchanging information and ideas.

Just as there are many definitions of communication, it also comes in many forms, including verbal, nonverbal, and written. Each of these forms of communication are discussed in this section.

4.1.0 Verbal Communication

Verbal communication refers to the spoken words exchanged between two or more people. Verbal communication consists of four distinct parts:

1. Sender
2. Message
3. Receiver
4. Feedback

Figure 5 depicts the relationship of these four parts within the communication process. In verbal communication, the focus is on feedback, which is used to verify that the sender's message was received as intended.

> **Did you know?**
>
> Research shows that the typical supervisor spends about 80 percent of his or her day communicating through writing, speaking, listening, or using body language. Of that time, studies suggest that approximately 20 percent of communication is written, and 80 percent involves speaking or listening.

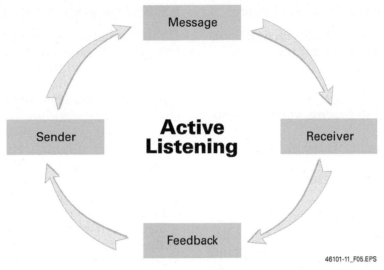

Figure 5 Communication process.

4.1.1 The Sender

The sender is the person who creates the message to be communicated. In verbal communication, the sender actually says the message aloud to the person(s) for whom it is intended.

The sender must be sure to speak in a clear and concise manner that can be easily understood by others. This is not an easy task; it takes practice. Some basic speaking tips are:

- Avoid talking with anything in your mouth (food, gum, etc.).
- Avoid swearing and acronyms.
- Don't speak too quickly or too slowly. In extreme cases, people tend to focus on the rate of speech rather than what is being communicated.
- Pronounce words carefully to prevent misunderstandings.
- Speak with enthusiasm. Avoid speaking in a harsh voice or in a monotone.

4.1.2 The Message

The message is what the sender is attempting to communicate to the audience. A message can be a set of directions, an opinion, or a feeling. Whatever its function, a message is an idea or fact that the sender wants the audience to know.

Before speaking, determine what must be communicated, then take the time to organize what to say, ensuring that the message is logical and complete. Taking the time to clarify your thoughts prevents rambling, not getting the message across effectively, or confusing the audience. It also permits the sender to get to the point quickly.

In delivering the message, the sender should assess the audience. It is important not to talk down to them. Remember that everyone, whether in a senior or junior position, deserves respect and courtesy. Therefore, the sender should use words and phrases that the audience can understand and avoid technical language or slang. In addition, the sender should use short sentences, which gives the audience time to understand and digest one point or fact at a time.

4.1.3 The Receiver

The receiver is the person to whom the message is communicated. For the communication process to be successful, it is important that the receiver understands the message as the sender intended. Therefore, the receiver must listen to what is being said.

There are many barriers to effective listening, particularly on a busy construction job site. Some of these obstacles include the following:

- Noise, visitors, cell phones, or other distractions
- Preoccupation, being under pressure, or daydreaming
- Reacting emotionally to what is being communicated
- Thinking about how to respond instead of listening
- Giving an answer before the message is complete
- Personal biases to the sender's communication style
- Finishing the sender's sentence

Some tips for overcoming these barriers are:

- Take steps to minimize or remove distractions; learn to tune out your surroundings
- Listen for key points
- Take notes
- Try not to take things personally
- Allow yourself time to process your thoughts before responding
- Let the sender communicate the message without interruption
- Be aware of your personal biases, and try to stay open-minded

There are many ways for a receiver to show that he or she is actively listening to what is being said. This can even be accomplished without saying a word. Examples include maintaining eye contact, nodding your head, and taking notes. It may also be accomplished through feedback.

4.1.4 Feedback

Feedback refers to the communication that occurs after the message has been sent by the sender and received by the receiver. It involves the receiver responding to the message.

Feedback is a very important part of the communication process because it allows the receiver to communicate how he or she interpreted the message. It also allows the sender to ensure that the message was understood as intended. In other words, feedback is a checkpoint to make sure the receiver and sender are on the same page.

The receiver can use the opportunity of providing feedback to paraphrase back what was heard. When paraphrasing what you heard, it is best to use your own words. That way, you can show the sender that you interpreted the message correctly and could explain it to others if needed.

In addition, the receiver can clarify the meaning of the message and request additional information when providing feedback. This is generally accomplished by asking questions.

One opportunity to provide feedback is in the performance of crew evaluations. Many companies have formal evaluation forms that are used on a yearly basis to evaluate workers for pay increases. These evaluations should not come as a once-a-year surprise. An effective crew leader provides constant performance feedback, which is ultimately reflected in the annual performance evaluation. It is also important to stress the importance of self-evaluation with your crew.

4.2.0 Nonverbal Communication

Unlike verbal or written communication, nonverbal communication does not involve the spoken or written word. Rather, non-verbal communication refers to things that you can actually see when communicating with others. Examples include facial expressions, body movements, hand gestures, and eye contact.

Nonverbal communication can provide an external signal of an individual's inner emotions. It occurs simultaneously with verbal communication; often, the sender of the nonverbal communication is not even aware of it.

Because it can be physically observed, nonverbal communication is just as important as the words used in conveying the message. Often, people are influenced more by nonverbal signals than by spoken words. Therefore, it is important to be conscious of nonverbal cues because you don't want the receiver to interpret your message incorrectly based on your posture or an expression on your face. After all, these things may have nothing to do with the communication exchange; instead, they may be carrying over from something else going on in your day.

4.3.0 Written or Visual Communication

Some communication will have to be written or visual. Written or visual communication refers to communication that is documented on paper or transmitted electronically using words or visuals.

Many messages on a job have to be communicated in text form. Examples include weekly reports, requests for changes, purchase orders, and correspondence on a specific subject. These items are written because they must be recorded for business and historical purposes. In addition, some communication on the job will have to be visual. Items that are difficult to explain verbally or by the written word can best be explained through diagrams or graphics. Examples include the plans or drawings used on a job.

When writing or creating a visual message, it is best to assess the reader or the audience before beginning. The reader must be able to read the message and understand the content; otherwise, the communication process will be unsuccessful. Therefore, the writer should consider the actual meaning of words or diagrams and how others might interpret them. In addition, the writer should make sure that all handwriting is legible if the message is being handwritten.

Here are some basic tips for writing:

- Avoid emotion-packed words or phrases.
- Be positive whenever possible.
- Avoid using technical language or jargon.
- Stick to the facts.
- Provide an adequate level of detail.
- Present the information in a logical manner.
- Avoid making judgments unless asked to do so.
- Proofread your work; check for spelling and grammatical errors.
- Make sure that the document is legible.
- Avoid using acronyms.
- Make sure the purpose of the message is clearly stated.
- Be prepared to provide a verbal or visual explanation, if needed.

Here are some basic tips for creating visuals:

- Provide an adequate level of detail.
- Ensure that the diagram is large enough to be seen.
- Avoid creating complex visuals; simplicity is better.
- Present the information in a logical order.
- Be prepared to provide a written or verbal explanation of the visual, if needed.

4.4.0 Communication Issues

It is important to note that each person communicates a little differently; that is what makes us unique as individuals. As the diversity of the workforce changes, communication will become even more challenging because the audience may include individuals from different ethnic groups, cultural backgrounds, educational levels, and economic status groups. Therefore, it is necessary to assess the audience in order to determine how to communicate effectively with each individual.

The key to effective communication is to acknowledge that people are different and to be able to adjust the communication style to meet the needs of the audience or the person on the receiving end of your message. This involves relaying the message in the simplest way possible, avoiding the use of words that people may find confusing. Be aware of how you use technical language, slang, jargon, and words that have multiple meanings. Present the information in a clear, concise manner. Avoid rambling and always speak clearly, using good grammar.

In addition, be prepared to communicate the message in multiple ways or adjust your level of detail or terminology to ensure that everyone understands the meaning as intended. For instance, a visual person who cannot comprehend directions in a verbal or written form may need a map. It may be necessary to overcome language barriers on the job site by using graphics or visual aids to relay the message.

Figure 6 shows how to tailor the message to the audience.

VERBAL INSTRUCTIONS Experienced Crew	VERBAL INSTRUCTIONS Newer Crew	WRITTEN INSTRUCTIONS	DIAGRAM/MAP
"Please drive to the supply shop to pick up our order."	"Please drive to the supply shop. Turn right here and left at Route 1. It's at 75th Street and Route 1. Tell them the company name and that you're there to pick up our order."	1. Turn right at exit. 2. Drive 2 miles to Route 1. Turn LEFT. 3. Drive 1 mile (pass the tire shop) to 75th Street. 4. Look for supply store on right. . . .	

Different people learn in different ways. Be sure to communicate so you can be understood.

46101-11_F06.EPS

Figure 6 Tailor your message.

Read the following verbal conversations, and identify any problems:

Conversation I:

Judy: Hey, Roger...
Roger: What's up?
Judy: Has the site been prepared for the job trailer yet?
Roger: Job trailer?
Judy: The job trailer—it's coming in today. What time will the job site be prepared?
Roger: The trailer will be here about 1:00 PM.
Judy: The job site! What time will the job site be prepared?

Conversation II:

John: Hey, Mike, I need your help.
Mike: What is it?
John: You and Joey go over and help Al's crew finish laying out the site.
Mike: Why me? I can't work with Joey. He can't understand a word I say.
John: Al's crew needs some help, and you and Joey are the most qualified to do the job.
Mike: I told you, I can't work with Joey.

Conversation III:

Ed: Hey, Jill.
Jill: Sir?
Ed: Have you received the latest DOL, EEO requirement to be sure the OFCP administrator finds our records up to date when he reviews them in August?
Jill: DOL, EEO, and OFCP?
Ed: Oh, and don't forget the MSHA, OSHA, and EPA reports are due this afternoon.
Jill: MSHA, OSHA, and EPA?

Conversation IV:

Susan: Hey, Bob, would you do me a favor?
Bob: Okay, Sue. What is it?
Susan: I was reading the concrete inspection report and found the concrete in Bays 4A, 3B, 6C, and 5D didn't meet the 3,000 psi strength requirements. Also, the concrete inspector on the job told me the two batches that came in today had to be refused because they didn't meet the slump requirements as noted on page 16 of the spec. I need to know if any placement problems happened on those bays, how long the ready mix trucks were waiting today, and what we plan to do to stop these problems in the future.

Participant Exercise A, Part II

Read the following written memos, and identify any problems:

Memo I:

Let's start with the transformer vault $285.00 due. For what you ask? Answer: practically nothing I admit, but here is the story. Paul the superintendent decided it was not the way good ole Comm Ed wanted it, we took out the ladder and part of the grading (as Paul instructed us to do) we brought it back here to change it. When Comm Ed the architect or Doe found out that everything would still work the way it was, Paul instructed us to reinstall the work. That is the whole story there is please add the $285.00 to my next payout.

Memo II:

Let's take rooms C 307-C-312 and C-313 we made the light track supports and took them to the job to erect them when we tried to put them in we found direct work in the way, my men spent all day trying to find out what to do so ask your Superintendent (Frank) he will verify seven hours pay for these men as he went back and forth while my men waited. Now the Architect has changed the system of hanging and has the gall to say that he has made my work easier, I can't see how. Anyway, we want an extra two (2) men for seven (7) hours for April 21 at $55.00 per hour or $385.00 on April 28th Doe Reference 197 finally resolved this problem. We will have no additional charges on Doe Reference 197, please note.

5.0.0 MOTIVATION

The ability to motivate others is a key skill that leaders must develop. Motivation is the ability to influence. It also describes the amount of effort that a person is willing to put forth to accomplish something. For example, a crew member who skips breaks and lunch in an effort to complete a job on time is thought to be highly motivated, but a crew member who does the bare minimum or just enough to keep his or her job is considered unmotivated.

Employee motivation has dimension because it can be measured. Examples of how motivation can be measured include determining the level of absenteeism, the percentage of employee turnover, and the number of complaints, as well as the quality and quantity of work produced.

5.1.0 Employee Motivators

Different things motivate different people in different ways. Consequently, there is no one-size-fits-all approach to motivating crew members. It is im-

portant to recognize that what motivates one crew member may not motivate another. In addition, what works to motivate a crew member once may not motivate that same person again in the future.

Frequently, the needs that motivate individuals are the same as those that create job satisfaction. They include the following:

- Recognition and praise
- Accomplishment
- Opportunity for advancement
- Job importance
- Change
- Personal growth
- Rewards

A crew leader's ability to satisfy these needs increases the likelihood of high morale within a crew. Morale refers to an individual's attitude toward the tasks he or she is expected to perform. High morale, in turn, means that employees will be motivated to work hard, and they will have a positive attitude about coming to work and doing their jobs.

5.1.1 Recognition and Praise

Recognition and praise refer to the need to have good work appreciated, applauded, and acknowledged by others. This can be accomplished by simply thanking employees for helping out on a project, or it can entail more formal praise, such as an award for Employee of the Month.

Some tips for giving recognition and praise include the following:

- Be available on the job site so that you have the opportunity to witness good work.
- Know good work and praise it when you see it.
- Look for good work and look for ways to praise it.
- Give recognition and praise only when truly deserved; otherwise, it will lose its meaning.
- Acknowledge satisfactory performance, and encourage improvement by showing confidence in the ability of the crew members to do above-average work.

5.1.2 Accomplishment

Accomplishment refers to a worker's need to set challenging goals and achieve them. There is nothing quite like the feeling of achieving a goal, particularly a goal one never expected to accomplish in the first place.

Crew leaders can help their crew members attain a sense of accomplishment by encouraging them to develop performance plans, such as goals for the year that will be used in performance evaluations. In addition, crew leaders can provide the support and tools (such as training and coaching) necessary to help their crew members achieve these goals.

5.1.3 Opportunity for Advancement

Opportunity for advancement refers to an employee's need to gain additional responsibility and develop new skills and abilities. It is important that employees know that they are not limited to their current jobs. Let them know that they have a chance to grow with the company and to be promoted as recognition for excelling in their work.

Effective leaders encourage their crew members to work to their full potentials. In addition, they share information and skills with their employees in an effort to help them to advance within the organization.

5.1.4 Job Importance

Job importance refers to an employee's need to feel that his or her skills and abilities are valued and make a difference. Employees who do not feel valued tend to have performance and attendance issues. Crew leaders should attempt to make every crew member feel like an important part of the team, as if the job wouldn't be possible without their help.

5.1.5 Change

Change refers to an employee's need to have variety in work assignments. Change is what keeps things interesting or challenging. It prevents the boredom that results from doing the same task day after day with no variety.

5.1.6 Personal Growth

Personal growth refers to an employee's need to learn new skills, enhance abilities, and grow as a person. It can be very rewarding to master a new competency on the job. Similar to change, personal growth prevents the boredom associated with doing the same thing day after day without developing any new skills.

Crew leaders should encourage the personal growth of their employees as well as themselves. Learning should be a two-way street on the job site; crew leaders should teach their crew members and learn from them as well. In addition, crew members should be encouraged to learn from each other.

5.1.7 Rewards

Rewards are compensation for hard work. Rewards can include a crew member's base salary or go beyond that to include bonuses or other incentives. They can be monetary in nature (salary raises, holiday bonuses, etc.), or they can be nonmonetary, such as free merchandise (shirts, coffee mugs, jackets, etc.) or other prizes. Attendance at training courses can be another form of reward.

5.2.0 Motivating Employees

To increase motivation in the workplace, crew leaders must individualize how they motivate different crew members. It is important that crew leaders get to know their crew members and determine what motivates them as individuals. Once again, as diversity increases in the workforce, this becomes even more challenging; therefore, effective communication skills are essential.

Education doesn't stop the day a person receives a diploma or certificate. It is a lifelong activity. Continuing education has long been recognized as a pathway to advancement, but in reality, it is essential to simply remaining in place. Regardless of what you do, new materials, methods, and processes are constantly emerging. Those who don't make the effort to keep up will fall behind.

Here is a list of some tips for motivating employees:

- Keep jobs challenging and interesting. Boredom is a guaranteed de-motivator.
- Communicate your expectations. People need clear goals in order to feel a sense of accomplishment when the goals are achieved.
- Involve the employees. Feeling that their opinions are valued leads to pride in ownership and active participation.
- Provide sufficient training. Give employees the skills and abilities they need to be motivated to perform.
- Mentor the employees. Coaching and supporting employees boosts their self-esteem, their self-confidence, and ultimately their motivation.
- Lead by example. Become the kind of leader employees admire and respect, and they will be motivated to work for you.
- Treat employees well. Be considerate, kind, caring, and respectful; treat employees the way that you want to be treated.
- Avoid using scare tactics. Threatening employees with negative consequences can backfire, resulting in employee turnover instead of motivation.
- Reward your crew for doing their best by giving them easier tasks from time to time. It is tempting to give your best employees the hardest or dirtiest jobs because you know they will do the jobs correctly.
- Reward employees for a job well done.

You are the crew leader of a masonry crew. Sam Williams is the person whom the company holds responsible for ensuring that equipment is operable and distributed to the jobs in a timely manner.

Occasionally, disagreements with Sam have resulted in tools and equipment arriving late. Sam, who has been with the company 15 years, resents having been placed in the job and feels that he outranks all the crew leaders.

Sam figured it was about time he talked with someone about the abuse certain tools and other items of equipment were receiving on some of the jobs. Saws were coming back with guards broken and blades chewed up, bits were being sheared in half, motor housings were bent or cracked, and a large number of tools were being returned covered with mud. Sam was out on your job when he observed a mason carrying a portable saw by the cord. As he watched, he saw the mason bump the swinging saw into a steel column. When the man arrived at his workstation, he dropped the saw into the mud.

You are the worker's crew leader. Sam approached as you were coming out of the work trailer. He described the incident. He insisted, as crew leader, you are responsible for both the work of the crew and how its members use company property. Sam concluded, "You'd better take care of this issue as soon as possible! The company is sick and tired of having your people mess up all the tools!"

You are aware that some members of your crew have been mistreating the company equipment.

1. How would you respond to Sam's accusations?

_____

_____

_____

_____

2. What action would you take regarding the misuse of the tools?

_____

_____

_____

_____

3. How can you motivate the crew to take better care of their tools? Explain.

_____

_____

_____

_____

6.0.0 Team Building

Organizations are making the shift from the traditional boss-worker mentality to one that promotes teamwork. The manager becomes the team leader, and the workers become team members. They all work together to achieve the common goals of the team.

There are a number of benefits associated with teamwork. They include the ability to complete complex projects more quickly and effectively, higher employee satisfaction, and a reduction in turnover.

6.1.0 Successful Teams

Successful teams are made up of individuals who are willing to share their time and talents in an effort to reach a common goal—the goal of the team. Members of successful teams possess an *Us* or *We* attitude rather than an *I* or *You* attitude; they consider what's best for the team and put their egos aside.

Some characteristics of successful teams include the following:

- Everyone participates and every team member counts.
- There is a sense of mutual trust and interdependence.
- Team members are empowered.
- They communicate.
- They are creative and willing to take risks.
- The team leader develops strong people skills and is committed to the team.

6.2.0 Building Successful Teams

To be successful in the team leadership role, the crew leader should contribute to a positive attitude within the team.

There are several ways in which the team leader can accomplish this. First, he or she can work with the team members to create a vision or purpose of what the team is to achieve. It is important that every team member is committed to the purpose of the team, and the team leader is instrumental in making this happen.

Team leaders within the construction industry are typically assigned a crew. However, it can be beneficial for the team leader to be involved in selecting the team members. Selection should be based on a willingness of people to work on the team and the resources that they are able to bring to the team.

When forming a new team, team leaders should do the following:

- Explain the purpose of the team. Team members need to know what they will be doing, how long they will be doing it (if they are temporary or permanent), and why they are needed.
- Help the team establish goals or targets. Teams need a purpose, and they need to know what it is they are responsible for accomplishing.
- Define team member roles and expectations. Team members need to know how they fit into the team and what is expected of them as members of the team.
- Plan to transfer responsibility to the team as appropriate. Teams should be responsible for the tasks to be accomplished.

7.0.0 Getting the Job Done

Crew leaders must implement policies and procedures to make sure that the work is done correctly. Construction jobs have crews of people with various experiences and skill levels available to perform the work. The crew leader's job is to draw from this expertise to get the job done well and in a timely manner.

7.1.0 Delegating

Once the various activities that make up the job have been determined, the crew leader must identify the person or persons who will be responsible for completing each activity. This requires that the crew leader be aware of the skills and abilities of the people on the crew. Then, the crew leader must put this knowledge to work in matching the crew's skills and abilities to specific tasks that must be accomplished to complete the job.

After matching crew members to specific activities, the crew leader must then delegate the assignments to the responsible person(s). Delegation is generally communicated verbally by the crew leader talking directly to the person who has been assigned the activity. However, there may be times when work is assigned indirectly through written instructions or verbally through someone other than the crew leader.

When delegating work, remember to:

- Delegate work to a crew member who can do the job properly. If it becomes evident that he or she does not perform to the standard desired, either teach the crew member to do the work correctly or turn it over to someone else who can.

- Make sure the crew member understands what to do and the level of responsibility. Make sure desired results are clear, specify the boundaries and deadlines for accomplishing the results, and note the available resources.
- Identify the standards and methods of measurement for progress and accomplishment, along with the consequences of not achieving the desired results. Discuss the task with the crew member and check for understanding by asking questions. Allow the crew member to contribute feedback or make suggestions about how the task should be performed in a safe and quality manner.
- Give the crew member the time and freedom to get started without feeling the pressure of too much supervision. When making the work assignment, be sure to tell the crew member how much time there is to complete it, and confirm that this time is consistent with the job schedule.
- Examine and evaluate the result once a task is complete. Then, give the crew member some feedback as to how well it has been done. Get the crew member's comments. The information obtained from this is valuable and will enable the crew leader to know what kind of work to assign that crew member in the future. It will also give the crew leader a means of measuring his or her own effectiveness in delegating work.

7.2.0 Implementing Policies and Procedures

Every company establishes policies and procedures that employees are expected to follow and the crew leaders are expected to implement. Company policies and procedures are essentially guides for how the organization does business. They can also reflect organizational philosophies such as putting safety first or making the customer the top priority. Examples of policies and procedures include safety guidelines, credit standards, and billing processes.

Here are some tips for implementing policies and procedures:

- Learn the purpose of each policy. That way, you can follow it and apply it properly and fairly.
- If you're not sure how to apply a company policy or procedure, check the company manual or ask your supervisor.
- Apply company policies and procedures. Remember that they combine what's best for the customer and the company. In addition, they provide direction on how to handle specific situations and answer questions.

- If you are uncertain how to apply a policy, check with your supervisor.

Crew leaders may need to issue orders to their crew members. Basically, an order initiates, changes, or stops an activity. Orders may be general or specific, written or oral, and formal or informal. The decision of how an order will be issued is up to the crew leader, but it is governed by the policies and procedures established by the company.

When issuing orders:

- Make them as specific as possible.
- Avoid being general or vague unless it is impossible to foresee all of the circumstances that could occur in carrying out the order.
- Recognize that it is not necessary to write orders for simple tasks unless the company requires that all orders be written.
- Write orders for more complex tasks that will take considerable time to complete or orders that are permanent.
- Consider what is being said, the audience to whom it applies, and the situation under which it will be implemented to determine the appropriate level of formality for the order.

8.0.0 PROBLEM SOLVING AND DECISION MAKING

Problem solving and decision making are a large part of every crew leader's daily work. There will always be problems to be resolved and decisions to be made, especially in fast-paced, deadline-oriented industries.

8.1.0 Decision Making vs. Problem Solving

Sometimes, the difference between decision making and problem solving is not clear. Decision making refers to the process of choosing an alternative course of action in a manner appropriate for the situation. Problem solving involves determining the difference between the way things are and the way things should be, and finding out how to bring the two together. The two activities are interrelated because in order to make a decision, you may also have to use problem-solving techniques.

8.2.0 Types of Decisions

Some decisions are routine or simple. Such decisions can be made based on past experiences. An example would be deciding how to get to and from work. If you've worked at the same place for a long time, you are already aware of the options for traveling to and from work (take the bus, drive a car, carpool with a co-worker, take a taxi, etc.). Based on past experiences with the options identified, you can make a decision about how best to get to and from work.

On the other hand, some decisions are more difficult. These decisions require more careful thought about how to carry out an activity by using a formal problem-solving technique. An example is planning a trip to a new vacation spot. If you are not sure how to get there, where to stay, what to see, etc., one option is to research the area to determine the possible routes, hotel accommodations, and attractions. Then, you will have to make a decision about which route to take, what hotel to choose, and what sites to visit, without the benefit of direct past experience.

8.3.0 Problem Solving

The ability to solve problems is an important skill in any workplace. It's especially important for craftworkers, whose workday is often not predictable or routine. In this section, you will learn a five-step process for solving problems, which you can apply to both workplace and personal issues. Review the following steps and then see how they can be applied to a job-related problem. Keep in mind that a problem will not be solved until everyone involved admits that there is a problem.

Step 1 *Define the problem.* This isn't as easy as it sounds. Thinking through the problem often uncovers additional problems.

Step 2 *Think about different ways to solve the problem.* There is often more than one solution to a problem, so you must think through each possible solution and pick the best one. The best solution might be taking parts of two different solutions and combining them to create a new solution.

Step 3 *Pick the solution that seems best and figure out an action plan.* It is best to receive input both from those most affected by the problem and from those who will be most affected by any potential solution.

Step 4 *Test the solution to determine whether it actually works.* Many solutions sound great in theory but in practice don't turn out to be effective. On the other hand, you might discover from trying to apply

a solution that it is acceptable with a little modification. If a solution does not work, think about how you could improve it, and then implement your new plan.

Step 5 *Evaluate the process.* Review the steps you took to discover and implement the solution. Could you have done anything better? If the solution turns out to be satisfactory, you can add the solution to your knowledge base.

Next, you will see how to apply the problem-solving process to a workplace problem. Read the following situation and apply the five-step problem-solving process to come up with a solution to the issues posed by the situation.

Situation:

You are part of a team of workers assigned to a new shopping mall project. The project will take about 18 months to complete. The only available parking is half a mile from the job site. The crew has to carry heavy toolboxes and safety equipment from their cars and trucks to the work area at the start of the day, and then carry them back at the end of their shifts.

Step 1 *Define the problem.* Workers are wasting time and energy hauling all their equipment to and from the work site.

Step 2 *Think about different ways to solve the problem.* Several solutions have been proposed:
- Install lockers for tools and equipment closer to the work site.
- Have workers drive up to the work site to drop off their tools and equipment before parking.
- Bring in another construction trailer where workers can store their tools and equipment for the duration of the project.
- Provide a round-trip shuttle service to ferry workers and their tools.

> NOTE
>
> Each solution will have pros and cons, so it is important to receive input from the workers affected by the problem. For example, workers will probably object to any plan (like the drop-off plan) that leaves their tools vulnerable to theft.

Step 3 *Pick the solution that seems best and figure out an action plan.* The workers decide that the shuttle service makes the most sense. It should solve the time and energy problem, and workers can keep their tools with them. To put the plan into effect, the project supervisor arranges for a large van and driver to provide the shuttle service.

Step 4 *Test the solution to determine whether it actually works.* The solution works, but there is a problem. All the workers are scheduled to start and leave at the same time, so there is not enough room in the van for all the workers and their equipment. To solve this problem, the supervisor schedules trips spaced 15 minutes apart. The supervisor also adjusts worker schedules to correspond with the trips. That way, all the workers will not try to get on the shuttle at the same time.

Step 5 *Evaluate the process.* This process gave both management and workers a chance to express an opinion and discuss the various solutions. Everyone feels pleased with the process and the solution.

8.4.0 Special Leadership Problems

Because they are responsible for leading others, it is inevitable that crew leaders will encounter problems and be forced to make decisions about how to respond to the problem. Some problems will be relatively simple to resolve, like covering for a sick crew member who has taken a day off from work. Other problems will be complex and much more difficult to handle.

Some complex problems are relatively common. A few of the major employee problems include:

- Inability to work with others
- Absenteeism and turnover
- Failure to comply with company policies and procedures

8.4.1 Inability to Work with Others

Crew leaders will sometimes encounter situations where an employee has a difficult time working with others on the crew. This could be a result of personality differences, an inability to communicate, or some other cause. Whatever the reason, the crew leader must address the issue and get the crew working as a team.

The best way to determine the reason for why individuals don't get along or work well together is to talk to the parties involved. The crew leader should speak openly with the employee, as well as the other individual(s) to uncover the source of the problem and discuss its resolution.

Once the reason for the conflict is found, the crew leader can determine how to respond. There may be a way to resolve the problem and get the workers communicating and working as a team again. On the other hand, there may be nothing that can be done that will lead to a harmonious solution. In this case, the crew leader would either have to transfer the employee to another crew or have the problem crew member terminated. This latter option should be used as a last measure and should be discussed with one's superiors or Human Resources Department.

8.4.2 Absenteeism and Turnover

Absenteeism and turnover are big problems. Without workers available to do the work, jobs are delayed, and money is lost.

Absenteeism refers to workers missing their scheduled work time on a job. Absenteeism has many causes, some of which are inevitable. For instance, people get sick, they have to take time off for family emergencies, and they have to attend family events such as funerals. However, there are some causes of absenteeism that can be prevented by the crew leader.

The most effective way to control absenteeism is to make the company's policy clear to all employees. Companies that do this find that chronic absenteeism is reduced. New employees should have the policy explained to them. This explanation should include the number of absences allowed and the reasons for which sick or personal days can be taken. In addition, all workers should know how to inform their crew leaders when they miss work and understand the consequences of exceeding the number of sick or personal days allowed.

Once the policy on absenteeism is explained to employees, crew leaders must be sure to implement it consistently and fairly. If the policy is administered equally, employees will likely follow it. However, if the policy is not administered equally and some employees are given exceptions, then it will not be effective. Consequently, the rate of absenteeism is likely to increase.

Despite having a policy on absenteeism, there will always be employees who are chronically late or miss work. In cases where an employee abuses the absenteeism policy, the crew leader should discuss the situation directly with the employee. The crew leader should confirm that the employee understands the company's policy and insist that the employee comply with it. If the employee's behavior continues, disciplinary action may be in order.

Turnover refers to the loss of an employee that is initiated by that employee. In other words, the employee quits and leaves the company to work elsewhere or is fired for cause.

Like absenteeism, there are some causes of turnover that cannot be prevented and others that can. For instance, it is unlikely that a crew leader could keep an employee who finds a job elsewhere earning twice as much money. However, crew leaders can prevent some employee turnover situations. They can work to ensure safe working conditions for their crew, treat their workers fairly and consistently, and help promote good working conditions. The key is communication. Crew leaders need to know the problems if they are going to be able to successfully resolve them.

Some of the major causes of turnover include the following:

- Unfair/inconsistent treatment by the immediate supervisor
- Unsafe project sites
- Lack of job security

For the most part, the actions described for absenteeism are also effective for reducing turnover. Past studies have shown that maintaining harmonious relationships on the job site goes a long way in reducing both turnover and absenteeism. This requires effective leadership on the part of the crew leader.

8.4.3 Failure to Comply With Company Policies and Procedures

Policies are the rules that define the relationship between the company, its employees, its clients, and its subcontractors. Procedures are the instructions for carrying out the policies. Some companies have dress codes that are reflected in their policies. The dress code may be partly to ensure safety, and partly to define the image a company wants to project to the outside world.

Companies develop procedures to ensure that everyone who performs a task does it safely and efficiently. Many procedures deal with safety. A lockout/tagout procedure is an example. In this procedure, the company defines who may perform a lockout, how it is done, and who has the authority to remove or override it. Workers who fail to follow the procedure endanger themselves, as well as their co-workers.

Among a typical company's policies is the policy on disciplinary action. This policy defines steps to be taken in the event that an employee violates the company's policies or procedures. The steps range from counseling by a supervisor for the first offense, to a written warning, to dismissal for repeat offenses. This will vary from one company to another. For example, some companies will fire an employee for any violation of safety procedures.

The crew leader has the first-line responsibility for enforcing company policies and procedures. The crew leader should take the time with a new crew member to discuss the policies and procedures and show the crew member how to access them. If a crew member shows a tendency to neglect a policy or procedure, it is up to the crew leader to counsel that individual. If the crew member continues to violate a policy or procedure, the crew leader has no choice but to refer that individual to the appropriate authority within the company for disciplinary action.

Case I:

On the way over to the job trailer, you look up and see a piece of falling scrap heading for one of the laborers. Before you can say anything, the scrap material hits the ground about five feet in front of the worker. You notice the scrap is a piece of conduit. You quickly pick it up, assuring the worker you will take care of this matter.

Looking up, you see your crew on the third floor in the area from which the material fell. You decide to have a talk with them. Once on the deck, you ask the crew if any of them dropped the scrap. The men look over at Bob, one of the electricians in your crew. Bob replies, "I guess it was mine. It slipped out of my hand."

It is a known fact that the Occupational Safety and Health Administration (OSHA) regulations state that an enclosed chute of wood shall be used for material waste transportation from heights of 20 feet or more. It is also known that Bob and the laborer who was almost hit have been seen arguing lately.

1. Assuming Bob's action was deliberate, what action would you take?

_____
_____
_____

2. Assuming the conduit accidentally slipped from Bob's hand, how can you motivate him to be more careful?

_____
_____
_____

3. What follow-up actions, if any, should be taken relative to the laborer who was almost hit?

_____
_____
_____

4. Should you discuss the apparent OSHA violation with the crew? Why or why not?

_____
_____
_____

5. What acts of leadership would be effective in this case? To what leadership traits are they related?

_____
_____
_____

Case II:

Mike has just been appointed crew leader of a tile-setting crew. Before his promotion into management, he had been a tile setter for five years. His work had been consistently of superior quality.

Except for a little good-natured kidding, Mike's co-workers had wished him well in his new job. During the first two weeks, most of them had been cooperative while Mike was adjusting to his supervisory role.

At the end of the second week, a disturbing incident took place. Having just completed some of his duties, Mike stopped by the job-site wash station. There he saw Steve and Ron, two of his old friends who were also in his crew, washing.

"Hey, Ron, Steve, you should not be cleaning up this soon. It's at least another thirty minutes until quitting time," said Mike. "Get back to your work station, and I'll forget I saw you here."

"Come off it, Mike," said Steve. "You used to slip up here early on Fridays. Just because you have a little rank now, don't think you can get tough with us." To this Mike replied, "Things are different now. Both of you get back to work, or I'll make trouble." Steve and Ron said nothing more, and they both returned to their work stations.

From that time on, Mike began to have trouble as a crew leader. Steve and Ron gave him the silent treatment. Mike's crew seemed to forget how to do the most basic activities. The amount of rework for the crew seemed to be increasing. By the end of the month, Mike's crew was behind schedule.

1. How do you think Mike should have handled the confrontation with Ron and Steve?

_____
_____

2. What do you suggest Mike could do about the silent treatment he got from Steve and Ron?

_____
_____

3. If you were Mike, what would you do to get your crew back on schedule?

_____
_____

4. What acts of leadership could be used to get the crew's willing cooperation?

_____
_____

5. To which leadership traits do they correspond?

_____
_____

1. A crew leader differs from a craftworker in that a _____.

   a. crew leader need not have direct experience in those job duties that a craftworker typically performs
   b. crew leader can expect to oversee one or more craftworkers in addition to performing some of the typical duties of the craftworker
   c. crew leader is exclusively in charge of overseeing, since performing technical work is not part of this role
   d. crew leader's responsibilities do not include being present on the job site

2. Among the many traits effective leaders should have is _____.

   a. the ability to communicate the goals of a project
   b. the drive necessary to carry the workload by themselves in order to achieve a goal
   c. a perfectionist nature, which ensures that they will not make useless mistakes
   d. the ability to make decisions without needing to listen to the opinions of others

3. Of the three styles of leadership, the _____ style would be effective in dealing with a craftworker's negative attitude.

   a. facilitator
   b. commander
   c. partner
   d. dictator

4. One way to overcome barriers to effective communication is to _____.

   a. avoid taking notes on the content of the message, since this can be distracting
   b. avoid reacting emotionally to the message
   c. anticipate the content of the message and interrupt if necessary in order to show interest
   d. think about how to respond to the message while listening

5. Feedback is important in verbal communication because it requires the _____.

   a. sender to repeat the message
   b. receiver to restate the message
   c. sender to avoid technical jargon
   d. sender to concentrate on the message

6. A good way to motivate employees is to use a one-size-fits-all approach, since employees are members of a team with a common goal.

   a. True
   b. False

7. A crew leader can effectively delegate responsibilities by _____.

   a. refraining from evaluating the employee's performance once the task is completed, since it is a new task for the employee
   b. doing the job for the employee to make sure the task is done correctly
   c. allowing the employee to give feedback and suggestions about the task
   d. communicating information to the employee, generally in written form

8. Problem solving differs from decision making in that _____.

   a. problem solving involves identifying discrepancies between the way a situation is and the way it should be
   b. decision making involves separating facts from non-facts
   c. decision making involves eliminating differences
   d. problem solving involves determining an alternative course of action for a given situation

SECTION THREE
SAFETY

Objectives

Upon completion of this section, you will be able to:

1. Explain the importance of safety.
2. Give examples of direct and indirect costs of workplace accidents.
3. Identify safety hazards of the construction industry.
4. Explain the purpose of OSHA.
5. Discuss OSHA inspection procedures.
6. Identify the key points of a safety program.
7. List steps to train employees on how to perform new tasks safely.
8. Identify a crew leader's safety responsibilities.
9. Explain the importance of having employees trained in first aid and cardiopulmonary resuscitation (CPR).
10. Describe the indications of substance abuse.
11. List the essential parts of an accident investigation.
12. Describe ways to maintain employee interest in safety. Distinguish between company policies and procedures.

1.0.0 SAFETY OVERVIEW

Businesses lose millions of dollars every year because of on-the-job accidents. Work-related injuries, sickness, and deaths have caused untold suffering for workers and their families. Project delays and budget overruns from injuries and fatalities result in huge losses for employers, and work-site accidents erode the overall morale of the crew.

Craftworkers are exposed to hazards as part of the job. Examples of these hazards include falls from heights, working on scaffolds, using cranes

---

Did you know?

When OSHA inspects a job site, they focus on the types of safety hazards that are most likely to cause fatal injuries. These hazards fall into the following classifications:

- Falls from elevations
- Struck-by hazards
- Caught in/between hazards
- Electrical shock hazards

---

in the presence of power lines, operating heavy machinery, and working on electrically-charged or pressurized equipment. Despite these hazards, experts believe that applying preventive safety measures could drastically reduce the number of accidents.

As a crew leader, one of your most important tasks is to enforce the company's safety program and make sure that all workers are performing their tasks safely. To be successful, the crew leader should:

- Be aware of the costs of accidents.
- Understand all federal, state, and local governmental safety regulations.
- Be involved in training workers in safe work methods.
- Conduct training sessions.
- Get involved in safety inspections, accident investigations, and fire protection and prevention.

Crew leaders are in the best position to ensure that all jobs are performed safely by their crew members. Providing employees with a safe working environment by preventing accidents and enforcing safety standards will go a long way towards maintaining the job schedule and enabling a job's completion on time and within budget.

1.1.0 Accident Statistics

Each day, workers in construction and industrial occupations face the risk of falls, machinery accidents, electrocutions, and other potentially fatal occupational hazards.

The National Institute of Occupational Safety and Health (NIOSH) statistics show that about 1,000 construction workers are killed on the job each year, more fatalities than in any other industry. Falls are the leading cause of deaths in the construction industry, accounting for over 40 percent of the fatalities. Nearly half of the fatal falls occurred from roofs, scaffolds, or ladders. Roofers, structural metal workers, and painters experienced the greatest number of fall fatalities.

In addition to the number of fatalities that occur each year, there are a staggering number of work-related injuries. In 2007, for example, more than 135,000 job-related injuries occurred in the construction industry. NIOSH reports that approximately 15 percent of all worker's compensation costs are spent on injured construction workers. The causes of injuries on construction sites include falls, coming into contact with electric current, fires, and mishandling of machinery or equipment. According to NIOSH, back injuries are the leading safety problem in workplaces.

2.0.0 COSTS OF ACCIDENTS

Occupational accidents are estimated to cost more than $100 billion every year. These costs affect the employee, the company, and the construction industry as a whole.

Organizations encounter both direct and indirect costs associated with workplace accidents. Examples of direct costs include workers' compensation claims and sick pay; indirect costs include increased absenteeism, loss of productivity, loss of job opportunities due to poor safety records, and negative employee morale attributed to workplace injuries. There are many other related costs involved with workplace accidents. A company can be insured against some of them, but not others. To compete and survive, companies must control these as well as all other employment-related costs.

2.1.0 Insured Costs

Insured costs are those costs either paid directly or reimbursed by insurance carriers. Insured costs related to injuries or deaths include the following:

- Compensation for lost earnings (known as worker's comp)
- Medical and hospital costs
- Monetary awards for permanent disabilities
- Rehabilitation costs
- Funeral charges
- Pensions for dependents

Insurance premiums or charges related to property damages include:

- Fire
- Loss and damage
- Use and occupancy
- Public liability
- Replacement cost of equipment, material, and structures

2.2.0 Uninsured Costs

The costs related to accidents can be compared to an iceberg, as shown in *Figure 7*. The tip of the iceberg represents direct costs, which are the visible costs. The more numerous indirect costs are not readily measureable, but they can represent a greater burden than the direct costs.

Uninsured costs related to injuries or deaths include the following:

- First aid expenses
- Transportation costs
- Costs of investigations
- Costs of processing reports
- Down time on the job site
- Costs to train replacement workers

Uninsured costs related to wage losses include:

- Idle time of workers whose work is interrupted
- Time spent cleaning the accident area
- Time spent repairing damaged equipment
- Time lost by workers receiving first aid
- Costs of training injured workers in a new career

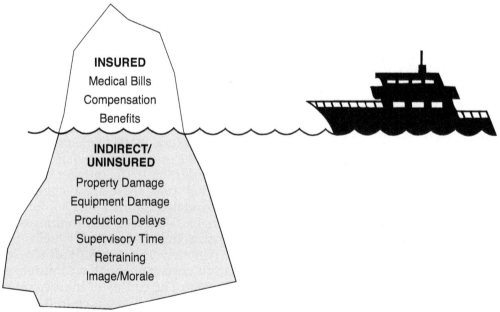

46101-11_F07.EPS

Figure 7 Costs associated with accidents.

Uninsured costs related to production losses include:

- Product spoiled by accident
- Loss of skill and experience
- Lowered production or worker replacement
- Idle machine time

Associated costs may include the following:

- Difference between actual losses and amount recovered
- Costs of rental equipment used to replace damaged equipment
- Costs of new workers used to replace injured workers
- Wages or other benefits paid to disabled workers
- Overhead costs while production is stopped
- Impact on schedule
- Loss of bonus or payment of forfeiture for delays

Uninsured costs related to off-the-job activities include:

- Time spent on injured workers' welfare
- Loss of skill and experience of injured workers
- Costs of training replacement workers

Uninsured costs related to intangibles include:

- Lowered employee morale
- Increased labor conflict
- Unfavorable public relations
- Loss of bid opportunities because of poor safety records
- Loss of client goodwill

3.0.0 SAFETY REGULATIONS

To reduce safety and health risks and the number of injuries and fatalities on the job, the federal government has enacted laws and regulations, including the *Occupational Safety and Health Act of 1970*. The purpose of OSHA is "to assure so far as possible every working man and woman in the Nation safe and healthful working conditions and to preserve our human resources."

To promote a safe and healthy work environment, OSHA issues standards and rules for working conditions, facilities, equipment, tools, and work processes. It does extensive research into occupational accidents, illnesses, injuries, and deaths in an effort to reduce the number of occurrences and adverse effects. In addition, OSHA regulatory agencies conduct workplace inspections to ensure that companies follow the standards and rules.

3.1.0 Workplace Inspections

To enforce OSHA regulations, the government has granted regulatory agencies the right to enter public and private properties to conduct workplace safety investigations. The agencies also have the right to take legal action if companies are not in compliance with the Act. These regulatory agencies employ OSHA Compliance Safety and Health Officers (CSHOs), who are chosen for their knowledge in the occupational safety and health field. The CSHOs are thoroughly trained in OSHA standards and in recognizing safety and health hazards.

States with their own occupational safety and health programs conduct inspections. To do so, they enlist the services of qualified state CSHOs.

Companies are inspected for a multitude of reasons. They may be randomly selected, or they may be chosen because of employee complaints, due to an imminent danger, or as a result of major accidents or fatalities.

OSHA can assess significant financial penalties for safety violations. In some cases, a superintendent or crew leader can be held criminally liable for repeat violations.

3.2.0 Penalties for Violations

OSHA has established monetary fines for the violation of their regulations. The penalties as of 2010 are shown in *Table 1*.

In addition to the fines, there are possible criminal charges for willful violations resulting in death or serious injury. There can also be personal liability for failure to comply with OSHA regulations. The attitude of the employer and their safety history can have a significant effect on the outcome of a case.

Table 1 OSHA Penalties for Violations

Violation	Penalty
Willful Violations	Maximum $70,000
Repeated Violations	Minimum $70,000
Serious, Other-than-Serious, Other Specific Violations	Minimum $7,000
OSHA Notice Violation	$1,000
Failure to Post *OSHA 300A Summary of Work-Related Injuries and Illnesses*	$1,000
Failure to Properly Maintain *OSHA 300 Log of Work-Related Injuries and Illnesses*	$1,000
Failure to Promptly and Properly Report Fatality/Catastrophe	$5,000
Failure to Permit Access to Records Under *OSHA 1904* Regulations	$1,000
Failure to Follow Advance Notification Requirements Under *OSHA 1903.6* Regulations	$2,000
Failure to Abate – for Each Calendar Day Beyond Abatement Date	$7,000
Retaliation Against Individual for Filing OSHA Complaint	$10,000

4.0.0 EMPLOYER SAFETY RESPONSIBILITIES

Each employer must set up a safety and health program to manage workplace safety and health and to reduce work-related injuries, illnesses, and fatalities. The program must be appropriate for the conditions of the workplace. It should consider the number of workers employed and the hazards to which they are exposed while at work.

To be successful, the safety and health program must have management leadership and employee participation. In addition, training and informational meetings play an important part in effective programs. Being consistent with safety policies is the key. Regardless of the employer's responsibility, however, the individual worker is ultimately responsible for his or her own safety.

4.1.0 Safety Program

The crew leader plays a key role in the successful implementation of the safety program. The crew leader's attitudes toward the program set the standard for how crew members view safety. Therefore, the crew leader should follow all program guidelines and require crew members to do the same.

Safety programs should consist of the following:

- Safety policies and procedures
- Hazard identification and assessment
- Safety information and training
- Safety record system
- Accident investigation procedures
- Appropriate discipline for not following safety procedures
- Posting of safety notices

4.1.1 Safety Policies and Procedures

Employers are responsible for following OSHA and state safety standards. Usually, they incorporate OSHA and state regulations into a safety policies and procedures manual. Such a manual is presented to employees when they are hired.

Basic safety requirements should be presented to new employees during their orientation to the company. If the company has a safety manual, the new employee should be required to read it and sign a statement indicating that it is understood. If the employee cannot read, the employer should have someone read it to the employee and answer any questions that arise. The employee should then sign a form stating that he or she understands the information.

It is not enough to tell employees about safety policies and procedures on the day they are hired and never mention them again. Rather, crew leaders should constantly emphasize and reinforce the importance of following all safety policies and procedures. In addition, employees should play an active role in determining job safety hazards and find ways that the hazards can be prevented and controlled.

4.1.2 Hazard Identification and Assessment

Safety policies and procedures should be specific to the company. They should clearly present the hazards of the job. Therefore, crew leaders should also identify and assess hazards to which employees are exposed. They must also assess compliance with OSHA and state standards.

To identify and assess hazards, OSHA recommends that employers conduct inspections of the workplace, monitor safety and health information logs, and evaluate new equipment, materials, and processes for potential hazards before they are used.

Crew leaders and employees play important roles in identifying hazards. It is the crew leader's responsibility to determine what working conditions are unsafe and to inform employees of hazards and their locations. In addition, they should encourage their crew members to tell them about hazardous conditions. To accomplish this, crew leaders must be present and available on the job site.

The crew leader also needs to help the employee be aware of and avoid the built-in hazards to which craftworkers are exposed. Examples include working at elevations, working in confined spaces such as tunnels and underground vaults, on caissons, in excavations with earthen walls, and other naturally dangerous projects.

In addition, the crew leader can take safety measures, such as installing protective railings to prevent workers from falling from buildings, as well as scaffolds, platforms, and shoring.

4.1.3 Safety Information and Training

The employer must provide periodic information and training to new and long-term employees. This must be done as often as necessary so that all employees are adequately trained. Special training and informational sessions must be provided when safety and health information changes or workplace conditions create new hazards. It is important to note that safety-related information must be presented in a manner that the employee will understand.

Whenever a crew leader assigns an experienced employee a new task, the crew leader must ensure that the employee is capable of doing the work in a safe manner. The crew leader can accomplish this by providing safety information or training for groups or individuals.

The crew leader should do the following:

- Define the task.
- Explain how to do the task safely.
- Explain what tools and equipment to use and how to use them safely.
- Identify the necessary personal protective equipment.
- Explain the nature of the hazards in the work and how to recognize them.
- Stress the importance of personal safety and the safety of others.
- Hold regular safety training sessions with the crew's input.
- Review material safety data sheets (MSDSs) that may be applicable.

4.1.4 Safety Record Systems

OSHA regulations (*CFR 29, Part 1904*) require that employers keep records of hazards identified and document the severity of the hazard. The information should include the likelihood of employee exposure to the hazard, the seriousness of the harm associated with the hazard, and the number of exposed employees.

In addition, the employer must document the actions taken or plans for action to control the hazards. While it is best to take corrective action immediately, it is sometimes necessary to develop a plan for the purpose of setting priorities and deadlines and tracking progress in controlling hazards.

Employers who are subject to the recordkeeping requirements of the *Occupational Safety and Health Act of 1970* must maintain a log of all recordable occupational injuries and illnesses. This is known as the *OSHA 300/300A* form.

An MSDS is designed to provide both workers and emergency personnel with the proper procedures for handling or working with a substance that may be dangerous. An MSDS will include information such as physical data (melting point, boiling point, flash point, etc.), toxicity, health effects, first aid, reactivity, storage, disposal, protective equipment, and spill/leak procedures. These sheets are of particular use if a spill or other accident occurs.

Any company with 11 or more employees must post an *OSHA 300A* form, *Log of Work-Related Injuries and Illnesses,* between February 1 and April 30 of each year. Employees have the right to review this form. Check your company's policies with regard to these reports.

OSHA's Form 300A (Rev. 01/2004)

Summary of Work-Related Injuries and Illnesses

Year 20____

U.S. Department of Labor
Occupational Safety and Health Administration

Form approved OMB no. 1218-0176

All establishments covered by Part 1904 must complete this Summary page, even if no work-related injuries or illnesses occurred during the year. Remember to review the Log to verify that the entries are complete and accurate before completing this summary.

Using the Log, count the individual entries you made for each category. Then write the totals below, making sure you've added the entries from every page of the Log. If you had no cases, write "0."

Employees, former employees, and their representatives have the right to review the OSHA Form 300 in its entirety. They also have limited access to the OSHA Form 301 or its equivalent. See 29 CFR Part 1904.35, in OSHA's recordkeeping rule, for further details on the access provisions for these forms.

Number of Cases

Total number of deaths	Total number of cases with days away from work	Total number of cases with job transfer or restriction	Total number of other recordable cases
(G)	(H)	(I)	(J)

Number of Days

Total number of days away from work	Total number of days of job transfer or restriction
(K)	(L)

Injury and Illness Types

Total number of . . .
(M)
(1) Injuries ____
(2) Skin disorders ____
(3) Respiratory conditions ____
(4) Poisonings ____
(5) Hearing loss ____
(6) All other illnesses ____

Post this Summary page from February 1 to April 30 of the year following the year covered by the form.

Public reporting burden for this collection of information is estimated to average 58 minutes per response, including time to review the instructions, search and gather the data needed, and complete and review the collection of information. Persons are not required to respond to the collection of information unless it displays a currently valid OMB control number. If you have any comments about these estimates or any other aspects of this data collection, contact: US Department of Labor, OSHA Office of Statistical Analysis, Room N-3644, 200 Constitution Avenue, NW, Washington, DC 20210. Do not send the completed forms to this office.

Establishment information

Your establishment name _____

Street _____
City _____ State ____ ZIP ____

Industry description (e.g., Manufacture of motor truck trailers) _____

Standard Industrial Classification (SIC), if known (e.g., 3715) _____

OR

North American Industrial Classification (NAICS), if known (e.g., 336212) _____

Employment information (If you don't have these figures, see the Worksheet on the back of this page to estimate.)

Annual average number of employees _____

Total hours worked by all employees last year _____

Sign here

Knowingly falsifying this document may result in a fine.

I certify that I have examined this document and that to the best of my knowledge the entries are true, accurate, and complete.

Company executive _____ Title _____

Phone (___) ___ - ___ Date ___ / ___ / ___

46101-11_SA01.EPS

Logs must be maintained and retained for five years following the end of the calendar year to which they relate. Logs must be available (normally at the establishment) for inspection and copying by representatives of the Department of Labor, the Department of Health and Human Services, or states accorded jurisdiction under the Act. Employees, former employees, and their representatives may also have access to these logs.

4.1.5 Accident Investigation

In the event of an accident, the employer is required to investigate the cause of the accident and determine how to avoid it in the future. According to OSHA, the employer must investigate each work-related death, serious injury or illness, or incident having the potential to cause death or serious physical harm. The employer should document any findings from the investigation, as well as the action plan to prevent future occurrences. This should be done immediately, with photos or video if possible. It is important that the investigation uncover the root cause of the accident so that it can be avoided in the future. In many cases, the root cause can be traced to a flaw in the system that failed to recognize the unsafe condition or the potential for an unsafe act (*Figure 8*).

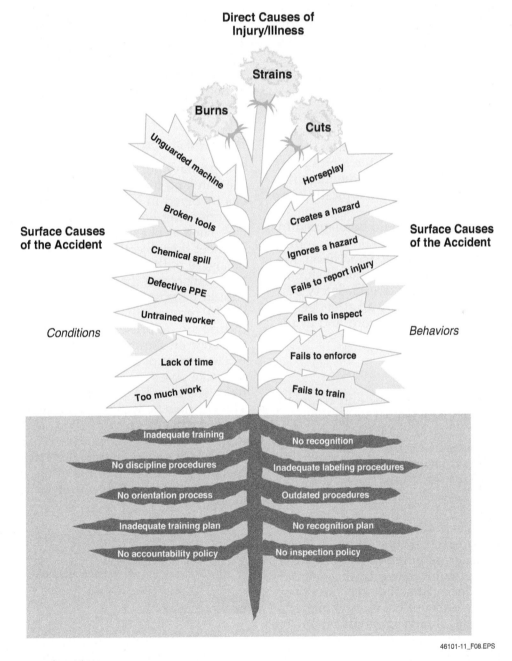

Figure 8 Root causes of accidents.

5.0.0 CREW LEADER INVOLVEMENT IN SAFETY

To be effective leaders, crew leaders must be actively involved in the safety program. Crew leader involvement includes conducting frequent safety training sessions and inspections; promoting first aid and fire protection and prevention; preventing substance abuse on the job; and investigating accidents.

5.1.0 Safety Training Sessions

A safety training session may be a brief, informal gathering of a few employees or a formal meeting with instructional films and talks by guest speakers. The size of the audience and the topics to be addressed determine the format of the meeting. Small, informal safety sessions are typically conducted weekly.

Safety training sessions should be planned in advance, and the information should be communicated to all employees affected. In addition, the topics covered in these training sessions should be timely and practical. A log of each safety session must be kept and signed by all attendees. It must then be maintained as a record and available for inspection.

5.2.0 Inspections

Crew leaders must make regular and frequent inspections to prevent accidents from happening. They must also take steps to avoid accidents. For that purpose, they need to inspect the job sites where their workers perform tasks. It is recommended that these inspections be done before the start of work each day and during the day at different times.

Crew leaders must protect workers from existing or potential hazards in their work areas. Crew leaders are sometimes required to work in areas controlled by other contractors. In these situations, the crew leader must maintain control over the safety exposure of his or her crew. If hazards exist, the crew leader should immediately bring the hazards to the attention of the contractor at fault, their superior, and the person responsible for the job site.

Crew leader inspections are only valuable if action is taken to correct potential hazards. Therefore, crew leaders must be alert for unsafe acts on their work sites. When an employee performs an unsafe action, the crew leader must explain to the employee why the act was unsafe, ask that the employee not do it again, and request cooperation in promoting a safe working environment. The crew leader must document what happened and what the employee was asked to do to correct the situation. It is then very important that crew leaders follow up to make certain the employee is complying with the safety procedures. Never allow a safety violation to go uncorrected. There are three courses of action that you, as a crew leader, can take in an unsafe situation:

- Get the appropriate party to correct the problem.
- Fix the problem yourself.
- Refuse to have the crew work in the area until the problem is corrected.

5.3.0 First Aid

The primary purpose of first aid is to provide immediate and temporary medical care to employees involved in accidents, as well as employees experiencing non-work-related health emergencies, such as chest pains or breathing difficulty. To meet this objective, every crew leader should be aware of the location and contents of first aid kits available on the job site. Emergency numbers should be posted in the job trailer. In addition, OSHA requires that at least one person trained in first aid be present at the job site at all times. Someone on site should also be trained in CPR.

The victim of an accident or sudden illness at a job site poses more problems than normal since he or she may be working in a remote location. The site may be located far from a rescue squad, fire department, or hospital, presenting a problem in the rescue and transportation of the victim to a hospital. The worker may also have been injured by falling rock or other materials, so special rescue equipment or first-aid techniques are often needed.

> **NOTE**
> CPR training must be renewed every two years.

The employer benefits by having personnel trained in first aid at each job site in the following ways:

- The immediate and proper treatment of minor injuries may prevent them from developing into more serious conditions. As a result, medical expenses, lost work time, and sick pay may be eliminated or reduced.
- It may be possible to determine if professional medical attention is needed.
- Valuable time can be saved when a trained individual prepares the patient for treatment when professional medical care arrives. As a result, lives can be saved.

The American Red Cross, Medic First Aid, and the United States Bureau of Mines provide basic and advanced first aid courses at nominal costs. These courses include both first aid and CPR. The local area offices of these organizations can provide further details regarding the training available.

5.4.0 Fire Protection and Prevention

Fires and explosions kill and injure many workers each year, so it is important that crew leaders understand and practice fire-prevention techniques as required by company policy.

The need for protection and prevention is increasing as new building materials are introduced. Some building materials are highly flammable. They produce great amounts of smoke and gases, which cause difficulties for fire fighters, and can quickly overcome anyone present. Other materials melt when ignited and may spread over floors, preventing fire-fighting personnel from entering areas where this occurs.

OSHA has specific standards for fire safety. They require that employers provide proper exits, fire-fighting equipment, and employee training on fire prevention and safety. For more information, consult OSHA guidelines.

5.5.0 Substance Abuse

Unfortunately, drug and alcohol abuse is a continuing problem in the workplace. Drug abuse means inappropriately using drugs, whether they are legal or illegal. Some people use illegal street drugs, such as cocaine or marijuana. Others use legal prescription drugs incorrectly by taking too many pills, using other people's medications, or self-medicating. Others consume alcohol to the point of intoxication.

It is essential that crew leaders enforce company policies and procedures regarding substance abuse. Crew leaders must work with management to deal with suspected drug and alcohol

abuse and should not handle these situations themselves. These cases are normally handled by the Human Resources Department or designated manager. There are legal consequences of drug and alcohol abuse and the associated safety implications. If you suspect that an employee is suffering from drug or alcohol abuse, immediately contact your supervisor and/or Human Resources Department for assistance. That way, the business and the employee's safety are protected.

It is the crew leader's responsibility to make sure that safety is maintained at all times. This may include removing workers from a work site where they may be endangering themselves or others.

For example, suppose several crew members go out and smoke marijuana or have a few beers during lunch. Then, they return to work to erect scaffolding for a concrete pour in the afternoon. If you can smell marijuana on the crew member's clothing or alcohol on their breath, you must step in and take action. Otherwise, they might cause an accident that could delay the project or cause serious injury or death to themselves or others.

It is often difficult to detect drug and alcohol abuse because the effects can be subtle. The best way is to look for identifiable effects, such as those mentioned above or sudden changes in behavior that are not typical of the employee. Some examples of such behaviors include the following:

- Unscheduled absences; failure to report to work on time
- Significant changes in the quality of work
- Unusual activity or lethargy
- Sudden and irrational temper flare-ups
- Significant changes in personal appearance, cleanliness, or health

There are other more specific signs that should arouse suspicion, especially if more than one is exhibited. Among them are:

- Slurring of speech or an inability to communicate effectively
- Shiftiness or sneaky behavior, such as an employee disappearing to wooded areas, storage areas, or other private locations
- Wearing sunglasses indoors or on overcast days to hide dilated or constricted pupils
- Wearing long-sleeved garments, particularly on hot days, to cover marks from needles used to inject drugs
- Attempting to borrow money from co-workers
- The loss of an employee's tools and company equipment

5.6.0 Job-Related Accident Investigations

There are two times when a crew leader may be involved with an accident investigation. The first time is when an accident, injury, or report of work-connected illness takes place. If present on site, the crew leader should proceed immediately to the accident location to ensure that proper first aid is being provided. He or she will also want to make sure that other safety and operational measures are taken to prevent another incident.

If mandated by company policy, the crew leader will need to make a formal investigation and submit a report after an incident. An investigation looks for the causes of the accident by examining the situation under which it occurred and talking to the people involved. Investigations are perhaps the most useful tool in the prevention of future accidents.

There are four major parts to an accident investigation. The crew leader will be concerned with each one. They are:

- Describing the accident
- Determining the cause of the accident
- Determining the parties involved and the part played by each
- Determining how to prevent re-occurrences

Case Study

For years, a prominent safety engineer was confused as to why sheet metal workers fractured their toes frequently. The crew leader had not performed thorough accident investigations, and the injured workers were embarrassed to admit how the accidents really occurred. It was later discovered they used the metal reinforced cap on their safety shoes as a "third hand" to hold the sheet metal vertically in place when they fastened it. The sheet metal was inclined to slip and fall behind the safety cap onto the toes, causing fractures. Several injuries could have been prevented by performing a proper investigation after the first accident.

6.0.0 PROMOTING SAFETY

The best way for crew leaders to encourage safety is through example. Crew leaders should be aware that their behavior sets standards for their crew members. If a crew leader cuts corners on safety, then the crew members may think that it is okay to do so as well.

The key to effectively promote safety is good communication. It is important to plan and coordinate activities and to follow through with safety programs. The most successful safety promotions occur when employees actively participate in planning and carrying out activities.

Some activities used by organizations to help motivate employees on safety and help promote safety awareness include:

- Safety training sessions
- Contests
- Recognition and awards
- Publicity

6.1.0 Safety Training Sessions

Safety training sessions can help keep workers focused on safety and give them the opportunity to discuss safety concerns with the crew. This topic was addressed in a previous section.

6.2.0 Safety Contests

Contests are a great way to promote safety in the workplace. Examples of safety-related contests include the following:

- Sponsoring housekeeping contests for the cleanest job site or work area
- Challenging employees to come up with a safety slogan for the company or department
- Having a poster contest that involves employees or their children creating safety-related posters
- Recording the number of accident-free workdays or person-hours
- Giving safety awards (hats, T-shirts, other promotional items or prizes)

One of the positive aspects of contests is their ability to encourage employee participation. It is important, however, to ensure that the contest has a valid purpose. For example, the posters or slogans created in a poster contest can be displayed throughout the organization as safety reminders.

6.3.0 Incentives and Awards

Incentives and awards serve several purposes. Among them are acknowledging and encouraging good performance, building goodwill, reminding employees of safety issues, and publicizing the importance of practicing safety standards. There are countless ways to recognize and award safety. Examples include the following:

- Supplying food at the job site when a certain goal is achieved
- Providing a reserved parking space to acknowledge someone for a special achievement
- Giving gift items such as T-shirts or gift certificates to reward employees
- Giving plaques to a department or an individual (*Figure 9*)
- Sending a letter of appreciation
- Publicly honoring a department or an individual for a job well done

Creativity can be used to determine how to recognize and award good safety on the work site. The only precautionary measure is that the award should be meaningful and not perceived as a bribe. It should be representative of the accomplishment.

6.4.0 Publicity

Publicizing safety is the best way to get the message out to employees. An important aspect of publicity is to keep the message accurate and current. Safety posters that are hung for years on end tend to lose effectiveness. It is important to keep ideas fresh.

Examples of promotional activities include posters or banners, advertisements or information on bulletin boards, payroll mailing stuffers, and employee newsletters. In addition, merchandise can be purchased that promotes safety, including buttons, hats, T-shirts, and mugs.

46101-11_F09.EPS

Figure 9 Examples of safety plaques.

Described here are three scenarios that reflect unsafe practices by craft workers. For each of these scenarios write down how you would deal with the situation, first as the crew leader of the craft worker, and then as the leader of another crew.

1. You observe a worker wearing his hard hat backwards and his safety glasses hanging around his neck. He is using a concrete saw.

   _____

   _____

   _____

   _____

   _____

   _____

2. As you are supervising your crew on the roof deck of a building under construction, you notice that a section of guard rail has been removed. Another contractor was responsible for installing the guard rail.

   _____

   _____

   _____

   _____

   _____

   _____

   _____

3. Your crew is part of plant shutdown at a power station. You observe that a worker is welding without a welding screen in an area where there are other workers.

   _____

   _____

   _____

   _____

   _____

   _____

   _____

1. One of a crew leader's responsibilities is to enforce company safety policies.

   a. True
   b. False

2. Which of the following is an indirect cost of an accident?

   a. Medical bills
   b. Production delays
   c. Compensation
   d. Employee benefits

3. A crew leader can be held criminally liable for repeat safety violations.

   a. True
   b. False

4. OSHA inspection of a business or job site _____.

   a. can be done only by invitation
   b. is done only after an accident
   c. can be conducted at random
   d. can be conducted only if a safety violation occurs

5. The *OSHA 300* form deals with _____.

   a. penalties for safety violations
   b. workplace illnesses and injuries
   c. hazardous material spills
   d. safety training sessions

6. A crew leader's responsibilities include all of the following, *except* _____.

   a. conducting safety training sessions
   b. developing a company safety program
   c. performing safety inspections
   d. participating in accident investigations

7. In order to ensure workplace safety, the crew leader should _____.

   a. hold formal safety training sessions
   b. have crew members conduct on-site safety inspections
   c. notify contractors and their supervisor of hazards in a situation where a job is being performed in an unsafe area controlled by other contractors
   d. hold crew members responsible for making a formal report and investigation following an accident

8. Prohibitions on the abuse of drugs deals only with illegal drugs such as cocaine and marijuana.

   a. True
   b. False

PROJECT CONTROL

Objectives

Upon completion of this section, you will be able to:

1. Describe the three phases of a construction project.
2. Define the three types of project delivery systems.
3. Define planning and describe what it involves.
4. Explain why it is important to plan.
5. Describe the two major stages of planning.
6. Explain the importance of documenting job site work.
7. Describe the estimating process.
8. Explain how schedules are developed and used.
9. Identify the two most common schedules.
10. Explain how the critical path method (CPM) of scheduling is used.
11. Describe the different costs associated with building a job.
12. Explain the crew leader's role in controlling costs.
13. Illustrate how to control the main resources of a job: materials, tools, equipment, and labor.
14. Explain the differences between production and productivity and the importance of each.

Performance Tasks

1. Develop and present a look-ahead schedule.
2. Develop an estimate for a given work activity

1.0.0 PROJECT CONTROL OVERVIEW

The contractor, project manager, superintendent, and crew leader each have management responsibilities for their assigned jobs. For example, the contractor's responsibility begins with obtaining the contract, and it does not end until the client takes ownership of the project. The project manager is generally the person with overall responsibility for coordinating the project. Finally, the superintendent and crew leader are responsible for coordinating the work of one or more workers, one or more crews of workers within the company and, on occasion, one or more crews of subcontractors. The crew leader directs a crew in the performance of work tasks.

This section describes methods of effective and efficient project control. It examines estimating, planning and scheduling, and resource and cost control. All the workers who participate in the job are responsible at some level for controlling cost and schedule performance and for ensuring that the project is completed according to plans and specifications.

> **NOTE**
>
> The material in this section is based largely on building-construction projects. However, the project control principles described here apply generally to all types of projects.

Construction projects are made up of three phases: the development phase, the planning phase, and the construction phase.

1.1.0 Development Phase

A new building project begins when an owner has decided to build a new facility or add to an existing facility. The development process is the first stage of planning for a new building project. This process involves land research and feasibility studies to ensure that the project has merit. Architects or engineers develop the conceptual drawings that define the project graphically. They then provide the owner with sketches of room layouts and elevations and make suggestions about what construction materials should be used.

During the development phase, an estimate for the proposed project is developed in order to establish a preliminary budget. Once that budget has been established, the financing of the project is discussed with lending institutions. The architects/engineers and/or the owner begins preliminary reviews with government agencies. These reviews include zoning, building restrictions, landscape requirements, and environmental impact studies.

Also during the development phase, the owner must analyze the project's cost and potential retun on investment (ROI) to ensure that its costs will not exceed its market value and that the project provides a good return on investment. If the project passes this test, the architect/engineer will proceed to the planning phase.

1.2.0 Planning Phase

When the architect/engineer begins to develop the project drawings and specifications, other design professionals such as structural, mechanical, and electrical engineers are brought in. They perform the calculations, make a detailed technical analysis, and check details of the project for accuracy.

The design professionals create drawings and specifications. These drawings and specifications are used to communicate the necessary information to the contractors, subcontractors, suppliers, and workers that contribute to a project.

During the planning phase, the owners hold many meetings to refine estimates, adjust plans to conform to regulations, and secure a construction loan. If the project is a condominium, an office building, or a shopping center, then a marketing program must be developed. In such cases, the selling of the project is often started before actual construction begins.

Next, a complete set of drawings, specifications, and bid documents is produced. Then the owner will select the method to obtain contractors. The owner may choose to negotiate with several contractors or select one through competitive bidding. Note that safety must also be considered as part of the planning process. A safety crew leader may walk through the site as part of the pre-bid process.

Contracts can take many forms. The three basic types from which all other types are derived are firm fixed price, cost reimbursable, and guaranteed maximum price.

- *Firm fixed price* – In this type of contract, the buyer generally provides detailed drawings and specifications, which the contractor uses to calculate the cost of materials and labor. To these costs, the contractor adds a percentage representing company overhead expenses such as office rent, insurance, and accounting/payroll costs. On top of all this, the contractor adds a profit factor. When submitting the bid, the contractor will state very specifically the conditions and assumptions on which the bid is based. These conditions and assumptions form the basis from which changes can be priced. Because the price is established in advance, any changes in the job requirements once the job is started will impact the contractor's profit margin. This is where the crew leader can play an important role by identifying problems that increase the amount of labor or material that was planned. By passing this information up the chain of command, the crew leader allows the company to determine if the change is outside the scope of the bid. If so, they can submit a change order request to cover the added cost.
- *Cost reimbursable* – In this type of contract, the buyer reimburses the contractor for labor, materials, and other costs encountered in the performance of the contract. Typically, the contractor and buyer agree in advance on hourly or daily labor rates for different categories of worker.

These rates include an amount representing the contractor's overhead expense. The buyer also reimburses the contractor for the cost of materials and equipment used on the job. The buyer and contractor also negotiate a profit margin. On this type of contract, the profit margin is likely to be lower than that of a fixed-price contract because the contractor's cost risk is significantly reduced. The profit margin is often subject to incentive or penalty clauses that make the amount of profit awarded subject to performance by the contractor. Performance is usually tied to project schedule milestones.

- *Guaranteed maximum price (GMP)* – This form of contract, also called a not-to-exceed contract, is most often used on projects that have been negotiated with the owner. Involvement in the process usually includes preconstruction, and the entire team develops the parameters that define the basis for the work. In some instances, the owner will require a competitively-bid GMP. In such cases, the scope of work has not been fully defined, but bids are taken for general conditions (direct costs) and fee based on an assumed volume of work. The advantages of the GMP contract vehicle are:
  - Reduced design time
  - Allows for phased construction
  - Uses a team approach to a project
  - Reduction in changes related to incomplete drawings

1.3.0 Construction Phase

The designated contractor enlists the help of mechanical, electrical, elevator, and other specialty subcontractors to complete the construction phase. The contractor may perform one or more parts of the construction, and rely on subcontractors for the remainder of the work. However, the general contractor is responsible for managing all the trades necessary to complete the project.

As construction nears completion, the architect/engineer, owner, and government agencies start their final inspections and acceptance of the project. If the project has been managed by the general contractor, the subcontractors have performed their work, and the architect/ engineers have regularly inspected the project to ensure that local codes have been followed, then the inspection procedure can be completed in a timely manner. This results in a satisfied client and a profitable project for all.

On the other hand, if the inspection reveals faulty workmanship, poor design or use of materials, or violation of codes, then the inspection and acceptance will become a lengthy battle and may result in a dissatisfied client and an unprofitable project.

Figure 10 shows the flow of a typical project.

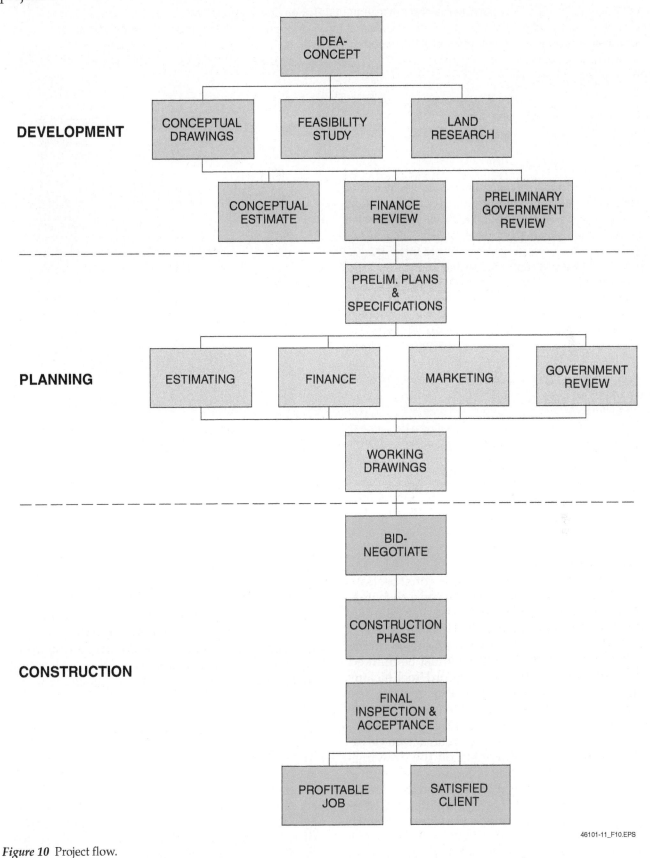

DEVELOPMENT

IDEA-CONCEPT

CONCEPTUAL DRAWINGS

FEASIBILITY STUDY

LAND RESEARCH

CONCEPTUAL ESTIMATE

FINANCE REVIEW

PRELIMINARY GOVERNMENT REVIEW

PLANNING

PRELIM. PLANS & SPECIFICATIONS

ESTIMATING

FINANCE

MARKETING

GOVERNMENT REVIEW

WORKING DRAWINGS

CONSTRUCTION

BID-NEGOTIATE

CONSTRUCTION PHASE

FINAL INSPECTION & ACCEPTANCE

PROFITABLE JOB

SATISFIED CLIENT

46101-11_F10.EPS

Figure 10 Project flow.

LEED stands for Leadership in Energy and Environmental Design. It is an initiative started by the U.S. Green Building Council (USGBC) to encourage and accelerate the adoption of sustainable construction standards worldwide through a Green Building Rating System™. USGBC is a non-government, not-for-profit group. Their rating system addresses six categories:

1. Sustainable Sites (SS)
2. Water Efficiency (WE)
3. Energy and Atmosphere (EA)
4. Materials and Resources (MR)
5. Indoor Environmental Quality (EQ)
6. Innovation in Design (ID)

LEED is a voluntary program that is driven by building owners. Construction crew leaders may not have input into the decision to seek LEED certification for a project, or what materials are used in the project's construction. However, these crew leaders can help to minimize material waste and support recycling efforts, both of which are factors in obtaining LEED certification.

An important question to ask is whether your project is seeking LEED certification. If the project is seeking certification, the next step is to ask what your role will be in getting the certification. If you are procuring materials, what information is needed and who should receive it? What specifications and requirements do the materials need to meet? If you are working outside the building or inside in a protected area, what do you need to do to protect the work area? How should waste be managed? Are there any other special requirements that will be your responsibility? Do you see any opportunities for improvement? LEED principles are described in more detail in the NCCER publications, *Your Role in the Green Environment* and *Sustainable Construction Supervisor*.

46101-11_SA02.EPS

1.3.1 As-Built Drawings

A set of drawings for a construction project reflects the completed project as conceived by the architect and engineers. During construction, changes usually are necessary because of factors unforeseen during the design phase. For example, if cabling or conduit is re-routed, or equipment is installed in a different location than shown on the original drawing, such changes must be marked on the drawings. Without this record, technicians called to perform maintenance or modify the equipment at a later date will have trouble locating all the cabling and equipment.

Any changes made during construction or installation must be documented on the drawings as the changes are made. Changes are usually made using a colored pen or pencil, so the change can be readily spotted. These drawings are commonly called redlines. When the drawings have been revised to reflect the redline changes, the final drawings are called as-builts, and are so marked. These become the drawings of record for the project.

2.0.0 PROJECT DELIVERY SYSTEMS

Project delivery systems refer to the process by which projects are delivered, from development through construction. Project delivery systems focus on three main systems: general contracting, design-build, and construction management (*Figure 11*).

2.1.0 General Contracting

The traditional project delivery system uses a general contractor. In this type of project, the owner determines the design of the project, and then solicits proposals from general contractors. After selecting a general contractor, the owner contracts directly with that contractor, who builds the project as the prime, or controlling, contractor.

	GENERAL CONTRACTING	DESIGN-BUILD	CONSTRUCTION MANAGEMENT
OWNER	Designs project (or hires architect)	Hires general contractor	Hires construction management company
GENERAL CONTRACTOR	Builds project (with owner's design)	Involved in project design, builds project	Builds, may design (hired by construction management company)
CONSTRUCTION MANAGEMENT COMPANY			Hires and manages general contractor and architect

46101-11_F11.EPS

Figure 11 Project delivery systems.

2.2.0 Design-Build

The design-build project delivery system is different from the general contracting delivery system. In the design-build system, both the design and construction of a project are managed by a single entity. GMP contracts are commonly used in these situations.

2.3.0 Construction Management

The construction management project delivery system uses a construction manager to facilitate the design and construction of a project. Construction managers are very involved in project control; their main concerns are controlling time, cost, and the quality of the project.

3.0.0 COST ESTIMATING AND BUDGETING

Before a project is built, an estimate must be prepared. An estimate is the process of calculating the cost of a project. There are two types of costs to consider, including direct and indirect costs. Direct costs, also known as general conditions, are those that can clearly be assigned to a budget. Indirect costs are overhead costs that are shared by all projects. These costs are generally applied as an overhead percentage to labor and material costs.

Direct costs include the following:

- Materials
- Labor
- Tools
- Equipment

Indirect costs refer to overhead items such as:

- Office rent
- Utilities
- Telecommunications
- Accounting
- Office supplies, signs

The bid price includes the estimated cost of the project as well as the profit. Profit refers to the amount of money that the contractor will make after all of the direct and indirect costs have been paid. If the direct and indirect costs exceed those estimated to perform the job, the difference between the actual and estimated costs must come out of the company's profit. This reduces what the contractor makes on the job.

Profit is the fuel that powers a business. It allows the business to invest in new equipment and facilities, provide training, and to maintain a reserve fund for times when business is slow. In the case of large companies, profitability attracts investors who provide the capital necessary for the business to grow. For these reasons, contractors cannot afford to lose money on a consistent basis. Those who cannot operate profitably are forced out of business. Crew leaders can help their companies remain profitable by managing budget, schedule, quality, and safety adhering to the drawings, specifications, and project schedule.

3.1.0 The Estimating Process

The cost estimate must consider a number of factors. Many companies employ professional cost estimators to do this work. They also maintain performance data for previous projects. This data can be used as a guide in estimating new projects. A complete estimate is developed as follows:

Step 1 Using the drawings and specifications, an estimator records the quantity of the materials needed to construct the job. This is called a quantity takeoff. The information is placed on a takeoff sheet like the one shown in *Figure 12*.

Step 2 Productivity rates are used to estimate the amount of labor required to complete the project. Most companies keep records of these rates for the type and size of the jobs that they perform. The company's estimating department keeps these records.

Step 3 The amount of work to be done is divided by the productivity rate to determine labor hours. For example, if the productivity rate for concrete finishing is 40 square feet per hour, and there are 10,000 square feet of concrete to be finished, then 250 hours of concrete finishing labor is required. This number would be multiplied by the hourly rate for concrete finishing to determine the cost of that labor category. If this work is subcontracted, then the subcontractor's cost estimate, raised by an overhead factor, would be used in place of direct labor cost.

Step 4 The total material quantities are taken from the quantity takeoff sheet and placed on a summary or pricing sheet, an example of which is shown in *Figure 13*. Material prices are obtained from local suppliers, and the total cost of materials is calculated.

Step 5 The cost of equipment needed for the project is determined. This number could reflect rental cost or a factor used by the company when their own equipment is to be used.

Step 6 The total cost of all resources—materials, equipment, tools, and labor—is then totaled on the summary sheet. The unit cost—the total cost divided by the total number of units of material to be put into place—can also be calculated.

Step 7 The cost of taxes, bonds, insurance, subcontractor work, and other indirect costs are added to the direct costs of the materials, equipment, tools, and labor.

Step 8 Direct and indirect costs are summed to obtain the total project cost. The contractor's expected profit is then added to that total.

> **NOTE**
>
> There are software programs available to simplify the cost estimating process. Many of them are tailored to specific industries such as construction and manufacturing, and to specific trades within the industries. For example, there are programs available for electrical and HVAC contractors. Estimating programs are typically set up to include a takeoff form and a form for estimating labor by category. Most of these programs include a data base that contains current prices for labor and materials, so they automatically price the job and produce a bid. Once the job is awarded, the programs can generate purchase orders for materials.

3.1.1 Estimating Material Quantities

The crew leader may be required to estimate quantities of materials.

A set of construction drawings and specifications is needed in order to estimate the amount of a certain type of material required to perform a job. The appropriate section of the technical specifications and page(s) of drawings should be carefully reviewed to determine the types and quantities of materials required. The quantities are then placed on the worksheet. For example, the specification section on finished carpentry should be reviewed along with the appropriate pages of drawings before taking off the linear feet of door and window trim.

If an estimate is required because not enough materials were ordered to complete the job, the estimator must also determine how much more work is necessary. Once this is known, the crew leader can then determine the materials needed. The construction drawings will also be used in this process.

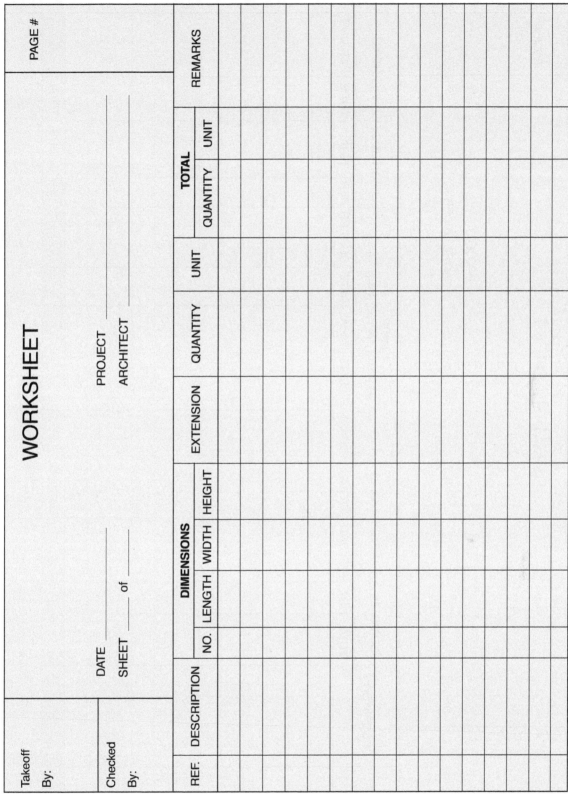

Figure 12 Quantity takeoff sheet.

46101-11_F12.EPS

SUMMARY SHEET

By:

DATE _____
SHEET _____ of _____

PROJECT _____
WORK ORDER # _____
TITLE: _____

PAGE # _____

| DESCRIPTION | QUANTITY | | MATERIAL COST | | | LABOR MAN HOURS FACTORS | | | | | | LABOR COST | | | ITEM COST | |
	TOTAL	UT	PER UNIT		TOTAL	CRAFT	PR UNIT	TOTAL	RATE	COST PR	PER	TOTAL		TOTAL	PER UNIT
	MATERIAL										LABOR		TOTAL		

46101-11_F13.EPS

Figure 13 Summary sheet.

Assume you are the leader of a crew building footing formwork for the construction shown in *Figure 14*. You have used all of the materials provided for the job, yet you have not completed it. You study the drawings and see that the formwork consists of two side forms, each 12" high. The total length of footing for the entire project is 115'-0". You have completed 88'-0" to date; therefore, you have 27'-0" remaining (115'-0" – 88'-0" = 27'-0"). Your job is to prepare an estimate of materials that you will need to complete the job. In this case, only the side forms will be estimated (the miscellaneous materials will not be considered here).

- Footing length to complete: 27'-0"
- Footing height: 1'-0"

Refer to the worksheet in *Figure 15* for a final tabulation of the side forms needed to complete the job.

1. Using the same footing as described in the example above, calculate the quantity (square feet) of formwork needed to finish 203 linear feet of the footing. Place this information directly on the worksheet.
2. You are the crew leader of a carpentry crew whose task is to side a warehouse with plywood sheathing. The wall height is 16 feet, and there is a total of 480 linear feet of wall to side. You have done 360 linear feet of wall and have run out of materials. Calculate how many more feet of plywood you will need to complete the job. If you are using 4' × 8' plywood panels, how many will you need to order, assuming no waste? Write your estimate on the worksheet.

Show your calculations to the instructor.

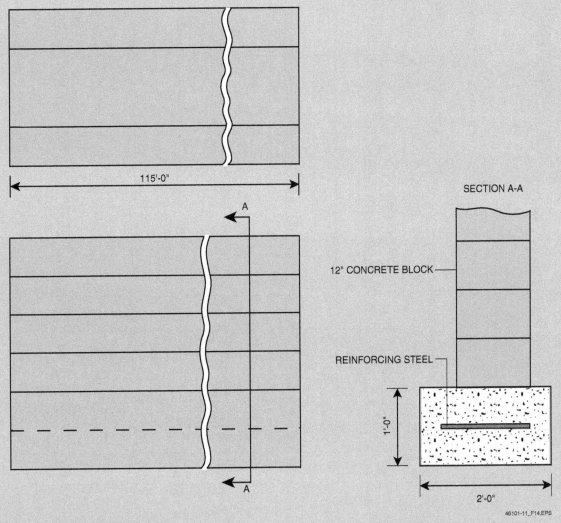

Figure 14 Footing formwork detail.

WORKSHEET

Takeoff
By: RWH

DATE _2/1/15_

Checked
By:

SHEET _01_ of _01_

PROJECT _Sam's Diner_

ARCHITECT _654b_

PAGE #1

| REF. | DESCRIPTION | DIMENSIONS | | | | | EXTENSION | QUANTITY | UNIT | TOTAL | | REMARKS |
		NO	LENGTH	WIDTH	HEIGHT					QUANTITY	UNIT	
	Footing Side Forms	2	27'0"		1'0"		2x27x1	54	SF	54	SF	

Figure 15 Worksheet with entries.

46101-11_F15.EPS

4.0.0 PLANNING

Planning can be defined as determining the method used to carry out the different tasks to complete a project. It involves deciding what needs to be done and coming up with an organized sequence of events or plan for doing the work.

Planning involves the following:

- Determining the best method for performing the job
- Identifying the responsibilities of each person on the work crew
- Determining the duration and sequence of each activity
- Identifying what tools and equipment are needed to complete a job
- Ensuring that the required materials are at the work site when needed
- Making sure that heavy construction equipment is available when required

- Working with other contractors in such a way as to avoid interruptions and delays

4.1.0 Why Plan?

With a plan, a crew leader can direct work efforts efficiently and can use resources such as personnel, materials, tools, equipment, work area, and work methods to their full potential.

Some reasons for planning include the following:

- Controlling the job in a safe manner so that it is built on time and within cost
- Lowering job costs through improved productivity
- Preparing for bad weather or unexpected occurrences
- Promoting and maintaining favorable employee morale
- Determining the best and safest methods for performing the job

---

Participant Exercise F

1. In your own words, define planning, and describe how a job can be done better if it is planned. Give an example.

   _____

   _____

   _____

   _____

2. Consider a job that you recently worked on to answer the following:
   a. List the material(s) used.
   b. List each member of the crew with whom you worked and what each person did.
   c. List the kinds of equipment used.

   _____

   _____

   _____

   _____

   _____

3. List some suggestions for how the job could have been done better, and describe how you would plan for each of the suggestions.

   _____

   _____

   _____

   _____

 Fundamentals of Crew Leadership

4.2.0 Stages of Planning

There are various times when planning is done for a construction job. The two most important occur in the pre-construction phase and during the construction work.

4.2.1 Pre-Construction Planning

The pre-construction stage of planning occurs before the start of construction. Except in a fairly small company or for a relatively small job, the crew leader usually does not get directly involved in the pre-construction planning process, but it is important to understand what it involves.

There are two phases of pre-construction planning. The first is when the proposal, bid, or negotiated price for the job is being developed. This is when the estimator, the project manager, and the field superintendent develop a preliminary plan for how the work will be done. This is accomplished by applying experience and knowledge from previous projects. It involves determining what methods, personnel, tools, and equipment will be used and what level of productivity they can achieve.

The second phase occurs after the contract is awarded. This phase requires a thorough knowledge of all project drawings and specifications. During this stage, the actual work methods and resources needed to perform the work are selected. Here, crew leaders might get involved, but their planning must adhere to work methods, production rates, and resources that fit within the estimate prepared before the contract was awarded. If the project requires a method of construction different from what is normal, the crew leader will usually be informed of what method to use.

4.2.2 Construction Planning

During construction, the crew leader is directly involved in planning on a daily basis. This planning consists of selecting methods for completing tasks before beginning work. Effective planning exposes likely difficulties, and enables the crew leader to minimize the unproductive use of personnel and equipment. Effective planning also provides a gauge to measure job progress. Effective crew leaders develop what is known as look-ahead (short-term) schedules. These schedules consider actual circumstances as well as projections two to three weeks into the future. Developing a look-ahead schedule helps ensure that all resources are available on the project when needed.

One of the characteristics of an effective crew leader is the ability to reduce each job to its simpler parts and organize a plan for handling each task.

Project planners establish time and cost limits for the project; the crew leader's planning must fit within those constraints. Therefore, it is important to consider the following factors that may affect the outcome:

- Site and local conditions, such as soil types, accessibility, or available staging areas
- Climate conditions that should be anticipated during the project
- Timing of all phases of work
- Types of materials to be installed and their availability
- Equipment and tools required and their availability
- Personnel requirements and availability
- Relationships with the other contractors and their representatives on the job

On a simple job, these items can be handled almost automatically. However, larger or more complex jobs require the planner to give these factors more formal consideration and study.

5.0.0 THE PLANNING PROCESS

The planning process consists of the following five steps:

Step 1 Establish a goal.

Step 2 Identify the work activities that must be completed in order to achieve the goal.

Step 3 Identify the tasks that must be done to accomplish those activities.

Step 4 Communicate responsibilities.

Step 5 Follow up to see that the goal is achieved.

5.1.0 Establish a Goal

The term *goal* has different meanings for different people. In general, a goal is a specific outcome that one works toward. For example, the project superintendent of a home construction project could establish the goal to have a house dried-in by a certain date. (Dried-in means ready for the application of roofing and siding.) In order to meet that goal, the leader of the framing crew and the superintendent would need to agree to a goal to have the framing completed by a given date. The crew leader would then establish sub-goals (objectives) for the crew to complete each element of the framing (floors, walls, roof) by a set time. The superintendent would need to set similar goals with the crews that install sheathing, building wrap, windows, and exterior doors. However, if the framing crew does not meet its goal, the other crews will be delayed.

5.2.0 Identify the Work to be Done

The second step in planning is to identify the work to be done to achieve the goal. In other words, it is a series of activities that must be done in a certain sequence. The topic of breaking down a job into activities is covered later in this section. At this point, the crew leader should know that, for each activity, one or more objectives must be set.

An objective is a statement of what is desired at a specific time. An objective must:

- Mean the same thing to everyone involved
- Be measurable, so that everyone knows when it has been reached
- Be achievable
- Have everyone's full support

Examples of practical objectives include the following:

- By 4:30 p.m. today, the crew will have completed installation of the floor joists.
- By closing time Friday, the roof framing will be complete.

Notice that both examples meet the first three requirements of an objective. In addition, it is assumed that everyone involved in completing the task is committed to achieving the objective. The advantage in developing objectives for each work activity is that it allows the crew leader to evaluate whether or not the plan and schedules are being followed. In addition, objectives serve as sub-goals that are usually under the crew leader's control.

Some construction work activities, such as installing 12"-deep footing forms, are done so often that they require little planning. However, other jobs, such as placing a new type of mechanical equipment, require substantial planning. This type of job requires that the crew leader set specific objectives.

Whenever faced with a new or complex activity, take the time to establish objectives that will serve as guides for accomplishing the job. These guides can be used in the current situation, as well as in similar situations in the future.

5.3.0 Identify Tasks to be Performed

To plan effectively, the crew leader must be able to break a work activity assignment down into smaller tasks. Large jobs include a greater number of tasks than small ones, but all jobs can be broken down into manageable components.

When breaking down an assignment into tasks, make each task identifiable and definable. A task is identifiable when the types and amounts of resources it requires are known. A task is definable if it has a specific duration. For purposes of efficiency, the job breakdown should not be too detailed or complex, unless the job has never been done before or must be performed with strictest efficiency.

For example, a suitable breakdown for the work activity to install 12" × 12" vinyl floor tile in a cafeteria might be the following:

Step 1 Prepare the floor.

Step 2 Lay out the tile.

Step 3 Spread the adhesive.

Step 4 Lay the tile.

Step 5 Clean the tile.

Step 6 Wax the floor.

The crew leader could create even more detail by breaking down any one of the tasks, such as lay the tile, into subtasks. In this case, however, that much detail is unnecessary and wastes the crew leader's time and the project's money. However, breaking tasks down further might be necessary in a case where the job is very complex or the analysis of the job needs to be very detailed.

Every work activity can be divided into three general parts:

- Preparing
- Performing
- Cleaning up

One of the most frequent mistakes made in the planning process is forgetting to prepare and to clean up. The crew leader must be certain that preparation and cleanup are not overlooked.

After identifying the various tasks that make up the job and developing an objective for each task, the crew leader must determine what resources the job requires. Resources include labor, equipment, materials, and tools. In most jobs, these resources are identified in the job estimate. The crew leader must make sure that these resources are available on the site when needed.

5.4.0 Communicating Responsibilities

A crew leader is unable to complete all of the activities within a job independently. Other people must be relied upon to get everything done. Therefore, most jobs have a crew of people with various experiences and skill levels to assist in the work. The crew leader's job is to draw from this expertise to get the job done well and in a safe and timely manner.

Once the various activities that make up the job have been determined, the crew leader must identify the person or persons responsible for completing each activity. This requires that the crew leader be aware of the skills and abilities of the people on the crew. Then, the crew leader must put this knowledge to work in matching the crew's skills and abilities to specific tasks that must be performed to complete the job.

After matching crew members to specific activities, the crew leader must then communicate the assignments to the crew. Communication of responsibilities is generally handled verbally; the crew leader often talks directly to the person to which the activity has been assigned. There

may be times when work is assigned indirectly through written instructions or verbally through someone other than the crew leader. Either way, the crew members should know what it is they are responsible for accomplishing on the job.

5.5.0 Follow-Up Activities

Once the activities have been delegated to the appropriate crew members, the crew leader must follow up to make sure that they are completed effectively and efficiently. Follow-up work involves being present on the job site to make sure all the resources are available to complete the work; ensuring that the crew members are working on their assigned activities; answering any questions; and helping to resolve any problems that occur while the work is being done. In short, follow-up activity means that the crew leader is aware of what's going on at the job site and is doing whatever is necessary to make sure that the work is completed on schedule.

Figure 16 reviews the planning steps.

The crew leader should carry a small note pad or electronic device to be used for planning and note taking. That way, thoughts about the project can be recorded as they occur, and pertinent details will not be forgotten. The crew leader may also choose to use a planning form such as the one illustrated in *Figure 17*.

As the job is being built, refer back to these plans and notes to see that the tasks are being done in sequence and according to plan. This is referred to as analyzing the job. Experience shows that jobs that are not built according to work plans usually end up costing more and taking more time; therefore, it is important that crew leaders refer back to the plans periodically.

The crew leader is involved with many activities on a day-to-day basis. As a result, it is easy to forget important events if they are not documented. To help keep track of events such as job changes, interruptions, and visits, the crew leader should keep a job diary.

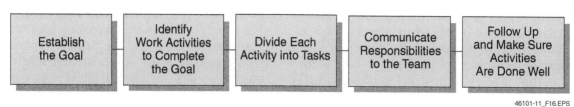

Figure 16 Steps to effective planning.

DAILY WORK PLAN

"PLAN YOUR WORK AND WORK YOUR PLAN = EFFICIENCY"

Plan of _____   Date _____

PRIORITY	DESCRIPTION	✓ When Completed ✗ Carried Forward

46101-11_F17.EPS

Figure 17 Planning form.

A job diary is a notebook in which the crew leader records activities or events that take place on the job site that may be important later. When recording in a job diary, make sure that the information is accurate, factual, complete, consistent, organized, and up to date. Follow company policy in determining which events should be recorded. However, if there is a doubt about what to include, it is better to have too much information than too little.

Figure 18 shows a sample page from a job diary.

6.0.0 PLANNING RESOURCES

Once a job has been broken down into its tasks or activities, the next step is to assign the various resources needed to perform them.

6.1.0 Safety Planning

Using the company safety manual as a guide, the crew leader must assess the safety issues associated with the job and take necessary measures to minimize any risk to the crew. This may involve working with the company or site safety officer and may require a formal hazard analysis.

6.2.0 Materials Planning

The materials required for the job are identified during pre-construction planning and are listed on the job estimate. The materials are usually ordered from suppliers who have previously provided quality materials on schedule and within estimated cost.

July 8, 2015

Weather: Hot and Humid

Project: Company XYZ Building

- The paving contractor crew arrived late (10 am).

- The owner representative inspected the footing foundation at approximately 1 pm.

- The concrete slump test did not pass. Two trucks had to be ordered to return to the plant, causing a delay.

- John Smith had an accident on the second floor. I sent him to the doctor for medical treatment. The cause of the accident is being investigated.

46101-11_F18.EPS

Figure 18 Sample page from a job diary.

The crew leader is usually not involved in the planning and selection of materials, since this is done in the pre-construction phase. The crew leader does, however, have a major role to play in the receipt, storage, and control of the materials after they reach the job site.

The crew leader is also involved in planning materials for tasks such as job-built formwork and scaffolding. In addition, the crew leader may run out of a specific material, such as fasteners, and need to order more. In such cases, a higher authority should be consulted, since most companies have specific purchasing policies and procedures.

6.3.0 Site Planning

There are many planning elements involved in site work. The following are some of the key elements:

- Emergency procedures
- Access roads
- Parking
- Stormwater runoff
- Sedimentation control
- Material and equipment storage
- Material staging
- Site security

6.4.0 Equipment Planning

Much of the planning for use of construction equipment is done during the pre-construction phase. This planning includes the types of equipment needed, the use of the equipment, and the length of time it will be on the site. The crew leader must work with the home office to make certain that the equipment reaches the job site on time. The crew leader must also ensure that sure equipment operators are properly trained.

Coordinating the use of the equipment is also very important. Some equipment is used in combination with other equipment. For example, dump trucks are generally required when loaders and excavators are used. The crew leader should also coordinate equipment with other contractors on the job. Sharing equipment can save time and money and avoid duplication of effort.

46101-11_SA03.EPS

The crew leader must designate time for equipment maintenance in order to prevent equipment failure. In the event of an equipment failure, the crew leader must know who to contact to resolve the problem. An alternate plan must be ready in case one piece of equipment breaks down, so that the other equipment does not sit idle. This planning should be done in conjunction with the home office or the crew leader's immediate superior.

6.5.0 Tool Planning

A crew leader is responsible for planning what tools will be used on the job. This task includes:

- Determining the tools required
- Informing the workers who will provide the tools (company or worker)
- Making sure the workers are qualified to use the tools safely and effectively
- Determining what controls to establish for tools

6.6.0 Labor Planning

All jobs require some sort of labor because the crew leader cannot complete all the work alone. When planning for labor, the crew leader must:

- Identify the skills needed to perform the work.
- Determine how many people having those specific skills are needed.
- Decide who will actually be on the crew.

In many companies, the project manager or job superintendent determines the size and makeup of the crew. Then, the crew leader is expected to accomplish the goals and objectives with the crew provided. Even though the crew leader may not be involved in staffing the crew, the crew leader is responsible for training the crew members to ensure that they have the skills needed to do the job.

In addition, the crew leader is responsible for keeping the crew adequately staffed at all times so that jobs are not delayed. This involves dealing with absenteeism and turnover, two common problems that affect industry today.

7.0.0 Scheduling

Planning and scheduling are closely related and are both very important to a successful job. Planning involves determining the activities that must be completed and how they should be accomplished. Scheduling involves establishing start and finish times or dates for each activity.

A schedule for a project typically shows:

- Operations listed in sequential order
- Units of construction
- Duration of activities
- Estimated date to start and complete each activity
- Quantity of materials to be installed

There are different types of schedules used today. They include the bar chart; the network schedule, which is sometimes called the critical path method (CPM) or precedence diagram; and the short-term, or look-ahead schedule.

7.1.0 The Scheduling Process

The following is a brief summary of the steps a crewleader must complete to develop a schedule.

Step 1 Make a list of all of the activities that will be performed to build the job, including individual work activities and special tasks, such as inspections or the delivery of materials.

At this point, the crew leader should just be concerned with generating a list, not with determining how the activities will be accomplished, who will perform them, how long they will take, or in what sequence they will be completed.

Step 2 Use the list of activities created in Step 1 to reorganize the work activities into a logical sequence.

When doing this, keep in mind that certain steps cannot happen until others have been completed. For example, footings must be excavated before concrete can be placed.

Step 3 Assign a duration or length of time that it will take to complete each activity and determine the start time for each. Each activity will then be placed into a schedule format. This step is important because it helps the crew leader ensure that the activities are being completed on schedule.

The crew leader must be able to read and interpret the job schedule. On some jobs, the beginning and expected end date for each activity, along with the expected crew or worker's production rate, is provided on the form. The crew leader can use this

information to plan work more effectively, set realistic goals, and measure whether or not they were accomplished within the scheduled time.

Before starting a job, the crew leader must:

- Determine the materials, tools, equipment, and labor needed to complete the job.
- Determine when the various resources are needed.
- Follow up to ensure that the resources are available on the job site when needed.

Availability of needed resources should be verified three to four working days before the start of the job. It should be done even earlier for larger jobs. This advance preparation will help avoid situations that could potentially delay the job or cause it to fall behind schedule.

7.2.0 Bar Chart Schedule

Bar chart schedules, also known as Gantt charts, can be used for both short-term and long-term jobs. However, they are especially helpful for jobs of short duration.

Bar charts provide management with the following:

- A visual concept of the overall time required to complete the job through the use of a logical method rather than a calculated guess
- A means to review each part of the job
- Coordination requirements between crafts
- Alternative methods of performing the work

A bar chart can be used as a control device to see whether the job is on schedule. If the job is not on schedule, immediate action can be taken in the office and the field to correct the problem and ensure that the activity is completed on schedule.

A bar chart is illustrated in *Figure 19*.

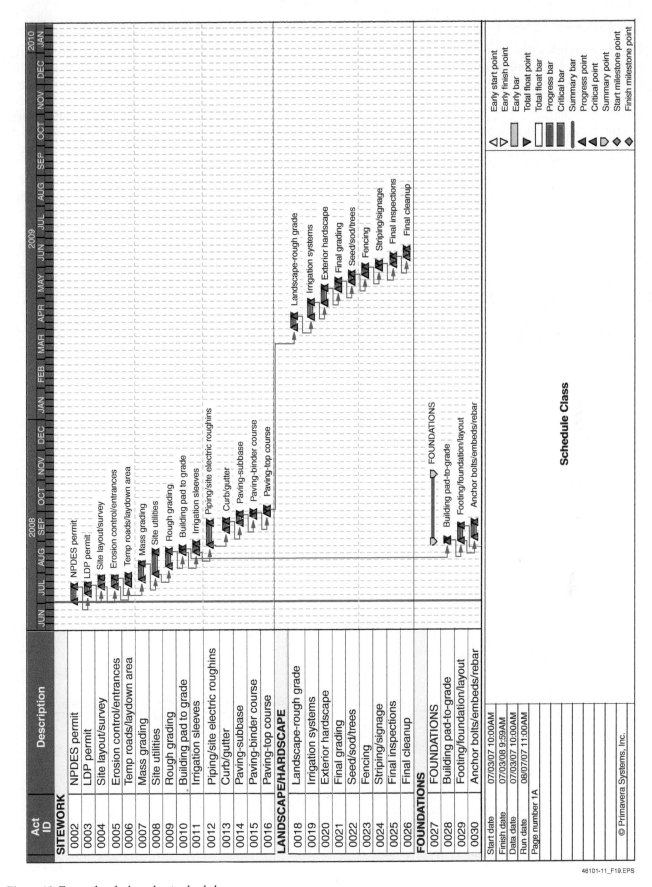

Figure 19 Example of a bar chart schedule.

7.3.0 Network Schedule

Network schedules are an effective project management tool because they show dependent (critical path) activities and activities that can be performed in parallel. In *Figure 20*, for example, reinforcing steel cannot be set until the concrete forms have been built and placed. Other activities are happening in parallel, but the forms are in the critical path. When building a house, drywall cannot be installed and finished until wiring, plumbing, and HVAC ductwork have been roughed-in. Because other activities, such as painting and trim work, depend on drywall completion, the drywall work is a critical-path function. That is, until it is complete, the other tasks cannot be started, and the project itself is likely to be delayed by the amount of delay in any dependent activity. Likewise, drywall work can't even be started until the rough-ins are complete. Therefore, the project superintendent is likely to focus on those activities when evaluating schedule performance.

The advantage of a network schedule is that it allows project leaders to see how a schedule change on one activity is likely to affect other activities and the project in general. A network schedule is laid out on a timeline and usually shows the estimated duration for each activity. Network schedules are generally used for complex jobs that take a long time to complete. The PERT (program evaluation and review technique) schedule is a form of network schedule.

7.4.0 Short-Term Scheduling

Since the crew leader needs to maintain the job schedule, he or she needs to be able to plan daily production. Short-term scheduling is a method used to do this. An example is shown in *Figure 21*.

The information to support short-term scheduling comes from the estimate or cost breakdown. The schedule helps to translate estimate data and the various job plans into a day-to-day schedule of events. The short-term schedule provides the crew leader with visibility over the project. If actual production begins to slip behind estimated production, the schedule will warn the crew leader that a problem lies ahead and that a schedule slippage is developing.

Short-term scheduling can be used to set production goals. It is generally agreed that production can be improved if workers:

- Know the amount of work to be accomplished
- Know the time they have to complete the work
- Can provide input when setting goals

Consider the following example:

Situation:

A carpentry crew on a retaining wall project is about to form and pour catch basins and put up wall forms. The crew has put in a number of catch basins, so the crew leader is sure that they can perform the work within the estimate. However, the crew leader is concerned about their production of the wall forms. The crew will work on both the basins and the wall forms at the same time.

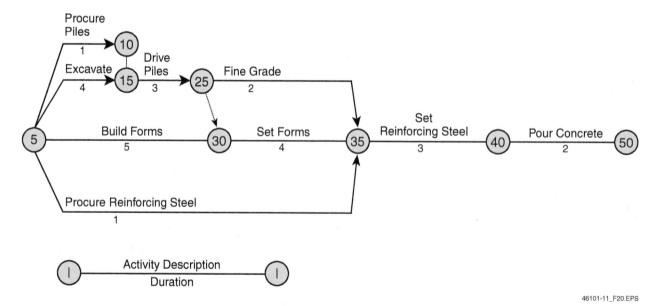

Figure 20 Example of a network schedule.

NCCER – *Masonry Level Three* 46101-11

Calendar Dates																
	7/1	7/2	7/3	7/7	7/8	7/9	7/10	7/11	7/14	7/15	7/16	7/17	7/18	7/21	7/22	7/23
ACTIVITY DESCRIPTION	Work Days															
	1	2	3	4	5	6	7	8	9	10	11	12	13	14	15	16
Process Piles	▓															
Excavate	▓	▓	▓	▓												
Build Forms	▓	▓	▓	▓	▓											
Process Reinforcing Steel	▓															
Drive Piles					▓	▓	▓									
Fine Grade								▓								
Set Forms								▓	▓	▓	▓					
Set Reinforced Steel												▓	▓	▓		
Pour Concrete															▓	▓

NOTES: The project start date is July 1st, which is a Tuesday.
Time placement of activity and duration may be done anytime through shaded portion.
Bottom portion of line available to show progress as activities are completed.

ADDITIONAL TASKS:

					▓											

46101-11_F21.EPS

Figure 21 Short-term schedule.

1. The crew leader notices the following in the estimate or cost breakdown:
   a. Production factor for wall forms = 16 work-hours per 100 square feet
   b. Work to be done by measurement = 800 square feet
   c. Total time = 128 work-hours (800 × 16 ÷ 100)
2. The carpenter crew consists of the following:
   a. One carpenter crew leader
   b. Four carpenters
   c. One laborer
3. The crew leader determines the goal for the job should be set at 128 work-hours (from the cost breakdown).
4. If the crew remains the same (six workers), the work should be completed in about 21 crew-hours (128 work hours ÷ 6 workers = 21.33 crew-hours).
5. The crew leader then discusses the production goal (completing 800 square feet in 21 crew-hours) with the crew and encourages them to work together to meet the goal of getting the forms erected within the estimated time.

The short-term schedule was used to translate production into work-hours or crew-hours and to schedule work so that the crew can accomplish it within the estimate. In addition, setting production targets provides the motivation to produce more than the estimate requires.

7.5.0 Updating a Schedule

No matter what type of schedule is used, it must be kept up to date to be useful to the crew leader. Inaccurate schedules are of no value.

The person responsible for scheduling in the office handles the updates. This person uses information gathered from job field reports to do the updates.

The crew leader is usually not directly involved in updating schedules. However, he or she may be responsible for completing field or progress reports used by the company. It is critical that the crew leader fill out any required forms or reports completely and accurately so that the schedule can be updated with the correct information.

8.0.0 COST CONTROL

Being aware of costs and controlling them is the responsibility of every employee on the job. It is the crew leader's job to ensure that employees uphold this responsibility. Control refers to the comparison of estimated performance against actual performance and following up with any needed corrective action. Crew leaders who use cost-control practices are more valuable to the company than those who do not.

On a typical job, many activities are going on at the same time. This can make it difficult to control the activities involved. The crew leader must be constantly aware of the costs of a project and effectively control the various resources used on the job.

When resources are not controlled, the cost of the job increases. For example, a plumbing crew of four people is installing soil pipe and does not have enough fittings. Three crew members wait while one crew member goes to the supply house for a part that costs only a few dollars. It takes the crew member an hour to get the part, so four hours of productive work have been lost. In addition, the travel cost for retrieving the supplies must be added.

8.1.0 Assessing Cost Performance

Cost performance on a project is determined by comparing actual costs to estimated costs. Regardless of whether the job is a contract bid project or an in-house project, a budget must first be established. In the case of a contract bid, the budget is generally the cost estimate used to bid the job. For an in-house job, participants will submit labor and material forecasts, and someone in authority will authorize a project budget.

It is common to estimate cost by either breaking the job into funded tasks or by forecasting labor and materials expenditures on a timeline. Many companies create a work breakdown structure (WBS) for each project. Within the WBS, each major task is assigned a discrete charge number. Anyone working on that task charges that number on their time sheet, so that project managers can readily track cost performance. However, knowing how much has been spent does not necessarily determine cost performance.

Although financial reports can show that actual expenses are tracking forecast expenses, they don't show if the work is being done at the required rate. Thus it is possible to have spent half the budget, but have less than half of the work compete. When the project is broken down into funded tasks related to schedule activities and events, there is far greater control over cost performance.

8.2.0 Field Reporting System

The total estimated cost comes from the job estimates, but the actual cost of doing the work is obtained from an effective field reporting system.

A field reporting system is made up of a series of forms, which are completed by the crew leader and others. Each company has its own forms and methods for obtaining information. The general information and the process of how they are used are described here.

First, the number of hours each person worked on each task must be known. This information comes from daily time cards. Once the accounting department knows how many hours each employee worked on an activity, it can calculate the total cost of the labor by multiplying the number of hours worked by the wage rate for each worker. The cost for the labor to do each task can be calculated as the job progresses. This cost will be compared with the estimated cost. This comparison will also be done at the end of the job.

When material is put in place, a designated person will measure the quantities from time to time and send this information to the home office and, possibly, the crew leader. This information, along with the actual cost of the material and the amount of hours it took the workers to install it, is compared to the estimated cost. If the cost is greater than the estimate, management and the crew leader will have to take action to reduce the cost.

A similar process is used to determine if the costs to operate equipment or the production rate are comparable to the estimated cost and production rate.

For this comparison process to be of use, the information obtained from field personnel must be correct. It is important that the crew leader be accurate in reporting. The crew leader is responsible for carrying out his or her role in the field reporting system. One of the best ways to do this is to maintain a daily diary, using a notebook or electronic device. In the event of a legal/contractual conflict with the client, such diaries are considered as evidence in court proceedings, and can be helpful in reaching a settlement.

Here is an example. You are running a crew of five concrete finishers for a subcontractor. When you and your crew show up to finish a slab, the GC says, "We're a day behind on setting the forms, so I need you and your crew to stand down until tomorrow." What do you do?

Of course, you would first call your office to let them know about the delay. Then, you would immediately record it in your job diary. A six-man crew for one day represents 48 labor hours. If your company charges $30 an hour, that's a potential loss of $1,440, which the company would want to recover from the GC. If there is a dispute, your entry in the job diary could result in a favorable decision for your employer.

8.3.0 Crew Leader's Role in Cost Control

The crew leader is often the company representative in the field, where the work takes place. Therefore, the crew leader has a great deal to do with determining job costs. When work is assigned to a crew, the crew leader should be given a budget and schedule for completing the job. It is then up to the crew leader to make sure the job is done on time and stays within budget. This is done by actively managing the use of labor, materials, tools, and equipment.

If the actual costs are at or below the estimated costs, the job is progressing as planned and scheduled, and the company will realize the expected profit. However, if the actual costs exceed the estimated costs, one or more problems may result in the company losing its expected profit, and maybe more. No company can remain in business if it continually loses money. One of the factors that can increase cost is client-related changes. The crew leaders must be able to assess the potential impact of such changes and, if necessary, confer with their employer to determine the course of action. If losses are occurring, the crew leader and superintendent will need to work together to get the costs back in line.

Noted below are some of the reasons why actual costs can exceed estimated costs and suggestions on what the crew leader can do to bring the costs back in line. Before starting any action, however, the crew leader should check with his or her superior to see that the action proposed is acceptable and within the company's policies and procedures.

- *Cause* – Late delivery of materials, tools, and/or equipment
  Corrective Action: Plan ahead to ensure that job resources will be available when needed
- *Cause* – Inclement weather
  Corrective Action: Work with the superintendent and have alternate plans ready
- *Cause* – Unmotivated workers
  Corrective Action: Counsel the workers
- *Cause* – Accidents
  Corrective Action: Enforce the existing safety program

There are many other methods to get the job done on time if it gets off schedule. Examples include working overtime, increasing the size of the crew, pre-fabricating assemblies, or working staggered shifts. However, these examples may increase the cost of the job, so they should not be done without the approval of the project manager.

9.0.0 RESOURCE CONTROL

The crew leader's job is to ensure that assigned tasks are completed safely according to the plans and specifications, on schedule, and within the scope of the estimate. To accomplish this, the crew leader must closely control how resources of materials, equipment, tools, and labor are used. Waste must be minimized whenever possible.

Control involves measuring performance and correcting deviations from plans and specifications to accomplish objectives. Control anticipates deviation from plans and specifications and takes measures to prevent it from occurring.

An effective control process can be broken down into the following steps:

Step 1   Establish standards and divide them into measurable units.

For example, a baseline can be created using experience gained on a typical job, where 2,000 LF of 1¼" copper water tube was installed in five days. Dividing 2,000 by 5 gives 400. Thus, the average installation rate in this case for 1¼" copper water tube was 400 LF/day.

Step 2   Measure performance against a standard.

On another job, 300 square feet of the same tube was placed during an average day. Thus, this average production of 300 LF/day did not meet the average rate of 400 LF/day.

Step 3   Adjust operations to ensure that the standard is met.

In Step 2 above, if the plan called for the job to be completed in five days, the crew leader would have to take action to ensure that this happens. If 300 LF/day is the actual average daily rate, it will have to be increased by 100 LF/day to meet the standard.

9.1.0 Materials Control

The crew leader's responsibility in materials control depends on the policies and procedures of the company. In general, the crew leader is responsible for ensuring on-time delivery, preventing waste, controlling delivery and storage, and preventing theft of materials.

9.1.1 Ensuring On-Time Delivery

It is essential that the materials required for each day's work be on the job site when needed. The crew leader should confirm in advance that all materials have been ordered and will be delivered on schedule. A week or so before the delivery date, follow-up is needed to make sure there will be no delayed deliveries or items on backorder.

To be effective in managing materials, the crew leader must be familiar with the plans and specifications to be used, as well as the activities to be performed. He or she can then determine how many and what types of materials are needed.

If other people are responsible for providing the materials for a job, the crew leader must follow up to make sure that the materials are available when they are needed. Otherwise, delays occur as crew members stand around waiting for the materials to be delivered.

9.1.2 Preventing Waste

Waste in construction can add up to loss of critical and costly materials and may result in job delays. The crew leader needs to ensure that every crew member knows how to use the materials efficiently. The crew should be monitored to make certain that no materials are wasted.

An example of waste is a carpenter who saws off a piece of lumber from a fresh piece, when the size needed could have been found in the scrap pile. Another example of waste involves installing a fixture or copper tube incorrectly. The time spent installing the item incorrectly is wasted because the task will need to be redone. In addition, the materials may need to be replaced if damaged during installation or removal.

Under LEED, waste control is very important. Credits are given for finding ways to reduce waste and for recycling waste products. Waste material should be separated for recycling if feasible (*Figure 22*).

Did you know?

Just-in-time (JIT) delivery is a strategy in which materials are delivered to the job site when they are needed. This means that the materials may be installed right off the truck. This method reduces the need for on-site storage and staging. It also reduces the risk of loss or damage as products are moved about the site. Other modern material management methods include the use of radio frequency identification tags (RFIDs) that make it easy to locate material in crowded staging areas.

9.1.3 Verifying Material Delivery

A crew leader may be responsible for the receipt of materials delivered to the work site. When this happens, the crew leader should require a copy of the shipping ticket and check each item on the shipping ticket against the actual materials to see that the correct amounts were received.

Figure 22 Waste material separated for recycling.

The crew leader should also check the condition of the materials to verify that nothing is defective before signing the shipping ticket. This can be difficult and time consuming because it means that cartons must be opened and their contents examined. However, it is necessary, because a signed shipping ticket indicates that all of the materials were received undamaged. If the crew leader signs for the materials without checking them, and then finds damage, no one will be able to prove that the materials came to the site in that condition.

Once the shipping ticket is checked and signed, the crew leader should give the original or a copy to the superintendent or project manager. The shipping ticket will then be filed for future reference because it serves as the only record the company has to check bills received from the supply house.

9.1.4 Controlling Delivery and Storage

Another very important element of materials control is where the materials will be stored on the job site. There are two factors in determining the appropriate storage location. The first is convenience. If possible, the materials should be stored near where they will be used. The time and effort saved by not having to carry the materials long distances will greatly reduce the installation costs.

Next, the materials must be stored in a secure area where they will not be damaged. It is important that the storage area suit the materials being stored. For instance, materials that are sensitive to temperature, such as chemicals or paints, should be stored in climate-controlled areas. Otherwise, waste may occur.

9.1.5 Preventing Theft and Vandalism

Theft and vandalism of construction materials increase costs because these materials are needed to complete the job. The replacement of materials and the time lost because the needed materials are missing can add significantly to the cost. In addition, the insurance that the contractor purchases will increase in cost as the theft and vandalism rate grows.

The best way to avoid theft and vandalism is a secure job site. At the end of each work day, store unused materials and tools in a secure location, such as a locked construction trailer. If the job site is fenced or the building can be locked, the materials can be stored within. Many sites have security cameras and/or intrusion alarms to help minimize theft and vandalism.

46101-11_F22.EPS

9.2.0 Equipment Control

The crew leader may not be responsible for long-term equipment control. However, the equipment required for a specific job is often the crew leader's responsibility. The first step is to identify when the required equipment must be transported from the shop or rental yard. The crew leader is responsible for informing the shop where it is being used and seeing that it is returned to the shop when the job is done.

It is common for equipment to lay idle at a job site because the job has not been properly planned and the equipment arrived early. For example, if wire-pulling equipment arrives at a job site before the conduit is in place, this equipment will be out of service while awaiting the conduit installation. In addition, it is possible that this unused equipment could be damaged, lost, or stolen.

The crew leader needs to control equipment use, ensure that the equipment is operated in accordance with its design, and that it is being used within time and cost guidelines. The crew leader must also ensure that equipment is maintained and repaired as indicated by the preventive maintenance schedule. Delaying maintenance and repairs can lead to costly equipment failures. The crew leader must also ensure that the equipment operators have the necessary credentials to operate the equipment, including applicable licenses.

The crew leader is responsible for the proper operation of all other equipment resources, including cars and trucks. Reckless or unsafe operation of vehicles will likely result in damaged equipment and a delayed or unproductive job. This, in turn, could affect the crew leader's job security.

The crew leader should also ensure that all equipment is secured at the close of each day's work in an effort to prevent theft. If the equipment is still being used for the job, the crew leader should ensure that it is locked in a safe place; otherwise, it should be returned to the shop.

9.3.0 Tool Control

Among companies, various policies govern who provides hand and power tools to employees. Some companies provide all the tools, while others furnish only the larger power tools. The crew leader should find out about and enforce any company policies related to tools.

Tool control is a twofold process. First, the crew leader must control the issue, use, and maintenance of all tools provided by the company. Next, the crew leader must control how the tools are being used to do the job. This applies to tools that are issued by the company as well as tools that belong to the workers.

Using the proper tools correctly saves time and energy. In addition, proper tool use reduces the chance of damage to the tool being used. Proper use also reduces injury to the worker using the tool, and to nearby workers.

Tools must be adequately maintained and properly stored. Making sure that tools are cleaned, dried, and lubricated prevents rust and ensures that the tools are in the proper working order.

In the event that tools are damaged, it is essential that they be repaired or replaced promptly. Otherwise, an accident or injury could result.

> **NOTE**
>
> Regardless of whether a tool is owned by a worker or the company, OSHA will hold the company responsible for it when it is used on a job site. The company can be held accountable if an employee is injured by a defective tool. Therefore, the crew leader needs to be aware of any defects in the tools the crew members are using.

Company-issued tools should be taken care of as if they are the property of the user. Workers should not abuse tools simply because they are not their own.

One of the major causes of time lost on a job is the time spent searching for a tool. To prevent this from occurring, a storage location for company-issued tools and equipment should be established. The crew leader should make sure that all company-issued tools and equipment are returned to this designated location after use. Similarly, workers should make sure that their personal toolboxes are organized so that they can readily find the appropriate tools and return their tools to their toolboxes when they are finished using them.

Studies have shown that the key to an effective tool control system lies in:

- Limiting the number of people allowed access to stored tools
- Limiting the number of people held responsible for tools
- Controlling the ways in which a tool can be returned to storage
- Making sure tools are available when needed

9.4.0 Labor Control

Labor typically represents more than half the cost of a project, and therefore has an enormous impact on profitability. For that reason, it is essential to manage a crew and their work environment in a way that maximizes their productivity. One of the ways to do that is to minimize the time spent on unproductive activities such as:

- Engaging in bull sessions
- Correcting drawing errors
- Retrieving tools, equipment, and materials
- Waiting for other workers to finish

If crew members are habitually goofing off, it is up to the crew leader to counsel those workers. The counseling should be documented in the crew leader's daily diary. Repeated violations will need to be referred to the attention of higher management as guided by company policy.

Errors will occur and will need to be corrected. Some errors, such as mistakes on drawings, may be outside of the crew leader's control. However, some drawing errors can be detected by carefully examining the drawings before work begins.

If crew members are making mistakes due to inexperience, the crew leader can help avoid these errors by providing on-the-spot training and by checking on inexperienced workers more often.

The availability and location of tools, equipment, and materials can have a profound effect on a crew's productivity. If the crew has to wait for these things, or travel a distance to get them, it reduces their productivity. The key to minimizing such problems is proactive management of these resources. As discussed earlier, practices such as checking in advance to be sure equipment and materials will be available when scheduled and placing materials close to the work site will help minimize unproductive time.

Delays caused by others can be avoided by carefully tracking the project schedule. By doing so, crew leaders can anticipate delays that will affect the work of their crews and either take action to prevent the delay or redirect the crew to another task.

Participant Exercise G

1. List the methods your company uses to minimize waste.

_____

_____

_____

_____

2. List the methods your company uses to control small tools on the job.

_____

_____

_____

_____

3. List five ways that you feel your company could control labor to maximize productivity.

_____

_____

_____

_____

_____

10.0.0 PRODUCTION AND PRODUCTIVITY

Production is the amount of construction put in place. It is the quantity of materials installed on a job, such as 1,000 linear feet of waste pipe installed in a given day. On the other hand, productivity depends on the level of efficiency of the work. It is the amount of work done per hour or day by one worker or a crew.

Production levels are set during the estimating stage. The estimator determines the total amount of materials to be put in place from the plans and specifications. After the job is complete, the actual amount of materials installed can be assessed, and the actual production can be compared to the estimated production.

Productivity relates to the amount of materials put in place by the crew over a certain time period. The estimator uses company records during the estimating stage to determine how much time and labor it will take to place a certain quantity of materials. From this information, the estimator calculates the productivity necessary to complete the job on time.

For example, it might take a crew of two people ten days to paint 5,000 square feet. The crew's productivity per day is obtained by dividing 5,000 square feet by ten days. The result is 500 square feet per day. The crew leader can compare the daily production of any crew of two painters doing similar work with this average, as discussed previously.

Planning is essential to productivity. The crew must be available to perform the work, and have all of the required materials, tools, and equipment in place when the job begins.

The time on the job should be for business, not for taking care of personal problems. Anything not work-related should be handled after hours, away from the job site. Planning after-work activities, arranging social functions, or running personal errands should be handled after work or during breaks.

Organizing field work can save time. The key to effectively using time is to work smarter, not necessarily harder. For example, most construction projects require that the contractor submit a set of as-built plans at the completion of the work. These plans describe how the materials were actually installed. The best way to prepare these plans is to mark a set of working plans as the work is in progress. That way, pertinent details will not be forgotten and time will not be wasted trying to remember how the work was done.

The amount of material actually used should not exceed the estimated amount. If it does, either the estimator has made a mistake, undocumented changes have occurred, or rework has caused the need for additional materials. Whatever the case, the crew leader should use effective control techniques to ensure the efficient use of materials.

When bidding a job, most companies calculate the cost per labor hour. For example, a ten-day job might convert to 160 labor hours (two painters for ten days at eight hours per day). If the company charges a labor rate of $30/hour, the labor cost would be $4,800. The estimator then adds the cost of materials, equipment, and tools, along with overhead costs and a profit factor, to determine the price of the job.

After a job has been completed, information gathered through field reporting allows the home office to calculate actual productivity and compare it to the estimated figures. This helps to identify productivity issues and improves the accuracy of future estimates.

The following labor-related practices can help to ensure productivity:

- Ensure that all workers have the required resources when needed.
- Ensure that all personnel know where to go and what to do after each task is completed.
- Make reassignments as needed.
- Ensure that all workers have completed their work properly.

1. Which of these activities occurs during the development phase of a project?
   a. Architect/engineer sketches are prepared and a preliminary budget is developed.
   b. Government agencies give a final inspection of the design, adherence to codes, and materials used.
   c. Project drawings and specifications are prepared.
   d. Contracts for the project are awarded.

2. The type of contract in which the client pays the contractor for their actual labor and material expenses they incur is known as a _____.
   a. firm fixed-price contract
   b. time-spent contract
   c. cost-reimbursable contract
   d. performance-based contract

3. On-site changes in the original design that are made during construction are recorded in the _____.
   a. as-built plans
   b. takeoff sheet
   c. project schedule
   d. job specifications

4. On a design-build project, _____.
   a. the owner is responsible for providing the design
   b. the architect does the design and the general contractor builds the project
   c. the same contractor is responsible for both design and construction
   d. a construction manager is hired to oversee the project

5. One example of a direct cost when bidding a job is _____.
   a. office rent
   b. labor
   c. accounting
   d. utilities

6. The control method that a crew leader uses to plan a few weeks in advance is a _____.
   a. network schedule
   b. bar chart schedule
   c. daily diary
   d. look-ahead schedule

7. A job diary should typically indicate _____.
   a. items such as job interruptions and visits
   b. changes needed to project drawings
   c. the estimated time for each job task related to a particular project
   d. the crew leader's ideas for improving employee morale

8. Gantt charts can help crew leaders in the field by _____.
   a. offering a comparison of actual production to estimated production
   b. providing short-term and long-term schedule information
   c. stating the equipment and materials necessary to complete a task
   d. providing the information needed to develop an estimate or an estimate breakdown

9. What is the crew leader's responsibility with regard to cost control?
   a. Cost control is outside the scope of a crew leader's responsibility.
   b. The crew leader is responsible only for minimizing material waste.
   c. The crew leader must ensure that all team members are aware of project costs and how to control them.
   d. The crew leader typically prepares the company's cost estimate and is therefore responsible for cost performance.

10. Which of the following is a correct statement regarding project cost?
   a. Cost is handled by the accounting department and is not a concern of the crew leader.
   b. A company's profit on a project is affected by the difference between the estimated cost and the actual cost.
   c. Wasted material is factored into the estimate, and is therefore not a concern.
   d. The contractor's overhead costs are not included in the cost estimate.

11. The crew leader is responsible for ensuring that equipment used by his or her crew is properly maintained.

    a. True
    b. False

12. Who is responsible if a defect in an employee's tool results in an accident?

    a. The employee
    b. The company
    c. The crew leader
    d. The tool manufacturer

13. Productivity is defined as the amount of work accomplished.

    a. True
    b. False

14. If a crew of masons is needed to lay 1,000 concrete blocks, and the estimator determined that two masons could complete the job in one eight-hour day, what is the estimated productivity rate?

    a. 125 blocks per hour
    b. 62.5 blocks per hour
    c. 31.25 blocks per hour
    d. 16 blocks per hour

Additional Resources

This module presents thorough resources for task training. The following resource material is suggested for further study.

Aging Workforce News, www.agingworkforce-news.com.

American Society for Training and Development (ASTD), www.astd.org.

Architecture, Engineering, and Construction Industry (AEC), www.aecinfo.com.

CIT Group, www.citgroup.com.

Equal Employment Opportunity Commission (EEOC), www.eeoc.gov.

National Association of Women in Construction (NAWIC), www.nawic.org.

National Census of Fatal Occupational Injuries (NCFOI), www.bls.gov.

National Center for Construction Education and Research, www.nccer.org.

National Institute of Occupational Safety and Health (NIOSH), www.cdc.gov/niosh.

National Safety Council, www.nsc.org.

NCCER Publications:

Your Role in the Green Environment

Sustainable Construction Supervisor

Occupational Safety and Health Administration (OSHA), www.osha.gov.

Society for Human Resources Management (SHRM), www.shrm.org.

United States Census Bureau, www.census.gov.

United States Department of Labor, www.dol.gov.

USA Today, www.usatoday.com.

Figure Credits

© 2010 Photos.com, a division of Getty Images. All rights reserved., Module opener

United States Department of Labor, Occupational Safety and Health Administration, SA01

plaquemaster.com, Figure 9

© U.S. Green Building Council, SA02

Sushil Shenoy, SA03 and Figure 22

John Ambrosia, Figure 19

NCCER CURRICULA — USER UPDATE

NCCER makes every effort to keep its textbooks up-to-date and free of technical errors. We appreciate your help in this process. If you find an error, a typographical mistake, or an inaccuracy in NCCER's curricula, please fill out this form (or a photocopy), or complete the online form at **www.nccer.org/olf**. Be sure to include the exact module ID number, page number, a detailed description, and your recommended correction. Your input will be brought to the attention of the Authoring Team. Thank you for your assistance.

Instructors – If you have an idea for improving this textbook, or have found that additional materials were necessary to teach this module effectively, please let us know so that we may present your suggestions to the Authoring Team.

NCCER Product Development and Revision

13614 Progress Blvd., Alachua, FL 32615

Email: curriculum@nccer.org
Online: www.nccer.org/olf

❏ Trainee Guide   ❏ Lesson Plans   ❏ Exam   ❏ PowerPoints   Other _____

Craft / Level: _____ Copyright Date: _____

Module ID Number / Title: _____

Section Number(s): _____

Description: _____

_____

_____

_____

_____

Recommended Correction: _____

_____

_____

_____

_____

Your Name: _____

Address: _____

Email: _____ Phone: _____

Glossary

Acid brick: Masonry units that are manufactured with special properties to withstand harsh chemical environments without disintegrating.

Arch: A form of construction in which a number of units span an opening by transferring vertical loads laterally to adjacent units and thus to the supports.

As-built drawing: A construction drawing that shows a project as it was completed, including all changes incorporated into the design during the construction process.

Ashlar: A square- or rectangular-cut stone masonry unit; or, a flat-faced surface having sawn or dressed bed and joint surfaces.

Backsight (BS): A reading taken on a leveling rod held on a point of known elevation to determine the height of the leveling instrument.

Basalt: A dark, durable form of igneous rock often used in walls and cobblestones.

Beam: Loadbearing horizontal framing element supported by walls or columns and girders.

Benchmark: A point established by the surveyor on or close to the building site and used as a reference for determining elevations during the construction of a building; a reference point established by the surveyor on or close to the property, usually at one corner of the lot.

Callout: Marking or identifying tag describing parts of a drawing on detail drawings, schedules, or other drawings.

Civil drawing: A drawing that shows the overall shape of the building site. Also called a site plan.

Cladding: A general term for a layer of non-bearing stone installed over a masonry wythe or structural backing.

Clamp: A mechanical lifting device that grips the sides of stone using friction.

Control point: A horizontal or vertical point established in the field to serve as part of a known framework for all points on the site.

Controlled access zone: A designated work area in which certain types of masonry work may take place without the use of conventional fall protection systems.

Crosshairs: A set of lines, typically horizontal and vertical, placed in a telescope used for sighting purposes.

Cut: A common term for a scaffold level; to remove soil or rock on site to achieve a required elevation.

Differential leveling: A method of leveling used to determine the difference in elevation between two points.

Dimension stone: A common term for stone that has been cut, shaped, and finished for use in masonry construction.

Dress: To square up the edges of stone to make them rectangular and to straighten any jagged edges.

Earthwork: All construction operations connected with excavating (cutting) or filling earth.

Easement: A legal right-of-way provision on another person's property (for example, the right of a neighbor to build a road or a public utility to install water and gas lines on the property). A property owner cannot build on an area where an easement has been identified.

Efflorescence: A deposit or crust of white powder on the surface of brickwork, resulting when soluble salts in the mortar or brick are drawn to the surface by moisture.

Elevation view: A drawing giving a view from the front or side of a structure.

Fabricator: A person who provides detailed shop drawings for the fabrication of components and who fabricates them in a shop for later installation at the job site.

Field notes: A permanent record of field measurement data and related information.

Fill: Adding soil or rock on site to achieve a required elevation.

Finishing: The process of honing, flame finishing, splitting, and polishing the exposed face or faces of stone.

Foresight (FS): A reading taken on a leveling rod held on a point in order to determine a new elevation.

Front setback: The distance from the property line to the front of the building.

Girder: Large steel or wooden beam supporting a building, usually around the perimeter.

Glazed block: Glazed masonry units made from concrete.

Granite: A very hard and durable form of igneous rock widely used in masonry for exterior and interior installations.

Grog: Burned, pulverized refractory material such as broken pottery or firebrick, utilized in the preparation of refractory bodies.

Guillotine: A hand tool consisting of two sharp spring-loaded segmental blades that split stone simultaneously from above and below.

Guyed derrick: An apparatus used for hoisting on high-rise buildings, consisting of a boom mounted on a column or mast that is held at the head by fixed-length supporting ropes or guys.

Hand clamp: A hand tool that is used to grip stone and to serve as a handle.

Height of instrument (HI): The elevation of the line of sight of the telescope relative to a known elevation. It is determined by adding the backsight elevation to the known elevation.

Honed: To be lightly polished.

Igneous: A type of stone that is formed when molten rock or volcanic lava cools and solidifies.

Isometric drawing: A three-dimensional drawing in which the object is tilted so that three faces are equally inclined to the picture plane.

Joist: Horizontal member of wood or steel supported by beams and holding up the planks of floors or the lathes of ceilings. Joists are laid edgewise to form the floor support.

Kerf: A slot cut into stone and designed to receive the end of a strap anchor.

Landscape drawing: A drawing that shows proposed plantings and other landscape features.

Lateral stress: Wind shear and other forces applying horizontal pressure to a wall or other structural unit.

Lath: Corrosion-resistant metal mesh attached to the substrate that serves as a base for mortar or plaster.

Leveling rod: A vertical measuring device that consists of two or more movable sections with graduated markings.

Limestone: A type of sedimentary stone consisting primarily of calcite that is widely used in loadbearing and veneer masonry applications, and is also a key ingredient in concrete.

Limited access zone: A restricted area alongside a masonry wall that is under construction.

Luster: The level of reflective shine on the exposed surface of stone.

Manufactured stone: Premade stone made from cast cementitious material with pigments and other added materials that give the appearance of natural stone.

MEP drawings: The set of construction drawings that consists of mechanical, electrical, and plumbing drawings.

Metamorphic: Igneous or sedimentary stone that has been subjected to extreme heat or pressure over a long period of time, causing it to change its physical or chemical structure.

Monument: A physical structure that marks the location of a survey point.

Mullion: A thin, vertical bar that divides lights in a window or panels in a door.

Offset: To position a stake at a specified distance and direction from the control point to allow that area to be worked in without disturbing the stake. Offset stakes include the distance from, and direction to, the control point.

Parallax: The apparent movement of the crosshairs in a surveying instrument caused by movement of the eyes.

Parapet: A low wall or railing.

Permeability: The extent to which a substance allows liquids and gases to pass through it.

Plan view: A drawing that represents a view looking down on an object.

Plasticity: A complex property of a material involving a combination of qualities of mobility and magnitude of yield value; a material's ability to be easily molded into various shapes.

Plugging chisel: A chisel with a tapered blade used for removing mortar from joints.

Point loading: An uneven distribution of structural load in a masonry structure that can cause cracking or buckling.

Portland cement paint: Cement-based paint. Type I, containing 65 percent portland cement by weight, is for general use; Type II, containing 80 percent portland cement by weight, is used where maximum durability is needed. Within each type there are two classes: Class A contains no aggregate filler and is for general use; Class B contains 20 to 40 percent sand filler and is used on open-textured surfaces.

Property line: The recorded legal boundary of a piece of property.

Pumice: A light-colored, powdery form of igneous stone rich in silica, often used to make concrete and cinder block and as an abrasive material.

Quarried: Mined or extracted.

Quarry sap: A brownish stain that forms on the surface of freshly quarried stone as water leaches from the stone.

Random rectangular stone: Stone of modular dimensions that has vertical and horizontal bed joints.

Reflected ceiling plan: A drawing that shows the details of the ceiling as though the ceiling were reflected by a mirror on the floor.

Refractory: A specialized masonry unit that can withstand high temperatures; used to form an insulating layer where extreme heat would damage other components of the structure. Refractories require special mortars that can also withstand high temperatures.

Reglet: A narrow molding used to separate two structural elements, usually roof and wall, to divert water.

Riser diagram: A type of isometric drawing that depicts the layout, components, and connections of a piping system.

Rubble: Small, irregular stone debris left over from the quarrying process.

Sandstone: A type of sedimentary stone consisting primarily of layers of quartz and feldspar; it is used in ornamental and decorative stone masonry installations and to make grindstones.

Scratch coat: A layer of mortar that has been scored or scratched with a raking tool in order to allow stone units to be adhered.

Season: To allow freshly quarried stone to dry out.

Sedimentary: A type of stone that is created by the gradual settling and compression of minerals and organic particles into layers.

Shop ticket: A document assigned to a piece of slabbed stone identifying the final dimensions and finish to be applied, and assigning the stone a number to aid in construction.

Slabbed: Sawn to a predetermined thickness from a larger quarried block.

Spall: A chip, fragment, or flake broken off from the edge or face of a stone masonry unit and having at least one thin edge.

Square-foot method: A method of estimating materials by calculating the area, in square feet, of a structural unit.

Station: An instrument setting location in differential leveling.

Strap: A mechanical lifting device that uses slings to support stone from underneath.

Strapmaster: A hand tool that uses a lever-operated ratchet to bend, cut, punch, twist, and shape metal sheet and rod.

Strata: Layers of sedimentary stone.

Structural glazed tile: Glazed masonry units made from burned clay or shale.

Substrate: The backing material to which natural and manufactured stone veneer is attached.

Takeoff: The process of measuring and counting individual items from a set of plans in order to estimate material quantities and associated items for construction projects.

Tape: A measuring tape, usually made of fiberglass, cloth, or stainless steel.

Taping: The process of making horizontal and vertical distance measurements.

Temporary benchmark: A point of known (reference) elevation determined from benchmarks through leveling, and permanent enough to last for the duration of a project.

Transit level: An optical instrument used in surveying.

Tuckpointing: The process of cutting away defective mortar and refilling the joints with fresh mortar.

Turning point (TP): A temporary point within an open or closed differential-leveling circuit whose elevation is determined by differential leveling. It is normally the leveling-rod location. Its elevation is determined by subtracting the foresight elevation from the height-of-the-instrument elevation.

Vacuum cup: A hand tool used to grip rough or irregularly shaped stone through the use of mechanically induced suction.

Vacuum lifter: A mechanical lifting device used to grip rough or irregularly shaped stone through the use of several mechanically or hydraulically operated suction surfaces.

Worm-drive saw: A power saw that uses a cylindrical gear to spin the blade slowly and with more torque than other types of power saws.

Index

Horizontal curtain walls, (28301):21
Horizontal lifelines, personal fall arrest systems, (28301):6
Hubs
    Marking, (28306):8–9
    Pilot holes for, (28306):9
Human Resources Department, (46101):7, 37
Hydrofluoric acid, (28303):18

I

Igneous, (28308):1, 2, 33
Ingalls building, (28301):19
Injury statistics, work-related, (46101):29
Inspection of existing structures
    Basements, (28303):6
    Checklist for, (28303):7–8
    Chimneys, (28303):7
    for condensation, (28303):5
    Exterior walls, (28303):7
    Frost damage, (28303):1
    Instruments used for, (28303):7
    Mortar joints, (28303):1
    Openings, (28303):7
    Stucco, (28303):7
    Weathering damage, (28303):1
Insulated concrete masonry units, (28304):21
Interior wall construction
    Elevated masonry systems, (28301):23, 25
    Maximum length- or height-to-thickness, by type, (28301):24
    Maximum spans, by type, (28301):24
International Building Code
    Height to thickness ratio for partition walls, (28301):23, 24
    Ladder support requirements, (28301):18
Intrados, (28302):6
Isometric drawing, (28304):1, 14, 46

J

Jack arch
    Bonded, (28302):11–12
    Common, (28302):11
    Example, (28302):4
    Term derivation, (28302):13
Jack arch construction
    Arch brick spacing marks, (28302):12–13
    Bonded arches, (28302):11–12
    Bonding patterns, (28302):10
    Common arches, (28302):11
    Embedding in mortar, (28302):13
    Layout from radial center point, (28302):12
Jamestown, New York Area Labor Management Committee (JALMC), (46101):4
Job datum. *See* Benchmarks
Job descriptions, (46101):8, 9
Job diary, (46101):57
Job importance, motivation and, (46101):20
Job satisfaction, (46101):19
Jointers, (28302):45
Joint reinforcement
    Pier-and-panel barrier walls, (28302):1
    Prefabricated, (28305):21
    Spalling, (28303):3–4
Joint reinforcement, continuous
    Estimating
        Conversion tables, (28305):20
        Linear-foot method, (28305):19–20
        Square-foot method, (28305):20–21
    Placement, (28305):19
Joint replacement
    Brick, (28303):24–25
    Foundation walls, (28303):29
Joints. *See also* Mortar joints
    Block parapet walls, (28301):23
    Collar joints, (28305):17
    Coping joints, (28301):23
    Pressure-relieving, (28301):19
Joist, (28304):18, 46
Just-in-time delivery, (46101):66

K

Kerf, (28308):6, 11, 20–21, 33
Keystone, (28302):6
Kiln jack installation method, refractory brick, (28302):27, 28

L

Labor control, (46101):69
Labor planning, (46101):59
Labor shortages, (46101):1
Ladder hoists, (28301):32–33
Ladders, (28301):16–17, 18
Landscape drawings, (28304):1, 2, 46
Language barriers, (46101):4
Lanyards, personal fall arrest systems, (28301):5
Laser leveling instruments, (28306):17
Lateral stresses, (28301):16, 18, 22, 40
Lath, (28306):8–9, (28308):20, 24, 33
Laws, chemical disposal, (28303):18
LAZ. *See* Limited access zone (LAZ)
Leaders
    Characteristics of effective, (46101):12–13
    Defined, (46101):11–12
    Ethical conduct in, (46101):14
    Expectations of, (46101):13
    Functions of, (46101):13
Leadership in Energy and Environmental Design (LEED), (46101):46, 59, 66
Leadership styles, (46101):13–14
Leaning Tower of Pisa, (28303):33
Learning retention, (46101):3
Learning styles, (46101):1, 2
LEED. *See* Leadership in Energy and Environmental Design (LEED)
Legal ethics, (46101):14
Legends, structural drawings, (28304):28
Leveling, differential. *See* Differential leveling
Leveling applications
    Cross-section leveling, (28306):34–35
    Grid leveling, (28306):35
    Profile leveling, (28306):34
    Transferring elevations up a structure, (28306):34, 35
Leveling instruments
    Adjusting, (28306):21, 23–24
    Automatic, (28306):17
    Builder's level, (28306):16
    Calibration testing, (28306):25–26
    Care and handling, (28306):19, 26
    Hand site levels, (28306):12
    Initial setup, (28306):21, 23–24
    Laser, (28306):17
    Leveling rod accessories, (28306):21, 23
    Leveling rods, (28306):11, 12, 20–21, 46
    Plumb bobs, (28306):12
    Total station, (28306):18–19
    Transit level, (28306):16, 17, 46
    Tripods, (28306):19–20
Leveling rod, (28306):11, 12, 20–21, 46
Leveling rod accessories, (28306):21, 23
Lewis pins, (28308):15
Lifelines, personal fall arrest systems, (28301):5–6
Lifting devices, stone masonry. *See also* Crane operations
    Box lewis, (28308):15
    Clamps, (28308):6, 12, 33
    Hand-powered, (28308):8
    Hoists, (28301):9, 32–33
    Inspection and storage, (28308):13–14
    Lewis pins, (28308):15
    Power driven, (28308):12–14
    Safety, (28308):13
    Shackles, (28308):13
    Straps, (28308):12, 33
    Vacuum lifter, (28308):13
Lime, estimating for mortar, (28305):16, (28308):22
Lime run, (28303):15–16
Limestone, (28308):1, 2–4, 33
Limited access zone (LAZ), (28301):1, 10, 40
Linear-foot method for estimating continuous joint reinforcement, (28305):19–20
Line-of-site test, (28306):25–26
Linings, acid brick masonry, (28302):15, 16
Lintels, (28305):4–6
Listening
    Active, (46101):15
    Effective, (46101):15–16
Loadbearing information on structural drawings, (28304):27–28
Long walls, cracks in, (28303):1
Luster, (28308):1, 4, 33